OT 33
Operator Theory: Advances and Applications
Vol. 33

Birkhäuser Verlag
Basel · Boston · Berlin

Topics in Interpolation Theory of Rational Matrix-valued Functions

Edited by

I. Gohberg

1988

Birkhäuser Verlag
Basel · Boston · Berlin

MATH
seplae

Volume Editorial Office:

Raymond and Beverly Sackler Faculty of Exact Sciences
School of Mathematical Sciences
Tel Aviv University
Tel Aviv, Israel

CIP-Titelaufnahme der Deutschen Bibliothek

Topics in interpolation theory of rational matrix valued functions / ed. by I. Gohberg. – Basel ; Boston ; Berlin : Birkhäuser, 1988
(Operator theory ; Vol. 33)
ISBN 3-7643-2233-0
NE: Gochberg, Izrail' [Hrsg.]; GT

sd
2/3/89
JJ

Printed in Germany
ISBN 3-7643-2233-0
ISBN 0-8176-2233-0

TABLE OF CONTENTS

EDITORIAL INTRODUCTION

In this volume several interpolation problems for rational matrix functions are treated systematically. Many of the papers are concerned with different aspects of the basic problem to construct a rational matrix function with a prescribed null (zero) and pole structure and its applications. The realization approach which comes from system theory and control is used in all papers and serves as a tool to reduce problems of rational matrix functions to problems of linear operators. This approach also enables the considered interpolation problems to be solved explicitly. The volume has a strong control theory flavour and all papers are connected with problems in system or network theory. The first paper, "Realization and Interpolation of Rational Matrix Functions," by J.A. Ball, I. Gohberg and L. Rodman, serves as a general introduction to the field. It contains terminology, description and analysis of the null (zero) and pole structure of rational matrix functions and solutions of the above mentioned basic interpolation problem. It also contains the solution of a tangential Nevanlinna-Pick problem and Nevanlinna-Pick–Takagi problem. The exposition continues with the paper of I. Gohberg, M.A. Kaashoek, A.C.M. Ran, "Interpolation Problems for Rational Matrix Functions with Incomplete Data and Wiener-Hopf Factorization," in which is solved the interpolation problem for rational matrix function when the given null and pole data are incomplete. This paper contains a new view of the Wiener-Hopf factorization. In both papers mentioned above is treated the case where the null and pole data is prescribed only in finite points, hence only analytic and invertible at infinity solutions are considered. This restriction is removed in the next short paper of I. Gohberg and M.A. Kaashoek, "Regular Rational Matrix Functions with Prescribed Pole and Zero Structure." Here the approach is based on the analysis of the appropriate Moebius transform. Another approach in solving different matrix interpolation problems without restrictions at infinity, based on earlier ideas of Ball-Helton, is used in the paper of J.A. Ball, N. Cohen, A.C.M. Ran, "Inverse Spectral Problems for Regular Improper Rational Matrix Functions." In the latter paper also can be found applications to the model reduction problem from linear system theory. In the final two papers in the volume different problems of interpolation for unitary on a line or on a circle, rational matrix function, are considered. In the paper, D. Alpay, I. Gohberg, "Unitary Rational Matrix Functions," are described unitary matrix functions in an indefinite metric in terms of their realizations, and the problem of multiplicative decomposition is studied. A new approach to inertia theorems is also presented. In the last paper, I. Gohberg, S. Rubinstein, "Proper Contractions and their Unitary Minimal Completions," is solved another problem of interpolation-completion which appears in network theory and especially in Darlington synthesis. The solution is based on a detailed description of rational matrix function, all values of which, on the real line, are contractions and strict contractions at infinity. The latter result is also used for the analysis of fractional decompositions of contractions.

The papers in this volume deal with related problems and are based on the same general approach. The volume is therefore more in the style of a monograph, where each paper forms a chapter.

Operator Theory:
Advances and Applications, Vol. 33

REALIZATION AND INTERPOLATION OF RATIONAL MATRIX FUNCTIONS[1]

Joseph A. Ball(*), Israel Gohberg(**) and Leiba Rodman(*)

In this paper we generalize for matrix valued functions a number of well known interpolation problems for scalar rational functions and obtain explicit formulas for the solutions. The realization approach toward the study of rational matrix functions from systems theory serves here as the main tool. The main results recently appeared in the literature; here we give a more systematic and transparent exposition based exclusively on analysis in finite dimensional spaces.

CONTENTS

[1]Research of the authors was partially supported by the (*) National Science Foundation and (**) Air Force Office of Scientific Research (Grant AF0SR-87-0287).

1. INTRODUCTION

One of the basic interpolation problems from our point of view is the problem of building a scalar rational function if its poles and zeros with their multiplicities are given. If one assumes that the function does not have a pole or a zero at infinity, the formula which solves this problem is

$$r(z) = \alpha \frac{(z-z_1)^{n_1}(z-z_2)^{n_2}\dots(z-z_l)^{n_l}}{(z-w_1)^{m_1}(z-w_2)^{m_2}\dots(z-w_p)^{m_p}} \tag{1}$$

where $z_1, \dots, z_l$ are the given zeros with given multiplicates $n_1, \dots, n_l$ and $w_1, \dots, w_p$ are the given poles with given multiplicities $m_1, \dots, m_p$, and α is an arbitrary nonzero number.

An obvious necessary and sufficient condition for solvability of this simplest interpolation problem is that $z_j \neq w_k (1 \leq j \leq l, \quad 1 \leq k \leq p)$ and $n_1 + \cdots + n_l = m_1 + \cdots + m_p$.

The second problem of interpolation in which we are interested is to build a rational matrix function via its zeros which on the imaginary line has modulus 1. In the case the function is scalar, the formula which solves this problem is a Blaschke product, namely

$$u(z) = \alpha \Pi_{j=1}^{n} \left(\frac{z - z_j}{z + \bar{z}_j} \right)^{m_j} \tag{2}$$

where $|\alpha| = 1$, and the z_j's are the given zeros with given multiplicities m_j. Here the necessary and sufficient condition for existence of such $u(z)$ is that $z_p \neq -\bar{z}_q$ for $1 \leq p, q \leq n$.

The third problem, the most sophisticated in this series, is the Nevanlinna-Pick interpolation problem: given points $z_1, \dots, z_n$ in the upper-half plane and complex numbers $w_1, \dots, w_m$ find a rational function $r(z)$ analytic with values of modulus less than 1 in the right half plane such that $r(z_j) = w_j, \quad 1 \leq j \leq n$. A necessary and sufficient condition for solutions to exist is that the nxn matrix

$$\Lambda = [(1 - w_i \bar{w}_j)(z_i + \bar{z}_j)^{-1}]_{1 \leq i,j \leq n}$$

be positive definite. A formula which describes all solutions of this problem is given in Chapter 4, section 1.

Our aim is to generalize these three basic problems and the formulas which give their solutions to the case of matrix-valued rational functions. Even the formulation of the problems leads to the necessity to define the null and pole structure of a rational matrix function. Simple examples show difficulties and new phenomena appearing in the matrix case. For instance, the matrix function

$$W_0(z) = \begin{bmatrix} 1 & z^{-1} \\ 0 & 1 \end{bmatrix}$$

clearly has a pole at $z = 0$, but unexpectedly it also has a zero at $z = 0$ because the inverse

$$W_0^{-1}(z) = \begin{bmatrix} 1 & -z^{-1} \\ 0 & 1 \end{bmatrix}$$

clearly also has a pole at $z = 0$.

In the second chapter of the paper we define what we mean by null and pole structure of a rational matrix function and we solve the matrix version of the first problem. The main tool in obtaining formulas for the solution consists of a result from systems theory which implies that any rational matrix function with no pole at infinity can be represented in the form

$$W(z) = D + C(zI - A)^{-1}B \tag{3}$$

where A, B, C are matrices of appropriate sizes. The necessary results about representations of type (3) which we need are also included in this paper.

The third chapter is a continuation of the first. Here we specify the properties of the null structure and pole structure and realization for a rational matrix function which is unitary in an indefinite metric on the imaginary line. We also obtain a generalization of the formula (2) for the matrix case.

The fourth chapter uses the results of the first two and presents a solution together with formulas of a matrix version of the Nevanlinna Pick problem.

Finally, in the last chapter we consider the Takagi problem (which is a generalization of Nevanlinna-Pick), where one allows the interpolating function to have a certain number of poles in the underlying domain. Here the presentation is sketchy (in contrast with the previous chapters).

The aim of this paper is to propose a systematic treatment of several important interpolation problems for rational matrix-valued functions, based on the unifying realization approach; a novelty here is the fact that all results are obtained with finite dimensional techniques only. The main results of this paper are not new. Some of them were published recently and some a while ago. More detailed remarks and references are included in the Notes at the end of each chapter. We would like to mention that this paper is essentially based on the following sources [AG], [BGK], [BH], [BR], [GKLR], [GS]. We use results, methods and valuable hints from the later papers. However, the exposition here is self-contained and all used results are included with their proofs.

Interpolation problems for rational matrix functions play an important role in systems and network theory and control. Because of space limitations we have excluded completely applications to engineering problems. Full analysis of a large number of interpolation problems for rational matrix functions together with engineering applications will appear in the forthcoming monograph [BGR2]. A large part of this paper was written during the last semester of 1987, while the first and the second named authors were visiting the University of Maryland at College Park and the third named author was visiting the College of William and Mary, Williamsburg, Virginia and for a short time at College Park also. We would like to use this opportunity to express our gratitude for hospitality of these institutions.

Throughout the paper we use the notation $diag(Z_j)_{j=1}^k$ or $diag(Z_1, \ldots, Z_k)$ to denote block diagonal matrix with the blocks $Z_1, \ldots, Z_k$ along the main diagonal. A block column matrix with blocks $Z_1, \ldots, Z_k$ is denoted $col(Z_j)_{j=1}^k$. The image of an mxn matrix A (considered as a linear transformation $\mathbb{C}^n \to \mathbb{C}^m$ in the standard orthonormal bases) is denoted ImA. We denote a left (right) inverse of a matrix X by X^{-L} (resp. X^{-R}).

2. NULL AND POLE STRUCTURE OF RATIONAL MATRIX FUNCTIONS AND RELATED INTERPOLATION PROBLEMS

Throughout this chapter we consider mxn matrices $A(z)$ whose entries are rational functions of the complex variable z with complex coefficients, in short rational matrix functions. It will be often assumed that $A(z)$ is *regular*, i.e., the size of $A(z)$ is square and $detA(z)$ is not identically zero. In this case the inverse matrix function $A(z)^{-1}$ is rational as well.

In this chapter we describe null and pole structures of a rational matrix function, using the basic notions of null and pole functions. Another description of these structures is given in terms of a minimal realization of the rational matrix function. We also state and solve the basic interpolation problem, namely, construction of a rational matrix function when its null and pole functions are given.

2.1 Null and pole functions: definitions and examples

Let $A(z)$ be nxn (regular) rational matrix function, and let $z_0 \in \mathbb{C}$. We say that an analytic $\mathbb{C}^n$-valued function $\varphi(z)$ defined in a neighborhood of z_0 is a *right null function* for $A(z)$ at z_0 of order $k > 0$ if $\varphi(z_0) \neq 0$ and if $A(z)\varphi(z)$ is analytic at z_0 with a zero of order k at z_0, i.e. k is the maximal positive integer for which

$$\lim_{z \to z_0} [(z - z_0)^{-k+1} A(z)\varphi(z)] = 0.$$

If $A(z)$ is analytic at z_0 and $det A(z_0) \neq 0$, then there is no right null function for $A(z)$ at z_0. Thus, a right null function for $A(z)$ can exist only for a finite number of points z_0; namely, for the poles of $A(z)$ and for those points z_0 of analyticity of $A(z)$ where $A(z_0)$ is not invertible. We say that z_0 is a *zero* of $A(z)$ if there is a right null function for $A(z)$ at z_0.

We illustrate that notion with several examples.

Example 1.1. Let $A(z) = \begin{bmatrix} 1 & z^{-1} \\ 0 & 1 \end{bmatrix}$ and suppose $\varphi(z)$ is a right null function for $A(z)$ at $z_0 = 0$. Write $\varphi(z)$ in component form $\begin{bmatrix} \varphi^{(1)}(z) \\ \varphi^{(2)}(z) \end{bmatrix}$. Then for φ to be a right null function of order k we must have $\varphi(z_0) \neq 0$ and $A(z)\varphi(z) = 0(|z|^k)$ for z near zero, i.e.

$$\varphi^{(1)}(z) + z^{-1}\varphi^{(2)}(z) = 0(|z|^k)$$

$$\varphi^{(2)}(z) = 0(|z|^k).$$

If $\varphi_j = \begin{bmatrix} \varphi_j^{(1)} \\ \varphi_j^{(2)} \end{bmatrix}$ is the j-th Taylor coefficient of $\varphi(z)$ at 0, this says

(1) $\varphi_0^{(2)} = 0$ and $\varphi_j^{(1)} + \varphi_{j+1}^{(2)} = 0$ for $j \leq k-1$

and

(2) $\varphi_j^{(2)} = 0$ for $j \leq k-1$.

If $j > 1$ then (2) forces $\varphi_0^{(2)} = \varphi_1^{(2)} = 0$ whence from (1) we get $\varphi_0^{(1)} = -\varphi_1^{(2)} = 0$. Thus $\varphi(0) = \begin{bmatrix} \varphi_0^{(1)} \\ \varphi_0^{(2)} \end{bmatrix} = \begin{bmatrix} 0 \\ 0 \end{bmatrix}$, a contradiction. On the other hand, for $k = 1$, the general solution is

$\varphi_0^{(2)} = 0, \varphi_1^{(2)} = -\varphi_0^{(1)}$=arbitrary, $\varphi_1^{(1)}$=arbitrary, $\varphi_j^{(1)}$ and $\varphi_j^{(2)}$ arbitrary, for $j > 2$. Thus the general right null function for this $A(z)$ at $z = 0$ is

$$\varphi(z) = \begin{bmatrix} c \\ 0 \end{bmatrix} + z \begin{bmatrix} d \\ -c \end{bmatrix} + 0(|z|^2)$$

where d and c are arbitrary complex numbers such that $c \neq 0$. Note that the second Taylor coefficient $\begin{bmatrix} d \\ -c \end{bmatrix}$ cannot be chosen independently of the first Taylor coefficient $\begin{bmatrix} c \\ 0 \end{bmatrix}$. Also $det A(z) \equiv 1$, so zeros of $det A(z)$ do not coincide with the zeros of $A(z)$.

Example 1.2. Consider the scalar case:

$$A(z) = \frac{\Pi_{j=1}^{p}(z - z_j)^{\alpha_j}}{\Pi_{j=1}^{q}(z - s_j)^{\beta_j}}$$

where $z_1, \ldots, z_p$ are the distinct zeros of the numerator of $A(z)$, and $s_1, \ldots, s_q$ are the distinct zeros of the denominator of $A(z)$. We assume that

$$\{z_1, \ldots, z_p\} \cap \{s_1, \ldots, s_q\} = \emptyset.$$

There is no right null function for $A(z)$ at z_0 for $z_0 \notin \{z_1, \ldots, z_p\}$. On the other hand, for each z_j we have a null function (namely, a non-zero constant function) of order α_j. This follows easily by considering the Taylor series for $A(z)$ in a neighborhood of z_j:

$$A(z) = \sum_{k=\alpha_j}^{\infty} (z - z_j)^k a_k, \qquad a_{\alpha_j} \neq 0.$$

It is also easy to see that α_j is the only order of any right null function of $A(z)$ at z_j. Thus, the notion of a zero of a rational matrix function, when applied to the scalar case, coincides with the usual notion of a zero of a scalar rational function.

Example 1.3. Let $A(z)$ have the form $A(z) = zI - A$ where A is the $n \times n$ Jordan cell upper triangular with eigenvalue z_0. We analyze the form of a right null function $\varphi(z) = \sum_{j=0}^{\infty} \varphi_j (z - z_0)^j$ for $A(z)$ at z_0. Write $A(z)$ in the form $A(z) = (z_0 I - A) + (z - z_0)I$ and multiply out power series in $(z - z_0)$ to get

$$A(z)\varphi(z) = \sum_{j=0}^{\infty} [(z_0 I - A)\varphi_j + \varphi_{j-1}](z - z_0)^j$$

(where we set $\varphi_{-1} = 0$). Thus, for $j = 0$ we must have

$$(z_0 I - A)\varphi_0 = 0$$

so $\varphi_0 = ce_1$ for a nonzero constant c, where $e_1, e_2, \ldots, e_n$ are the standard basis vectors for $\mathbb{C}^n$. For φ to be a right null function of order at least two, we must have in addition

$$(z_0 I - A)\varphi_1 + \varphi_0 = 0$$

This then forces $\varphi_1 = ce_2$. To achieve a right null function of maximum possible order we choose

$$\varphi_j = ce_{j+1} \text{ for } 0 \leq j \leq n-1.$$

The resulting vector function

$$\varphi(z) = c[e_1 + (z - z_0)e_2 + \cdots + (z - z_0)^{n-1} e_n] + 0(z - z_0)^n$$

is a right null function at z_0 of order n. As $\varphi_{n-1} = ce_n$ is not in the image of $z_0 I - A$, there is no way to define φ_n so as to make $\varphi(z)$ a right null function at z_0 of order larger than n. Thus n is the largest possible order for a right null function $\varphi(z)$, and we have shown that every right null function of order n has the form (1) for a $c \neq 0$. Note that for this example the Taylor coefficients $\varphi_0, \varphi_1, \ldots, \varphi_{n-1}$ for the right null function φ coincide with a Jordan chain of vectors for the matrix A. This gives a connection between right null functions for analytic matrix functions and the theory of Jordan canonical form for finite matrices. ■

It is easy to see that if $\varphi^{(1)}(z), \ldots, \varphi^{(m)}(z)$ are right null functions for $A(z)$ at z_0 and $c_1, \ldots, c_m$ are complex numbers, then the linear combination $\varphi(z) = c_1\varphi^{(1)}(z) + \cdots + c_m\varphi^{(m)}(z)$ is again a right null function as long as $\varphi(z_0) \neq 0$; if no $c_j = 0$, then the order of $\varphi(z)$ as a null function is at least the minimum of the orders of the null function $\varphi^{(1)}(z), \ldots, \varphi^{(m)}(z)$. If φ is a right null function and $s(z)$ is an arbitrary scalar analytic function with $s(z_0) \neq 0$, then $\tilde{\varphi}(z) = s(z)\varphi(z)$ is also a right null function of the same order.

If $0 \neq x$ is a column vector such that $x = \varphi(z_0)$ for some right null function for the rational matrix function $A(z)$ at z_0, we say that x is a (right) eigenvector for $A(z)$ at z_0. The set of all eigenvectors for $A(z)$ at z_0 together with 0 is a linear subspace of $\mathbb{C}^n$ which we call the (right) eigenspace for $A(z)$ at z_0; note that we cannot identify the eigenspace as $Ker\ A(z_0)$, since $A(z_0)$ is not defined as a finite matrix if A has a pole at z_0. For example, from Example 1.1, we see that the right eigenspace for the matrix function $A(z) = \begin{bmatrix} 1 & z^{-1} \\ 0 & 1 \end{bmatrix}$ at the point $z_0 = 0$ is $\{[c\ 0]^T : c \in \mathbb{C}\}$.

There is a completely analogous notion of pole functions. We say that the analytic vector function $\psi(z)$ with $\psi(z_0) \neq 0$ is a *right pole function* for the rational matrix function $A(z)$ at the point z_0 of order $k > 0$ if there is an analytic vector function $\varphi(z)$ such that $A(z)\varphi(z) - (z - z_0)^{-k}\psi(z)$ is analytic at z_0. As $A(z)$ is regular, the inverse function $A(z)^{-1}$ is rational as well.

It is easy to see that $y(z)$ is a right pole function for $A(z)$ at z_0 if and only if $y(z)$ is a right null function for $A^{-1}(z)$ at z_0. Thus the analysis of right pole functions for a given rational matrix function reduced to the study of right null functions for the inverse function. We illustrate with an example.

Example 1.4. As in Example 1.1, let $A(z) = \begin{bmatrix} 1 & z^{-1} \\ 0 & 1 \end{bmatrix}$. Then if $\varphi(z) = \begin{bmatrix} \varphi_0^{(1)} \\ \varphi_0^{(2)} \end{bmatrix} + \begin{bmatrix} \varphi_1^{(1)} \\ \varphi_1^{(2)} \end{bmatrix} (z - z_0) + 0(|z - z_0|^2)$, we see that near z_0,

$$A(z)\varphi(z) = z^{-1} \begin{bmatrix} \varphi_0^{(2)} \\ 0 \end{bmatrix} + \begin{bmatrix} \varphi_0^{(1)} + \varphi_1^{(2)} \\ \varphi_0^{(2)} \end{bmatrix} + 0(|z|).$$

Thus a pole function for $A(z)$ at 0 can have order at most 1, in which case

$$\psi(z) = \begin{bmatrix} c \\ 0 \end{bmatrix} + z \begin{bmatrix} d \\ c \end{bmatrix} + 0(|z|^2) \tag{1}$$

where $0 \neq c$ and d are any complex numbers, is a pole function for $A(z)$ for order 1. Alternatively, we may note that $A^{-1}(z) = \begin{bmatrix} 1 & -z^{-1} \\ 0 & 1 \end{bmatrix}$ and calculate right null functions for $A^{-1}(z)$. ■

Finally, we introduce the notions of left null and left pole functions. By a *left null function* $\varphi^*(z) = \varphi_0^* + \varphi_1^*(z - z_0) + \ldots$ for the rational matrix function $A(z)$ at the point z_0, we mean a row vector function $\varphi^*(z)$ analytic at z_0 with $\varphi^*(z_0) \neq 0$ for which the analytic row vector function $\varphi^*(z)A(z)$ has value 0 at z_0; the order of the zero of $\varphi^*(z)A(z)$ at z_0 is then the order of $\varphi^*(z)$ as a left null function. Note that the row vector function $\varphi^*(z)$ is a left null function for $A(z)$ at z_0 if and only if $\varphi(z) = \varphi^*(z)^T$ is right null function for $A(z)^T$ at z_0 of the same order. In this way all the results concerning right null functions can be adapted to the setting of left null functions. Analogously one can introduce left pole functions. A *left pole function* of order k for a rational matrix function $A(z)$ at z_0 is a row vector function $\psi^*(z)$ analytic at z_0 with $\psi^*(z_0) \neq 0$ such that $(z-z_0)^{-k}\psi^*(z) = \varphi^*(z)A(z)$ for some analytic row vector function $\varphi^*(z)$. Equivalently, $\psi^*(z)$ is a left null function for $A^{-1}(z)$ of order k at z_0.

2.2 Canonical sets of null and pole functions

Fix an nxn rational matrix function $A(z)$ and a zero at the point z_0 in $\mathbb{C}$. To organize the collection of all right null functions of $A(z)$ at z_0 we now define the notion of a *canonical set of right null functions.* We shall see later (see Corollary 2.4) that the set of possible orders for a right null function for $A(z)$ at z_0 is bounded. Let $\varphi^{(1)}$ be a right null function having this maximum order. If the eigenvector $\varphi^{(1)}(z_0)$ already spans the right eigenspace of $A(z)$ at z_0, then the single right null function $\{\varphi^{(1)}(z)\}$ forms a canonical set of right null functions. Otherwise we choose a direct complement K_1 of $span\{\varphi^1(z_0)\}$ in the right eigenspace and let $\varphi^{(2)}(z)$ be a right null function of maximum order such that $\varphi^{(2)}(z_0) \in K_1$. Continuing in this way we construct right null functions $\varphi^{(1)}(z), \varphi^{(2)}(z), \ldots, \varphi^{(m)}(z)$ with nonincreasing orders $k_1 \geq k_2 \geq \cdots \geq k_m$ such that $\{\varphi^{(1)}(z_0), \varphi^{(2)}(z_0), \ldots, \varphi^{(m)}(z_0)\}$ is a basis for the eigenspace. A collection of right null functions $\{\varphi^{(1)}(z), \varphi^{(2)}(z), \ldots, \varphi^{(m)}(z)\}$ with these properties we call a *canonical set of right null functions* for $A(z)$ at z_0.

For instance, in Example 1.1 the function

$$\varphi(z) = \begin{bmatrix} 1 \\ 0 \end{bmatrix} + z \begin{bmatrix} 0 \\ -1 \end{bmatrix}$$

by itself forms a canonical set of right null functions of $A(z)$ at $z_0 = 0$. In the scalar case the function $\varphi(z) = 1$ forms a canonical set of null functions at each zero z_j of $A(z)$.

Analogously the notions of canonical sets of left null functions, of right pole functions, and of left pole functions are defined. So the row functions $\psi_r^*(z), \ldots, \psi_1^*(z)$ form a canonical set of left pole functions for $A(z)$ at z_0 if and only if the transposed functions $\psi_1^*(z)^T, \ldots, \psi_r^*(z)^T$ form a canonical set of right null functions for $(A(z)^{-1})^T$ at z_0.

Our next goal is to compute a canonical set of null and pole functions at a point z_0 for a general rational matrix function $A(z)$. As we shall see, the construction reduces to piecing together the scalar case, the behavior under direct sums and the behavior under multiplication.

We start with the behavior under direct sums.

Example 2.1. We suppose that $A(z)$ has a block diagonal form

$$A(z) = diag.(A_j(z))_{j=1}^m$$

where (say) each $A_j(z)$ has size $n_j \times n_j$. Then a right null function $\varphi(z)$ for $A(z)$ can be partitioned conformally with respect to the block diagonal structure of $A(z)$ as $\varphi(z) = col.(\varphi^{(j)}(z))_{j=1}^m$ where $\varphi^{(j)}(z)$ is C^{n_j}-valued. We first observe that if $\varphi^{(k)}(z)$ is a null function for $A_k(z)$ at z_0, then $\varphi(z) = col(\delta_{kj}\varphi^{(k)}(z))_{j=1}^m(\delta_{kj} = 1$ for $k = j$ and 0 otherwise) is a right null function for $A(z)$ of the same order. Conversely, if $\varphi(z) = col(\varphi^{(j)}(z))_{j=1}^m$ is a right null function for $A(z)$ of order (say) k, then $\varphi^{(j)}(z)$ is a right null function for $A_j(z)$ of order at least k for each j such that $\varphi^{(j)}(z_0) \neq 0$. Thus the maximum order for a right null function for $A(z)$ is equal to the maximum of the maximum orders of the right null functions for the $A_j(z)'s$. In this way it is possible to construct a canonical set of right null functions for $A(z)$ from canonical sets of right null functions for all the blocks $A_j(z)$. Analogously one considers canonical sets of left null functions and of right and left pole functions. ∎

Next, to build the canonical set we need the local Smith form of a rational matrix function which is given in the following theorem.

THEOREM 2.1. *Let $A(z)$ be an nxn rational matrix function with $det A(z)$ not identically zero, and let $z_0 \in \mathbb{C}$. Then $A(z)$ admits the following representation:*

$$A(z) = E(z) diag((z - z_0)^{n_1}, \dots, (z - z_0)^{n_n}) F(z), \tag{2.1}$$

where $E(z)$ and $F(z)$ are nxn rational matrix functions analytic and invertible at z_0, and

$$n_1 \geq \cdots \geq n_n \tag{2.2}$$

are integers. Moreover, the integers $n_1, \dots, n_n$ are uniquely determined by $A(z)$ and z_0 (subject to condition (2.2)), and one of $E(z)$ and $F(z)$ in (2.1) can be chosen to be a matrix polynomial with constant non-zero determinant.

The representation (2.1) is called *local Smith form* of $A(z)$ at z_0. It should be noted that an analogous result holds also for rectangular rational matrix functions, or for square rational matrix functions with determinant identically zero. However, the framework of regular rational matrix functions is adequate for the purposes of this paper.

Theorem 2.1 is well-known; its proof can be obtained by repeating the arguments in a proof of the Smith form for matrix polynomials (see, e.g., proof of Theorem S1.1 in [GLR1]).

The integers $n_1 \geq \cdots \geq n_n$ will be called the *partial multiplicities* of $A(z)$ at z_0.

Theorem 2.1 suggests a way to examine null and pole functions of a rational matrix function $A(z)$. As a first step, study the transformation of null and pole functions at z_0 if the rational matrix is pre- and post- multiplied by rational matrix functions analytic and invertible at z_0. Secondly, study the null and pole functions of $diag((z-z_0)^{n_1}, \ldots, (z-z_0)^{n_n})$. The first problem is settled as follows.

THEOREM 2.2. *Suppose A_1, E, F are nxn rational matrix functions, and E and F are analytic at z_0 with $detE(z_0) \neq 0, detF(z_0) \neq 0$. Let $A_2(z) = E(z)A_1(z)F(z)$. Then $\varphi(z)$ is a right null function for $A_2(z)$ at z_0 if and only if $F(z)\varphi(z)$ is a right null function for $A_1(z)$ at z_0 of the same order. Moreover, if $\{\varphi^{(1)}(z), \ldots, \varphi^{(m)}(z)\}$ is a canonical set of right null function for $A_2(z)$, then $\{F(z)\varphi^{(1)}(z), \ldots, F(z)\varphi^{(m)}(z)\}$ is a canonical set of right null functions for $A_1(z)$.*

PROOF: If $\varphi(z_0) \neq 0$ then $F(z_0)\varphi(z_0) \neq 0$ since $detF(z_0) \neq 0$. If $A_2(z)\varphi(z)$ has a zero of order k at z_0, then the same holds for $A_1(z) \cdot F(z)\varphi(z) = E(z)^{(-1)} \cdot A_2(z)\varphi(z)$ since $E(z)^{-1}$ is analytic at z_0 with $detE(z_0)^{-1} \neq 0$. This essentially proves the first assertion, and the second assertion is an easy consequence, again by using $detF(z_0) \neq 0$. ∎

We leave it to the reader the formulation and proof of statements analogous to Theorem 2.2 concerning left null functions, right pole functions and left pole functions.

Consider now the null functions for the local Smith form. Let $A(z) = diag.((z-z_0)^{\kappa_2}, \ldots, (z-z_0)^{\kappa_n})$ where $\kappa_1 \geq \kappa_2 \geq \cdots \geq \kappa_n$ are integers. Then the vector function $\varphi(z) = col.(\varphi^{(i)}(z))_{i=1}^{n}$ is a right null function for $A(z)$ of order k if and only if $\varphi(z_0) \neq 0$ and $\varphi^{(i)}(z)$ has a zero at

z_0 of order at least $k - \kappa_i$ for $1 \leq i \leq n$. When $k > \kappa_1$ the second condition is incompatible with the first; thus κ_1 is the maximum possible order of a null function (assuming $\kappa_1 > 0$). If some $\kappa_i' s$ are negative, then the second condition places restriction on some Taylor coefficients $\varphi_j := \frac{1}{j!}\varphi^{[j]}(z)$ of φ at z_0 for $j > k$. Assume that $k_1 > k_2 > \cdots > k_n > 0 > k_{n+1} > \cdots > k_m$ are the distinct nonzero partial multiplicities of $A(z)$, each with multiplicity $n_1, n_2, \ldots, n_r, n_{r+1}, \ldots, n_m$ (so $k_1 = \kappa_1 = \cdots = \kappa_{n_1} > k_2 = \kappa_{n_1+1} = \cdots = \kappa_{n+n_2} > \cdots > k_m = \kappa_{n+1-n_m} = \kappa_{n+2-n_m} = \cdots = \kappa_n$). We wish to describe an arbitrary canonical set of right null function for $A(z)$:

THEOREM 2.3. *In the above notation,* $\{\varphi^{(1)}, \ldots, \varphi^{(\hat{r})}\}$ *is a canonical set of right null functions for* $A(z)$ *at* z_0 *if and only if* $\hat{r} = r$, $\{\varphi_0^{(1)}, \ldots, \varphi_0^{(r)}\}$ *is linearly independent, and*

$$\varphi_\alpha^{(i)} \in span\{e_1, \ldots, e_{n_1}, \ldots, e_{n_1+n_2+\cdots+n_j}\}$$

where $j = max\{m : k_m + \alpha \geq \kappa_i\}$, *and in this case* $\varphi^{(i)}$ *is a right null function of precise order* κ_i.

PROOF: By definition, $\varphi^{(1)}(z_0) \neq 0$ and $A(z)\varphi^{(1)}(z)$ has a zero at z_0 of maximum possible order subject to this constraint. By the diagonal form of $A(z)$ it is easy to check that this forces $0 \neq \varphi^{(1)}(z_0) \in span\{e_1, e_2, \ldots, e_{n_1}\}$ and more generally $\varphi_\alpha^{(1)} \in span\{e_1, \ldots, e_{n_1}, e_{n_1+1}, \ldots, e_{n_1+n_2+\cdots+n_j}\}$ where $j = max\{m : k_m + \alpha \geq \kappa_1\}$, and then the order is precisely κ_1. Conversely, any such $\varphi^{(1)}(z)$ is a right null function of the maximum possible order $\kappa_1 = k_1$.

In general we claim: $\{\varphi^{(1)}, \ldots, \varphi^{(K)}\}$ satisfies

(i) for $1 \leq i \leq K$, $\varphi^{(i)}(z)$ *is a right null function for* $A(z)$ *at* z_0 *of maximum possible order such that* $\varphi_0^{(i)}$ *is not in span* $\{\varphi_0^{(1)}, \ldots, \varphi_0^{(i-1)}\}$

if and only if

(ii) $\{\varphi_0^{(1)}, \ldots, \varphi_0^{K}\}$ *is linearly independent*

and

(iii) for $1 \leq i \leq K$, $\varphi_\alpha^{(i)} \in span\{e_1, \ldots, e_{n_1}, \ldots, e_{n_1+n_2+\cdots+n_j}\}$ *where* $j = max\{m : k_m + \alpha \geq \kappa_i\}$, *and in this case* $\varphi^{(i)}$ *is a right null function of precise order* κ_i.

We prove the claim by induction on K. The case $K = 1$ was verified above. Assume that the claim holds for $i \leq K$. We wish to prove that it holds for $i = K + 1$.

Case 1. $K = n_1 + n_2 + \cdots + n_\beta$. By condition (ii), $\{\varphi_0^{(i)} : 1 \le i \le K\}$ is linearly independent. Since $\kappa_1 \ge \kappa_2 \ge \cdots \ge \kappa_K$, a particular consequence of (iii) is that

$$\varphi^{(i)} \in span\{e_1, \ldots, e_{n_1}, \ldots, e_{n_1+n_2+\cdots+n_j}\}$$

for $1 \le i \le K$ if $j = max\{m : k_m \ge \kappa_K\}$. Since $K = n_1 + n_2 + \cdots + n_\beta$, $\kappa_K = k_\beta$ and hence $j = \beta$. We have now established

$$\varphi_0^{(i)} \in span\{e_1, \ldots, e_{n_1}, \ldots, e_{n_1+n_2+\cdots+n_\beta}\}$$

for $1 \le i \le K$. By dimension count

$$span\{\varphi_0^{(i)} : 1 \le i \le K\} = span\{e_i : 1 \le i \le K\}.$$

By (ii) we must therefore choose $\varphi^{(K+1)}$ so that $\varphi_0^{(K+1)}$ is linearly independent of $span\{e_i : 1 \le i \le K\}$. By the diagonal form of $A(z)$ we read off that the maximal possible order for a right null function $\varphi^{(K+1)}$ for $A(z)$ subject to this linear independence constraint is $k_{\beta+1}$, and this is achieved as long as

$$\varphi_\alpha^{(K+1)} \in span\{e_1, \ldots, e_{n_1}, \ldots, e_{n_1+n_2+\cdots+n_j}\}$$

where

$$j = max\{m : k_m + \alpha \ge \kappa_{K+1} = k_{\beta+1}\}.$$

Conversely, any such $\varphi^{(K+1)}(z)$ may serve as the $(K+1)$-st function in a set of $K+1$ null functions for $A(z)$ satisfying (i).

Case 2. $n_1 + \cdots + n_\alpha < K < n_1 + \cdots + n_{\alpha+1}$. In this case also $\varphi_0^{(i)} \in span\{e_i : 1 \le i \le n_j\}$ for $1 \le i \le k$ where $j = max\{m : k_m \ge \kappa_K\}$. For this case $\kappa_K = k_{\alpha+1}$, so $j = \alpha + 1$. In this case by dimension count $span\{\varphi_0^{(i)} : 1 \le i \le K\}$ does not fill out $span\{e_i : 1 \le i \le n_1 + \cdots + n_{\alpha+1}\}$. The maximal possible order for a right null function $\varphi^{(K+1)}$, subject to the restriction that $\varphi_0^{(K+1)}$ is linearly independent of $span\{\varphi_0^{(i)} : 1 \le i \le K\}$ is therefore $k_{\alpha+1} = \kappa_{K+1}$, and this is achieved as long as $\varphi_0^{(K+1)}$ is chosen linearly independent of $span\{\varphi_0^{(i)} : 1 \le i \le K\}$ but still in $span\{e_i : 1 \le i \le n_1 + \cdots + n_{\beta+1}\}$, and also $\varphi_\alpha^{(K+1)} \in span\{e_i : 1 \le i \le n_1 + \cdots + n_j\}$ where

$j = max\{m : k_m + \alpha \geq k_{\beta+1} = \kappa_{(K+1)}\}$. Converely, any right null function $\varphi^{(K+1)}$ of maximal possible order $k_{\beta+1}$ such that $\varphi_0^{(K+1)}$ is linearly independent of $span\{\varphi_0^{(i)} : 1 \leq i \leq K\}$ must arise in this way.

Next note that the inductive construction continues until $K = n_1 + n_2 + \cdots + n_j = r$. If $\varphi(z_0) \notin span\{e_1, e_2, \ldots, e_r\}$, then $A(z)\varphi(z)$ is not analytic at z_0, or it is but $A(z_0)\varphi(z_0) \neq 0$; thus the set of right null functions $\{\varphi^{(i)}(z) : 1 \leq i \leq r\}$ cannot be enlarged to a larger set and still satisfy (i). Thus the number of right null functions in a canonical set is precisely r and all canonical sets are of the form in the theorem. ∎

COROLLARY 2.4. *(a) The orders of right (resp. left) null functions in a canonical set of right (resp. left) null functions of $A(z)$ at z_0 coincide with the positive partial multiplicities of $A(z)$ at z_0.*

(b) The orders of the right (resp. left) pole functions in a canonical set of right (resp. left) pole functions of $A(z)$ at z_0 coincide with the absolute values of negative partial multiplicities of $A(z)$ at z_0.

For the proof combine Theorems 2.2 and 2.3. We remark that the results of this and the previous section apply to meromorphic matrix functions as well.

2.3 Null and pole data and the problem of interpolation

In this section we introduce an alternative language for dealing with the local zero and pole structure of a rational matrix function.

Suppose $A(z)$ is a regular rational matrix function with Laurent expansion in a deleted neighborhood of z_0 $A(z) = \sum_{j=-q}^{\infty}(z - z_0)^j A_j$. Here it is assumed that $q \geq 0$. Suppose that the analytic vector function $\varphi(z)$ with Taylor series expansion $\varphi(z) = \sum_{j=0}^{\infty} \frac{1}{j!}\varphi^{[j]}(z_0)(z - z_0)^j$ is a right null function for $A(z)$ at z_0. Thus $A(z)\varphi(z)$ has analytic continuation to z_0 which has a zero of order k at z_0, where $A(z)\varphi(z)$ has Laurent series expansion

$$A(z)\varphi(z) = \sum_{i=1}^{\infty}\left[\sum_{l=0}^{i} A_{-q+l}\frac{1}{(i-l)!}\varphi^{[i-l]}(z_0)\right](z - z_0)^{i-q}.$$

Thus, if φ is a right null function of order k for $A(z)$ at z_0 we have

$$A_{-q}\frac{1}{i!}\varphi^{[i]}(z_0) + \cdots + A_{-q+i}\varphi(z_0) = 0, \quad i = 1, \ldots, q + k - 1. \tag{3.1}$$

Let us now say that our ordered set of column vectors $(x_0, \ldots, x_{k-1})$ $(k \geq 0)$ in $\mathbb{C}^n$ is a right null chain for the rational matrix function $A(z)$ at z_0 if there exists a right null function $\varphi(z)$ for $A(z)$ at z_0 of order k having its first k Taylor coefficients satisfy

$$\frac{1}{i!}\varphi^{[i]}(z_0) = x_i, \qquad 0 \leq i \leq k-1.$$

Equivalently, from (3) we see that $(x_0, \ldots, x_{k-1})$ is a null chain for $A(z)$ at z_0 if there exist vectors $x_k, \ldots, x_{q+k-1}$ in $\mathbb{C}^n$ such that

$$A_{-q}x_i + \cdots + A_{-q+i}x_0 = 0, i = 0, \ldots, q+k-1. \tag{3.2}$$

The number k is called the length of the chain. The leading vector x_0 in the chain is the value $\varphi(z_0)$ of the associated right null function for $A(z)$ at z_0, i.e. an eigenvector for $A(z)$ at z_0. Given an eigenvector x_0 for W at z_0, of course, there may in general be many right null chains for $A(z)$ at z_0 which have x_0 as a first vector.

Note that condition (32) for a right null chain does not depend on the choice of q, i.e. writing

$$W(z) = \sum_{j=-r}^{\infty} (z - z_0)^j A_j \tag{3.3}$$

with $r > q$ and $A_{-r} = \cdots = A_{-q+1} = 0$, we obtain precisely the same right null chains by using (3.3). Note that if $(x_0, x_1, \ldots, x_{k-1})$ is a right null chain then there exist $x_k, \ldots, x_{q+k-1}$ such that

$$\varphi(z) = \sum_{i=0}^{q+k-1} (z - z_0)^i x_i \tag{3.4}$$

is a right null function for $A(z)$ at z_0; if $A(z)$ happens to be analytic at z_0, then for any choice of $x_k, \ldots, x_{q+k-1}$ the function (3.4) is a right null function for $A(z)$ at z_0. Generally, however, there may exist other choices of $x_k, x_{k+1}, \ldots, x_{q+k-1}$ for which (3.4) is not a right null function of order k.

We emphasize that right null functions are not determined by the right null chains (as is the case for analytic functions). Indeed, if $A(z)$ has a pole and a zero at z_0, a number of Taylor coefficients beyond the right null chain appear in the equations that define a right null function, and hence generally cannot be arbitrarily chosen.

By a *canonical set of right null chains* we mean an ordered set of vectors

$$x_0^{(1)}, \dots, x_{r_1-1}^{(1)}; x_0^{(2)}, \dots, x_{r_2-1}^{(2)}; \dots; x_0^{(p)}, \dots, x_{r_p-1}^{(p)} \tag{3.5}$$

such that a corresponding set of right null functions

$$\{\varphi^{(j)}(z) = \sum_{i=0}^{r+q-1} (z - z_0)^i x_i^{(j)} : j = 1, \dots, p\}$$

is a canonical set of right null functions for $A(z)$ at z_0.

The information contained in (3.6) can be put into a pair of matrices. In order to do this let X_i be the $m \times r_i$ matrix whose $j-th$ column is $x_{j-1}^{(i)}$. Let J_i be the $r_i \times r_i$ Jordan block with z_0 on the main diagonal. Put

$$X = [X_1, \dots, X_p], \qquad J_1 \oplus \cdots \oplus J_p. \tag{3.6}$$

So J is a Jordan matrix with a single eigenvalue. Note that $dim\ Ker\ (z_0 I - J) = dim$ (right eigenspace of $A(z)$ at z_0).

The pair (X, J) is called a right null pair of A corresponding to the zero z_0. Note that X is the list of vectors in a canonical set of right null chains organized as a matrix, and the Jordan matrix J can be thought of as a bookkeeping device which encodes where one chain ends and the next begins in X and which point z_0 we are considering.

Also, any pair (Z, T) *similar* to (X, J) (i.e. such that $Z = XS, T = S^{-1}JS$ for some invertible matrix S) will be called the *right null pair* of A at z_0.

A pair of matrices (Z, T), where Z has size $n \times p$, and T has square size $p \times p$, is said to be a *null kernel pair* if $\cap_{i=0}^{l-1} ker(ZT^i) = \{0\}$ for some integer $l \geq 1$. The size p of T is said to be the *order* of the pair (Z, T).

THEOREM 3.1. *If (Z, T) is a right null pair for a regular rational matrix function $A(z)$ at z_0, then (Z, T) is a null kernel pair with order equal to the sum of the positive partial multiplicities for $A(z)$ at z_0.*

PROOF: The statement concerning the order of (Z, T) follows from Corollary 2.4(a).

To prove that (Z,T) is a null kernel pair we can assume that $Z = X, T = J$ are given by (3.6). It is elementary to see that a vector $v \in \mathbb{C}^p$ is in $Ker\ col(XJ^i)_{i=0}^{l-1}$ if and only if v is in $Ker\ col(X(J - z_0I^i)_{i=0}^{l-1}$, and thus $rank\ col(XJ^i)_{i=0}^{l-1} = rank\ col(X(J - z_0I)^i)_{i=0}^{l-1}$. As $J - z_0I$ is the shift matrix, we see that XJ^i has the form

$$XJ^i = [\underbrace{0,\ldots,0}_{i}, x_0^{(1)},\ldots,x_{r_1-i-1}^{(1)}; \underbrace{0,\ldots,0}_{i}, x_0^{(2)},\ldots,x_{r_2-i-1}^{(2)};\ldots;\underbrace{0,\ldots,0}_{i}, x_0^{(j)},\ldots,x_{r_j-i-1}^{(j)}]$$

Using the linear independence of the leading vectors $\{x_0^{(1)}, x_0^{(2)}, \ldots, x_0^{(j)}\}$ in the chains, one can now deduce that $col(X(J - z_0I)^i)_{i=0}^{r_1-1}$ has $r_1 + r_2 + \cdots + r_j = p$ linearly independent columns, and thus has full rank p. ∎

It turns out that the null pair of $A(z)$ at z_0 is determined uniquely up to similarity:

THEOREM 3.2. *If* (Z_1, T_1) *and* (Z_1, T_1) *are two right null pairs of* $A(z)$ *at* z_0, *then*

$$Z_1 = Z_2S, \qquad T_1 = S^{-1}T_2S$$

for some unique invertible matrix S. *Moreover, the sizes of the blocks in the Jordan form of* T *coincide with the positive partial multiplicities for* $A(z)$ *at* z_0.

PROOF: The second statement follows again from Corollary 2.4(a). Because of this we can assume that $T_1 = T_2 = J = J_1 \oplus \cdots \oplus J_p$ is given as in (3.6), and the matrices Z_1 and Z_2 are constructed as X in (3.6) using (possibly different) canonical sets of right null functions of $A(z)$ at z_0.

Let us show that

$$Im\ col(Z_j(T - \lambda_0 I)^i)_{i=0}^{l-1} = Span\{(x_k, x_{k-1},\ldots,x_0,0,\ldots,0)^T | x_0,\ldots,x_k \quad (3.7)$$

$$\text{is a right null chain of } A(z) \text{ at } z_0\}, j = 1,2.$$

Here l is such that the columns of both matrices $col(Z_2T^i)_{i=0}^{l-1}$ and $col(Z_1T^i)_{i=0}^{l-1}$ are linearly independent.

Indeed, by using Theorem 2.3, (3.7) is easily verified when

$$A(z) = diag((z - z_0)^{\kappa_1}, \ldots (z - z_0)^{\kappa_n})$$

for some integers $\kappa_1, \ldots, \kappa_n$.

Consider now the general case, and let $A(z) = E(z) \cdot diag((z-z_0)^{k_1}, \ldots, (z-z_0)^{k_n})F(z)$ be a local Smith form of $A(z)$. Assume

$$Z_1 = [x_0^{(1)} \ldots x_{r_1-1} x_0^{(2)} \ldots x_{r_2-1}^{(2)} \ldots x_0^{(p)} \ldots x_{r_p-1}^{(p)}] \tag{3.8}$$

and put

$$y_r^{(i)} = \sum_{j=0}^{r} \frac{1}{j!}(F)^{(i)}(z_0) x_{r-j}^{(i)}; \tag{3.9}$$

let $\tilde{Z}_1$ be the matrix obtained from Z_1 by replacing $x_r^{(i)}$ by $y_r^{(i)}$. Then (cf. Theorem 2.2) $(\tilde{Z}_1, T)$ is a right null pair for $D(z) = diag((z-z_0)^{K_1}, \ldots, (z-z_0)^{K_n})$ at z_0. Since (3.7) is already proved for $D(z)$

$$Im\ col(\tilde{Z}_1(T - \lambda_0 I)^i)_{i=0}^{l-1} = span\{(y_k, y_{k-1}, \ldots, y_0, 0, \ldots, 0)^T | y_0, \ldots, y_k \tag{3.10}$$

$$\textit{is a right null chain of } D(z) \textit{ corresponding to } z_0\}$$

Applying the matrix

$$\begin{bmatrix} F^{-1}(z_0) & (F^{-1}(z_0))^{(1)} & \ldots & (F^{-1}(z_0))^{(l-1)} & \\ 0 & F^{-1}(z_0) & \ldots & (F^{-1}(z_0))^{(l-2)} & \\ & \ldots & \ldots & \ldots & \ldots \\ 0 & 0 & \ldots & F^{-1}(z_0) & \end{bmatrix}$$

to both sides of (3.10), we obtain (in view of (3.9), the equality (3.7) for $j = 1$ (and analogously for $j = 2$).

In particular, $Im\ col(Z_1(T - z_0 I)^j)_{j=0}^{l-1} = Im\ col(Z_2(T - \lambda_0 I)^j)_{j=0}^{l-1}$. So there exists a $p \times p$ matrix S (where p is the size of T) such that

$$col(Z_1(T - z_0 I)^j)_{j=0}^{l-1} S = col(Z_2(T - \lambda_0 I)^j)_{j=0}^{l-1}. \tag{3.11}$$

As the columns of $col(Z_q(T - z_0 I)^j)_{j=0}^{l-1}, q = 1, 2$ are linearly independent, it is easily seen that S is invertible and uniquely determined. In fact,

$$S = [col(Z_1(T - z_0 I)^j)_{j=0}^{l-1}]^{-L} [col(Z_2(T - z_0 I)^j)_{j=0}^{l-1}],$$

and S^{-1} is given by a similar formula with the roles of Z_1 and Z_2 interchanged.

The first block row in (3.11) implies that $Z_1S = Z_2$. To prove that $TS = ST$ let us consider the matrices $Q_i := col(Z_i(T - z_0I)^j)_{j=0}^{l}, i = 1, 2$. As above, we have $Q_1\tilde{S} = Q_2$ for some $p \times p$ matrix $\tilde{S}$. By uniqueness of S we obtain that in fact $\tilde{S} = S$. Then

$$col(Z_1(T - z_0I)^j)_{j=0}^{l-1} \cdot (T - z_0I)S = col(Z_2(T - z_0I)^j)_{j=0}^{l-1}(T - z_0I)$$

$$= col(Z_1(T - z_0I)^j)_{j=0}^{l-1}S(T - z_0I).$$

Since the columns of $col(Z_1(T - z_0I)^j)_{j=0}^{l-1}$ are linearly independent, the equality $(T - z_0I)S = S(T - z_0I)$, or $TS = ST$, follows. So $Z_1S = Z_2$ and $TS = ST$, i.e. the pairs (Z_1, T) and (Z_2, T) are similar. ∎

We need also the notion of right null pair with respect to several zeros. Let σ be a non-empty set in the complex plane, and let $z_1, \ldots, z_r$ be all the distinct zeros (if any) of $A(z)$ in σ. A *right null pair* of $A(z)$ *with respect to* σ is, by definition, any pair of matrices (C_ζ, A_ζ) which is similar to $([Z_1Z_2 \ldots Z_r], T_1 \oplus \cdots \oplus T_r)$, where (Z_j, T_j) is a right null pair of $A(z)$ corresponding to z_j, for $j = 1, \ldots, r$.

In particular, a right null pair of $A(z)$ with respect to the whole complex plane will be called a *global right null pair* for $A(z)$.

THEOREM 3.3. *A right null pair (C_ζ, A_ζ) of $A(z)$ with respect to σ is a null kernel pair, and is unique up to similarity.*

PROOF: We can assume that

$$C_\zeta = [Z_1Z_2 \ldots Z_r]; \; A_\zeta = T_1 \oplus \cdots \oplus T_r, \tag{3.12}$$

where (Z_j, T_j) is a right null pair for z_j, and $z_1, \ldots, z_r$ are all the distinct zeros of $A(z)$ in σ. By Theorem 3.1,

$$\cap_{q=0}^{l} ker(Z_jT_j^q) = \{0\}$$

for some l.

Let

$$x \in \cap_{q=0}^{\infty} ker(C_\zeta A_\zeta^q),$$

and partition x conformally with (3.12):$x = col(x_j)_{j=1}^r$

Then

$$\sum_{j=1}^{r} Z_j T_j^q x_j = 0; \qquad q = 0, 1, \dots . \tag{3.13}$$

Using the fact that

$$\sigma(T_j) = \{z_j\}, j = 1, \dots, r$$

and hence the spectra of $T_1, \dots, T_r$ are mutually disjoint, for each s there is a polynomial $\varphi_\zeta(z)$ such that $\varphi_\zeta(T_j) = 0$ for $j \neq s$ and $\varphi_\zeta(T_\zeta) = I$. Now (3.13) implies

$$0 = \sum_{j=1}^{r} Z_j T_j^q \varphi_\zeta(T_j) x_j = Z_\zeta T_\zeta^q x_\zeta$$

for $q = 0, 1, \dots$, and since (Z_ζ, T_ζ) is a null kernel pair we obtain $x_\zeta = 0$. It follows that $x = 0$, and so (C_ζ, A_ζ) is a null kernel pair.

The uniqueness of (C_ζ, A_ζ) up to similarity follows from the corresponding uniqueness property for right null pairs of $A(z)$ corresponding to each $z_j, j = 1, \dots, r$. ∎

We introduce now the notions of left null chains and left null pair. A chain of row vectors $y_0, y_1, \dots, y_{r-1}$ is a *left null chain* for the rational matrix function $A(z)$ at z_0 if there exists a *left null function* $\psi(z)$ of order r having $\frac{1}{j!}\psi^{[j]}(z_0) = y_j$ for $0 \leq j \leq r-1$; equivalently, the chain of column vectors $y_0^T, y_1^T, \dots, y_{r-1}^T$ is a right null chain for $A^T(z) := A(z)^T$ at z_0. A system of left null chains

$$y_{10}, y_{11}, \dots, y_{1,r_1-1}; \dots ; y_{p0}, \dots, y_{p,r_p-1} \tag{3.14}$$

is said to be *canonical* if there is an associated set of left null functions

$$\{\psi^{(1)}(z), \dots, \psi^{(p)}(z)\}$$

with $\frac{1}{\alpha!}\psi^{(j)[\alpha]}(z_0) = y_{j\alpha}$ for $1 \leq j \leq p$, $0 \leq \alpha \leq r_j - 1$ which is canonical. Now let (3.14) be a canonical set of left null chains. For $1 \leq i \leq p$ let $R_i = col(y_{i,r_i-1-\alpha})_{\alpha=0}^{r_i-1}$. Let J_i be the $r_i \times r_i$ Jordan block with z_0 on the main diagonal and put $R_0 = col(R_j)_{j=1}^p$, $J_0 = diag(J_1, \dots, J_p)$. Then the pair (J_0, R_0) will be called a *left null pair* for $A(z)$ at z_0. Any pair of the form (S, Y), where $S = M^{-1}J_0M, y = M^{-1}R_0$ for some invertible matrix M, will also be called a *left null pair* for

$A(z)$ at z_0. It follows from the definition that (S,Y) is a left null pair of $A(z)$ at z_0 if and only if (Y^T, S^T) is a right null pair of $A(z)^T$ at z_0. Using this fact properties analogous to those described above for right null pairs can be verified for left null pairs.

Observe that if (Z_1, T_1) and (T_2, Y_2) are right and left null pairs of $A(z)$ at z_0, then T_1 and T_2 are similar. Indeed, $\sigma(T_1) = \sigma(T_2) = \{z_0\}$, and the sizes of Jordan blocks in each T_1 and T_2 are just the positive partial multiplicities of $A(z)$ at z_0.

Given a non-empty set $\sigma \subset \mathbb{C}$ with the distinct zeros $z_1, \ldots, z_r$ of $A(z)$ in σ, a *left null pair* of $A(z)$ *with respect to* σ is defined as any pair of matrices (A_ζ, B_ζ) which is similar to

$$(S_1 \oplus \cdots \oplus S_r,\ col(Y_j)_{j=1}^r),$$

where (S_j, Y_j) is a left null pair for $A(z)$ at $z_j, j = 1, \ldots, r$. The similarity here is understood in the following sense: there exists an invertible matrix T such that

$$S_1 \oplus \cdots \oplus S_r = T^{-1} A_\zeta T;\ col(Y_j)_{j=1}^r = T^{-1} B_\zeta.$$

The left null pair of $A(z)$ with respect to $\mathbb{C}$ will be called *global left null pair* of $A(z)$.

To state the analogue of Theorem 3.3, we introduce the concept of full range pair which is dual to the notion of null kernel pair. A pair of matrices (S,Y), where S is $p \times p$ and Y is $p \times n$ is called *full range pair* if

$$\sum_{j=0}^{l-1} Im(S^j Y) = \mathbb{C}^p$$

for some positive integer l.

THEOREM 3.4. *A left null pair of $A(z)$ with respect to σ is a full range pair and is unique up to similarity.*

The proof is reduced to Theoremm 3.3 by observing that (A_ζ, B_ζ) is a left null pair for $A(z)$ with respect to σ if and only (B_ζ^T, A_ζ^T) is a right null pair for $A(z)^T$ with respect to the same σ ∎.

By definition, a chain of column vectors $(y_1, \ldots, y_r)$ with $y_1 \neq 0$ is a *right pole chain* for $A(z)$ at z_0 if there exists a right pole function $\varphi(z)$ for $A(z)$ at z_0 of order r such that $\frac{1}{\alpha!}\varphi^{[\alpha]}(z_0) = y_{\alpha+1}$ for $0 \leq \alpha \leq r-1$. Equivalently, $(y_1, \ldots, y_r)$ with $y_1 \neq 0$ is a right pole chain of $A(z)$ at z_0 if this chain is a right null chain of $A(z)^{-1}$ at z_0, that is, if there exist additional vectors $y_{r+1},\dots,y_{r+q}$ (q being the order of z_0 as a pole of $A(z)^{-1}$) such that $A(z)^{-1}y(z)$ is analytic at z_0 with a zero of order r at z_0, where $y(z)=\sum_{j=1}^{r+q}(z-z_0)^{j-1}y_j$. As for the null chains, we introduce the *right pole pair* (Z,T) and the *left pole pair* of (S,Y) of $A(z)$ at z_0. So (Z,T) (resp. (S,Y)) is a right (resp. left) null pair of $A(z)^{-1}$ at z_0.

In analogy with the null pairs with respect to a set $\sigma\subset\mathbb{C}$, we introduce a left pole pair (A_π,B_π) and a right pole pair (C_π,A_π) of $A(z)$ with respect to σ (the matrices A_π in both pairs are similar, so the usage of the same notation for these matrices is justified). A left (resp. right) pole pair of $A(z)$ with respect to $\mathbb{C}$ will be called a *global left* (resp. *right) pole pair* for $A(z)$. The analogues of Theorems 3.3 and 3.4 hold for pole pairs. We leave their statements and proofs to the reader.

From now on we will use the description of zero and pole data of a rational matrix function in terms of the corresponding pairs of matrices.

We can formulate now the basic interpolation problems: Given some or all of the four pairs of matrices corresponding to a set $\sigma\subset\mathbb{C}$ (namely, right null pair, right pole pair, left null pair, left pole pair) construct a rational matrix function with these pairs, or, better yet, describe all rational matrix functions with the given pairs.

In this paper we will study later (in Section 2.6) only one such problem and some of its applications: namely, when a global right pole pair and a global left null pair are given.

2.4 Realizations of rational matrix functions and systems theory.

This section reviews the results from systems theory that we shall need, for more information, see, e.g., [K,KFA].

A very important role for the next development is a special type of representation for a rational matrix function which appears in systems theory and is called a realization. Namely, any rational matrix function which is analytic at infinity can be represented in the form

$$R(z)=D+C(zI-A)^{-1}B \tag{4.1}$$

where A,B,C,D are matrices with compatible sizes and A is square This representation is natural as the representation of a *transfer function*: namely, if we have a system then the system is characterized by a rational matrix function $R(z)$ which transforms the input $\hat{u}(z)$ into the output $\hat{y}(z)$

$$\hat{y}(z) = R(z)\hat{u}(z). \tag{4.2}$$

This is the representation of the system in the frequency domain and $R(z)$ is called the transfer function. If the transfer function has a realization (4.1) then the system can be rewritten in the time domain in the following way. Define $\hat{x}(z) = (zI - A)^{-1}B\hat{u}(z)$. Then

$$\begin{aligned} z\hat{x}(z) &= A\hat{x}(z) + B\hat{u}(z) \\ \hat{y}(z) &= C\hat{x}(z) + D\hat{u}(z) \end{aligned} \tag{4.3}$$

If one applies to the system the inverse Laplace transform, we get the system in the time domain

$$\begin{aligned} \frac{dx}{dt} &= Ax(t) + Bu(t) \\ y(t) &= Cx(t) + Du(t), \qquad x(0) = 0. \end{aligned} \tag{4.4}$$

Then $x(t)$ is called the state space variable and the space where it is varying is called the state space. In the realization formula (4.1) it is clear that $D = R(\infty)$. One way of finding A, B, C for a given rational matrix function is via the following lemma.

LEMMA 4.1. *Let $H(z) = \sum_{j=0}^{l-1} z^j H_j$ and $L(z) = z^l I + \sum_{j=0}^{l-1} z^j A_j$ be $r \times n$ and $n \times n$ matrix polynomials, respectively. Put*

$$B = \begin{bmatrix} 0 \\ \vdots \\ 0 \\ I \end{bmatrix}, \; A = \begin{bmatrix} 0 & I & \dots & 0 \\ \vdots & & \ddots & \\ 0 & 0 & \dots & I \\ -A_0 & -A_0 & \dots & -A_{l-1} \end{bmatrix}, \; C = [H_0 \dots H_{l-1}].$$

Then

$$H(z)L(z)^{-1} = C(zI - A)^{-1}B.$$

PROOF: Let

$$Q = [I0 \dots 0].$$

We verify first that

$$L(z)^{-1} = Q(zI - A)^{-1}B. \tag{4.5}$$

Indeed, by straightforward multiplication one verifies the formula

$$\begin{bmatrix} E_1(z) & \dots & & E_{l-1}(z) & I \\ -I & 0 & \dots & 0 & 0 \\ 0 & -I & \dots & 0 & \vdots \\ \vdots & \vdots & \vdots & & \\ 0 & 0 & \dots & -I & 0 \end{bmatrix} (zI - A) = \begin{bmatrix} L(z) & 0 \\ 0 & I_{n(l-1)} \end{bmatrix} \begin{bmatrix} I & & & & 0 \\ -zI & I & & & \\ & -zI & \ddots & & \\ & & \ddots & & \\ 0 & & & & \\ & & & -zI & I \end{bmatrix}. \tag{4.6}$$

where the matrix polynomials $E_1, \dots, E_{l-1}$ are defined by

$$E_j(z) = z^{l-1}I + \sum_{k=0}^{l-j-1} z^k A_{k+j}, \qquad j = 1, \dots, l-1.$$

Formula (4.6) implies

$$\begin{bmatrix} L(z)^{-1} & 0 \\ 0 & I \end{bmatrix} = \begin{bmatrix} I & & & 0 \\ -zI & I & & \\ & & \ddots & \\ & \ddots & & \\ 0 & & -zI & I \end{bmatrix} (zI - A)^{-1} \begin{bmatrix} \begin{matrix} 0 \\ \vdots \\ 0 \\ I \end{matrix} & * \end{bmatrix}$$

and premultiplication by Q and postmultiplication by B gives (4.5).

Now define $C_1(z), \dots, C_l(z)$ for all $z \notin \sigma(A)$ by

$$col[C_j(z)]_{j=1}^l = (zI - A)^{-1}B.$$

From (4.5) we see that $C_1(z) = L(z)^{-1}$. As $(zI - A)[col(C_j(z))_{j=1}^l] = B$, the special form of A yields

$$C_i(z) = z^{i-l}C_1(z), \qquad 1 \le i \le i.$$

It follows that $C(z - A)^{-1}B = \sum_{j=0}^{l-1} H_j C_{j+1}(z) = H(z)L(z)^{-1}$, and the proof is complete. ∎

THEOREM 4.2. *Every $r \times n$ rational matrix function which is analytic at infinity has a realization.*

PROOF: Let $R(z)$ be an $r \times n$ rational matrix function with finite value at infinity. There exists a monic scalar polynomial $l(z)$ such that $l(z)R(z)$ is a (matrix) polynomial. For instance, take $l(z)$ to be a least common multiple of the denominators of entries in $R(z)$. Put $H(z) = l(z)(R(z) - R(\infty))$.

Then $H(z)$ is an $r \times n$ matrix polynomial. Clearly, $L(z) = l(z)I_n$ is monic and $R(z) = R(\infty) + H(z)L(z)^{-1}$. Furthermore,

$$\lim_{z \to \infty} H(z)L(z)^{-1} = \lim_{z \to \infty} [R(z) - R(\infty)] = 0.$$

So the degree of $H(z)$ is strictly less than the degree of $L(z)$. We can apply Lemma 4.1 to find A, B, C for which

$$R(z) = R(\infty) + H(z)L(z)^{-1} = R(\infty) + C(zI - A)^{-1}B,$$

i.e. this is a realization of $R(z)$. ■

A realization for $R(z)$ is far from being unique. This can be seen from our construction of a realization because there are many choices for $l(z)$.

Among all the realizations of $R(z)$ those with the properties that (C, A) is a *null kernel* pair and (A, B) is a *full range* pair will be of special interest. That is, for which

$$\cap_{j=0}^{\infty} Ker\ CA^j = \{0\} \tag{4.7}$$

$$\sum_{j=0}^{\infty} Im\ A^j B = \mathbb{C}^m. \tag{4.8}$$

Observe that conditions (4.7) and (4.8) are equivalent to

$$\cap_{j=0}^{p-1} Ker\ CA^j = \{0\} \tag{4.9}$$

for some positive integer p, and to

$$\sum_{j=0}^{p-1} Im A^j B = \mathbb{C}^m \tag{4.10}$$

for some p, respectively. In turn, if (4.9) holds for some p then it holds for any p not smaller than the degree of the minimal polynomial of A. An analogous remark applies to (4.10).

It turns out that a realization (A, B, C) for which conditions (4.7), (4.8) are satisfied is essentially unique.

THEOREM 4.3. *Let (A_1, B_2, C_2) be realizations for a rational matrix function $R(z)$ for which (C_1, A_1) and (C_2, A_2) are null kernel pairs and (A_1, B_1), (A_2, B_2) are full range pairs. Then the sizes of A_1 and A_2 coincide, and there exists a nonsingular matrix S such that*

$$A_1 = S^{-1}A_2S, \qquad B_1 = S^{-1}B_2, \qquad C_1 = C_2S. \tag{4.11}$$

Moreover, the matrix S is unique and is given by

$$S = [col(C_2A_2^j)_{j=0}^{p-1}]^{-L}[col(C_1A_1^j)_{j=0}^{p-1}] = [B_2, A_2B_2, \ldots, A_2^{p-1}B_2][B_1, A_1B_1, \ldots, A_1^{p-1}B_1]^{-R}. \tag{4.12}$$

Here p is any integer greater than or equal to the maximum of the degrees of minimal polynomials for A_1 and A_2, and the superscript $-L$ (resp. $-R$) indicates left (resp. right) inverse.

Note that the hypotheses on the pairs (C_j, A_j) and (A_j, B_j) ensure that the one-sided inverses that appear in (4.12) indeed exist.

PROOF: We have

$$R(z) = R(\infty) + C_1(zI - A_1)^{-1}B_1 = R(\infty) + C_2(zI - A_2)^{-1}B_2.$$

For $|z| > max\{||A_1||, ||A_2||\}$, the matrices $zI - A_1$ and $zI - A_2$ are nonsingular.
Consequently, we have

$$C_1(\sum_{j=0}^{\infty} z^{-j-1}A_1^j)B_1 = C_2(\sum_{j=0}^{\infty} z^{-j-1}A_2^j)B_2$$

for any z with $|z| > max\{||A_1||, ||A_2||\}$. Comparing coefficients, we see that $C_1A_1^jB_1 = C_2A_2^jB_2, j = 0, 1, \ldots$. This implies $\Omega_1\Delta_1 = \Omega_2\Delta_2$, where, for $k = 1, 2$ we write

$$\Omega_k = col[C_kA_k^j]_{j=0}^{p-1}, \qquad \Delta_k = [B_k, A_kB_k \ldots, A_k^{p-1}B_k\}.$$

Premultiplying by a left inverse of Ω_2 and postmultiplying by a right inverse of Δ_2 we find that the second equality in (4.12) holds. Now define S as in (4.12). Let us check first that S is (two-sided) invertible. Indeed, we shall verify the relations

$$(\Omega_1^{-L}\Omega_2)S = I, \qquad S(\Delta_1\Delta_2^{-R}) = I.$$

Since $C_1A_1^jB_1 = C_2A_2^jB_2$, for $j = 0, 1, \dots$, we have $\Omega_2^{-L}\Omega_1\Delta_1\Delta_2^{-R} = \Omega_2^{-L}\Omega_2\Delta_2\Delta_2^{-R} = I$. Similarly, one checks that $\Omega_1^{-L}\Omega_2\Delta_2\Delta_1^{-R} = I$. Because S is invertible the sizes of A_1 and A_2 must coincide.

It remains to check the equations (4.11). Write

$$\Omega_2A_2\Delta_2 = \Omega_1A_1\Delta_1 = \Omega_1\Delta_1\Delta_1^{-R}A_1\Delta_1 = \Omega_2\Delta_2\Delta_1^{-R}A_1\Delta_1.$$

Premultiply by Ω_2^{-L} and postmultiply by Δ_2^{-R} to obtain $A_2S = SA_1$. Now

$$SB_1 = \Omega_2^{-L}\Omega_1B_1 = \Omega_2^{-L}\Omega_2B_2 = B_2$$

and

$$C_2S = C_2\Delta_2\Delta_1^{-R} = C_1\Delta_1\Delta_1^{-R} = C_1. \blacksquare$$

THEOREM 4.4. *In a realization* (A, B, C) *of* $R(z), (C, A)$ *and* (A, B) *are null kernel pairs and full range pairs, respectively, if and only if the size of* A *is minimal among all possible realizations of* $R(z)$.

We omit the proof of Theorem 4.4. A transparent proof (with full details) is given in [GLR2] (see Theorems 7.13 and 7.14).

Realizations of the kind described in Theorem 4.4 are, naturally, called minimal realizations of $R(z)$. That is, they are those realizations for which the dimension of the space on which A acts is as small as possible.

COROLLARY 4.5. *Any two minimal realizations* $(A_j, B_j, C_j), j = 1, 2$ *of* $R(z)$ *are similar, i.e.*

$$A_1 = S^{-1}A_2S, B_1 = S^{-1}B_2, C_1 = C_2S$$

for some invertible matrix S. *The invertible matrix* S *is unique.*

Assume now in addition that $R(z)$ is of square size and invertible at infinity. Then $R(z)^{-1}$ is again a rational matrix function which is analytic and invertible at infinity. It is not difficult to find a realization for $R(z)^{-1}$ if a realization for $R(z)$ is given. To this end consider the system (4.4) with the transfer function

$$R(z) = D + C(zI - A)^{-1}B. \tag{4.13}$$

Then $R(z)^{-1}$ is the transfer function of the system obtained from (4.4) by interchanging of the roles of $\hat{y}(z)$ and $\hat{u}(z)$, i.e. by assuming that $\hat{y}(z)$ is input and $\hat{u}(z)$ is output. To write down this system, rearrange the second equation in (4.4) in the form

$$\hat{u}(z) = -D^{-1}C\hat{x}(z) + D^{-1}\hat{y}(z)$$

and substitute in the first equation in (4.2):

$$z\hat{x}(z) = (A - BD^{-1}C)\hat{x}(z) + BD^{-1}\hat{y}(z).$$

Now it is clear that

$$R(z)^{-1} = D^{-1} - D^{-1}C(zI - (A - BD^{-1}C))^{-1}BD^{-1}. \tag{4.14}$$

Of course, one can verify that (4.14) is indeed the inverse of (4.13) by purely algebraic manipulations, writing out the product of the right-hand sides of (4.13) and (4.14).

It turns out that this passage from a realization of $R(z)$ to a realization of $R(z)^{-1}$ preserves minimality:

THEOREM 4.6. *If (4.13) is minimal and D is invertible, then (4.14) is minimal as well.*

PROOF: By Theorem 4.4 (C, A) is a null kernel pair and (A, B) is a full range pair. We have to prove (using the same Theorem 4.4) that $(-D^{-1}C, A - BD^{-1}C)$ is a null kernel pair and $(A - BD^{-1}C, BD^{-1})$ is a full range pair. Using induction on j one can easily verify that $C(A - BD^{-1}C)^{l_x} = 0$ for $0 \leq l \leq j$ is equivalent to $CA^{l_x} = 0$ for $0 \leq l \leq j$.

$$\cap_{j=0}^{\infty} ker C(A - BD^{-1}C)^j = \cap_{j=0}^{\infty} ker CA^j,$$

and since (C, A) is null kernel, so is $(C, A - BD^{-1}C)$, and hence also $(-D^{-1}C, A - BD^{-1}C)$. A dual argument works for $(A - BD^{-1}C, B)$. Namely, use the fact that for every $x \in \mathbb{C}^n$ the vector A^jBx is contained in the subspace

$$\sum_{k=0}^{j} Im((A - BD^{-1}C)^k B). \blacksquare$$

2.5 Realization and null-pole data

In this section we show how the null and pole pairs of a rational matrix function (as introduced and studied in Section 3) can be recovered from its minimal realization. The results of this section (with the exception of Theorem 5.1) are taken from [BGK].

We start with an introduction of the notion of pole and null triples. Let $R(z)$ be rational $n \times n$ matrix function with $detA(z)$ not identically zero, and let z_0 be a pole of $R(z)$. A triple of matrices (Z, T, Y) is called a *pole triple* of $R(z)$ at z_0, if (Z, T) is a right pole pair of $R(z)$ at $z_0, (T, Y)$ is a left pole pair of $R(z)$ at z_0 and $R(z) - Z(zI - T)^{-1}Y$, is analytic at z_0. In particular, (Z, T) is a null kernel pair and (T, Y) is a full range pair. The definition of a null triple is analogous: (Z, T, Y) is a null triple for $R(z)$ at its zero z_0 if (Z, T) is a right null pair at $z_0, (T, Y)$ is a left null pair of $R(z)$ at z_0, and $R(z)^{-1} - Z(zI - T)^{-1}Y$ is analytic at z_0.

THEOREM 5.1. *A null triple (Z, T, Y) of $R(z)$ at z_0 exists and is unique up to similarity. More precisely, if $(Z_j, T_j, Y_j), j = 1, 2$ are two null triples of $R(z)$ at z_0, then there exists a nonsingular matrix S such that*

$$Z_2 = Z_1 S, T_2 = S^{-1} T_1 S, Y_2 = S^{-1} Y_1.$$

An analogous statement holds for pole triples.

PROOF: We start with the proof of uniqueness. Let (Z_1, T_1, Y_1) and (Z_2, T_2, Y_2) be two null triples of $R(z)$ at z_0. Then

$$Z_1(zI - T_1)^{-1}Y_1 - Z_2(zI - T_2)^{-1}Y_2 = [Z_1(zI - T_1)^{-1}Y_1 - R(z)^{-1}]$$

$$+[R(z)^{-1} - Z_2(zI - T_2)^{-1}Y_2]$$

is analytic at z_0. Since $\sigma(T_1) = \sigma(T_2) = \{z_0\}$, by Liouville's theorem we obtain

$$Z_1(zI - T_1)^{-1}Y_1 \equiv Z_2(zI - T_2)^{-1}Y_2,$$

i.e.

$$Z_1 T_1^j Y_1 = Z_2 T_2^j Y_2, j = 0, 1, \ldots.$$

Now the existence of an invertible S such that (5.1) holds is proved as in the proof of Theorem 4.3.

For the proof of existence consider first the case when $R(z)$ in the local Smith form:

$$R(z) = diag((z-z_0)^{k_1}, ..., (z-z_0)^{k_n}),$$

where $k_1 \geq \cdots \geq k_p > 0 \geq k_{p+1} \geq \cdots \geq k_n$. In this case the existence of a null triple is easily seen. For instance, assuming for notational simplicity that $p = 1$, one such null triple is given by

$$Z = [e_1 0 \ldots 0]; T = \begin{bmatrix} z_0 & 1 & & 0 \\ & z_0 & 1 & \\ & & \ddots & \\ & & & 1 \\ 0 & & & z_0 \end{bmatrix}; y = \begin{bmatrix} 0 \\ \vdots \\ 0 \\ e_1^T \end{bmatrix},$$

where the size of T is $k_1 \times k_1$.

The general case is reduced to the local Smith form, as follows. Assume (Z, T, Y) is a null triple of $R(z)$ at z_0, and let $F(z)$ be a rational matrix function analytic and invertible at z_0. As (Z, T) is a right null pair of $R(z)$ at z_0, without loss of generality we can assume that

$$Z = [Z_1, \ldots Z_r], \quad Z_j = [x_0^{(j)} \ldots x_{k_j-1}^{(j)}],$$

$$T = diag(T_1, T_2, \ldots, T_r),$$

where the columns of Z_j form a right null chain of $R(z)$ at z_0 and T_j is the corresponding Jordan block with eigenvalue z_0. By Theorem 2.2 the pair $(\tilde{Z} = [\tilde{Z}_1 \ldots \tilde{Z}_r], T)$ is a right null pair of $\tilde{R}(z) := R(z)F^{-1}(z)$ corresponding to z_0 where

$$\tilde{Z}_j = [y_0^{(j)} \ldots y_{k_j-1}^{(j)}]$$

and

$$y_q^{(j)} = \sum_{p=0}^{q} \frac{1}{p!} F^{(p)}(z_0) x_{q-p}^{(j)} \tag{5.2}$$

We verify now that the function

$$G(z) := (F(z)Z - \tilde{Z})(zI - T)^{-1}$$

is analytic at z_0. We have

$$G(z) = \sum_{j=1}^{r} (F(z)Z_j - \tilde{Z}_j)(zI - T_j)^{-1} =$$

$$\sum_{j=1}^{r}\{F(z)[x_0^{(j)}\dots x_{k_j-1}^{(j)}]-[y_0^{(j)}\dots y_{k_j-1}^{(j)}]\}\begin{bmatrix}(z-z_0)^{-1} & (z-z_0)^{-2} & \cdots & (z-z_0)^{-k_j}\\ 0 & (z-z_0)^{-1} & & \vdots\\ \vdots & \vdots & & \\ 0 & 0 & \cdots & (z-z_0)^{-1}\end{bmatrix}.$$

For $p=1,2\dots$ the coefficient of $(z-z_0)^{-p}$ in the right-hand side is

$$\sum_{j=1}^{r}\Big\{F(z_0)[0\dots 0x_0^{(j)}\dots x_{k_j-p}^{(j)}]+F'(z_0)[0\dots 0x_0^{(j)}\dots x_{k_j-p-1}^{(j)}]+\dots$$

$$+\frac{1}{(k_j-p)!}F^{(k_j-p)}(z_0)[0\dots 0x_0^{(j)}]-[0\dots 0y_0^{(j)}\dots y_{k_j-p}^{(j)}]\Big\}$$

which is zero in view of (5.2). As

$$F(z)R(z)^{-1}-\tilde{Z}(zI-T)^{-1}Y=$$

$$=F(z)(R(z)^{-1}-Z(zI-T)^{-1}Y)+(F(z)Z-\tilde{Z})(zI-T)^{-1}Y,$$

it follows that $(\tilde{Z},T,Y)$ is a null triple of $\tilde{R}(z)$ at z_0.

Analogously one verifies that the existence of a null triple for $R(z)$ at z_0 implies existence of a null triple for $E(z)R(z)$ at z_0, where $E(z)$ is a rational matrix function analytic and invertible at z_0. It remains to use Theorem 2.1 and the existence of a null triple for a local Smith form verified above. ∎

Using Theorem 5.1 we can analyze the connections between minimal realizations and global null and pole pairs.

THEOREM 5.2. *Let $R(z)=D+C(zI-A)^{-1}B$ be a minimal realization of the rational matrix function R with invertible $R(\infty)=D$. Let $A^\times=A-BD^{-1}C$. Then (C,A) and (A,B) are respectively right and left global pole pairs of $R(z)$ and $(D^{-1}C,A^\times)$ and $(A^\times,BD^{-1})$ are respectively right and left global null pairs of $R(z)$.*

PROOF: Let $z_1,\dots,z_r$ be the poles of $R(z)$. By Theorem 5.1 there exists a pole triple (Z_j,T_j,Y_j) for $R(z)$ at z_j for $1\le j\le r$. Set

$$Z=[Z_1\cdots Z_r],T=diag(T_1,T_2,\cdots,T_r)$$

$$Y=col.(Y_j)_{j=1}^{r}.$$

Since (Z_j, T_j) in particular is a right pole pair for $R(z)$ at z_j and $z_1, \cdots, z_r$ are all poles of $R(z)$, we see that (Z, T) is a global right pole pair for $R(z)$. Similarly, (T, Y) is a global left pole pair for $R(z)$. By the definition of a pole triple for $R(z)$ at z_j, $Z_j(zI - T_j)^{-1}Y_j$ has a pole only at z_j and $R(z) - Z_j(zI - T_j)^{-1}Y_j$ is analytic at z_j. From this we see that

$$R(z) - Z(zI - T)^{-1}Y = R(z) - \sum_{j=1}^{r} Z_j(zI - T_j)^{-1}Y_j$$

has no poles in the complex plane. Since the value at infinity is D, by Liouville's Theorem we get $R(z) = D + Z(zI - T)^{-1}Y$, i.e. (T, Y, Z, D) is a realization of $R(z)$. By the analogues of Theorems 3.1 and 3.3 for pole pairs, (Z, T) is a null-kernel pair and (T, Y) is a full range pair, so (T, Y, Z, D) is a minimal realization for $R(z)$. Then by Theorem 4.3, this realization must be similar to the given realization (A, B, C, D). In particular, (C, A) is a global right pole pair for $R(z)$ since its is similar to the right pole pair (Z, T) and (A, B) is a global left pole pair for $R(z)$ since it is similar to (T, Y).

Finally, to see that $(D^{-1}C, A^{\times})$ and $(A^{\times}, BD^{-1})$ are right and left null pairs for $R(z)$, use that a right (resp. left) null pair for $R(z)$ is just a right (resp. left) pole pair for $R^{-1}(z)$ and apply the result above for pole pairs to the minimal realization

$$R^{-1}(z) = D^{-1} - D^{-1}C(zI - A^{\times})^{-1}BD^{-1}$$

for $R^{-1}(z)$ is given by (4.14). ∎

2.6. A global interpolation problem for rational matrix functions

A scalar rational function is uniquely determined up to a nonzero constant factor by its zeros and poles; this is essentially a consequence of the fundamental theorem of algebra. For the matrix case we have seen that a canonical set of null chains at each point plays the role of the zeros and similarly a canonical set of pole chains at each pole gives more complete pole information. In the scalar case, multiplying by a nonzero constant does not affect the poles and the zeros; for the matrix case let us suppose we are interested only in left-sided information concerning a rational matrix function, i.e. in pole-zero structure that is unaffected by multiplying by a constant invertible matrix on the right. It is easy to see that a canonical set of left null chains at each point (or equivalently, a global left null pair (A_ζ, B_ζ)) and a canonical set of right pole chains at each point (i.e. a global right pole pair (C_π, A_π) have these properties. We restrict ourselves in this section to rational matrix functions $R(z)$ which are analytic and invertible at infinity. Then a natural question to ask is: does a global left null pair (A_ζ, B_ζ) and a global right pole pair (C_π, A_π) for a rational matrix function $R(z)$ determine $W(z)$ uniquely up to a right invertible constant factor? Also, given a full range pair (A_ζ, B_ζ) and a null kernel pair (C_π, A_π), when is there a rational matrix function $R(z)$ (invertible at ∞) having (A_ζ, B_ζ) as a global left null pair and (C_π, A_π) as a global right pole pair? For the scalar case we know that the answer to the first question is yes while the answer to the second question is yes if and only if A_ζ and A_π have the same size and disjoint spectra (i.e. the number of zeros is the same as the number of poles (including multiplicities) and the set of zeros is disjoint from the set of poles). For the matrix case the answer is more complicated. We have already seen (eg. via the local Smith form) that it is possible in the matrix case for a point z_0 to be a pole and a zero for a function $R(z)$, so the answer to the second question is certainly more subtle in the matrix case. As the following example illustrates, in general the answer to the first question is no in the matrix case.

Example 6.1. Consider the function

$$R_\alpha(z) = \begin{bmatrix} 1 & \alpha z^{-1} \\ 0 & 1 \end{bmatrix}$$

where $\alpha \neq 0$ is a complex parameter. Then for all $\alpha \neq 0$, R_2 has a pole only at 0, and a canonical set of right pole chains at 0 consists of the single vector $\left\{\begin{bmatrix} 1 \\ 0 \end{bmatrix}\right\}$. Thus $(C_\pi, A_\pi) = \left(\begin{bmatrix} 1 \\ 0 \end{bmatrix}, 0\right)$ is a right pole pair for $R_\alpha(z)$ for all $\alpha \neq 0$. As $R_\alpha^{-1}(z) = \begin{bmatrix} 1 & -\alpha z^{-1} \\ 0 & 1 \end{bmatrix}$, we see similarly that a left pole pair for R_α^{-1} (i.e. a left null pair for R_α) is $(A_\zeta, B_\zeta) = (0, [0, 1])$. Thus each of the infinitely many functions $R_\alpha (\alpha \neq 0)$ has the same left null pair (A_ζ, B_ζ) and right pole pair (C_π, A_π). As $R_\alpha(\infty) = I$ for all α, any two of these functions must differ from each other by more than a constant right factor. ∎

The answer for the matrix case can be derived by using the connection between a left null pair and right pole pair for a rational matrix function $R(z)$ and a minimal realization for $R(z)$. We assume that $R(z)$ is analytic and invertible at infinity; by considering $R(z)R(\infty)^{-1}$ in place of $R(z)$ we may assume that the value at ∞ is I. Then if (C_π, A_π) is a right pole pair for $R(z)$, we know by Corollary 5.3 that there exists a matrix $\tilde{B}$ such that

$$R(z) = I + C_\pi(zI - A_\pi)^{-1}\tilde{B} \tag{6.1}$$

is a minimal realization for $R(z)$. Then by Theorem 4.7 we know that

$$R^{-1}(z) = I - C\pi(zI - (A_\pi - \tilde{B}C_\pi))^{-1}\tilde{B} \tag{6.2}$$

is a minimal realization for R^{-1}. On the other hand, if (A_ζ, B_ζ) is a left null pair for $R(z)$, i.e. a left pole pair for R^{-1}, then by the same Corollary 5.3, there exists a matrix $\tilde{C}$ such that $R^{-1}(z)$ has the minimal realization

$$R^{-1}(z) = I - \tilde{C}(zI - A_\zeta)^{-1}B_\zeta \tag{6.3}$$

Thus the realization (6.2) and (6.3) for R^{-1} must be similar to each other (see Corollary 4.5), i.e. there must exist an invertible matrix S such that

$$\tilde{C}S = C_\pi, S^{-1}A_\zeta S = A_\pi - \tilde{B}C_\pi, S^{-1}B_\zeta = \tilde{B}. \tag{6.4}$$

From the second equation we get $SA_\pi - A_\zeta S = S\tilde{B}C_\pi$ while from the last equation we can replace $S\tilde{B}$ with B_ζ. Thus the Sylvester equation

$$SA_\pi - A_\zeta S = B_\zeta C_\pi$$

must have an invertible solution S whenever there exists a rational matrix function $R(z)$ invertible at ∞ having (C_π, A_π) is a right pole pair and (A_ζ, B_ζ) as a left null pair. This necessary condition turns out to be also sufficient. The following result was proved in [GKLR].

THEOREM 6.1. *Suppose (C_π, A_π) is a null kernel pair and (A_ζ, B_ζ) is a full range pair of matrices. Then a necessary and sufficient condition for the existence of a rational matrix function $R(z)$ invertible at ∞ such that* *(i)* (C_π, A_π) *is a right pole pair for* $R(z)$ *and*

(ii) (A_ζ, B_ζ) *is a left null pair for* $R(z)$

is that there exist an invertible solution S *of the Sylvester equation*

$$SA_\pi - A_\zeta S = B_\zeta C_\pi \tag{6.5}$$

In this case, solutions $R(z)$ *of (i) and (ii) having* $R(\infty) = I$ *are in one-to-one correspondence with invertible solutions* S *of (6.5) according to the formula*

$$R(z) = I + C_\pi(zI - A_\pi)^{-1}S^{-1}B_\zeta \tag{6.6}$$

with inverse given by

$$R^{-1}(z) = I - C_\pi S^{-1}(zI - A_\zeta)^{-1}B_\zeta \tag{6.7}$$

PROOF: We have already noted in the discussion immediately preceding the Theorem that whenever there exists a $R(z)$ satisfying (i) and (ii) then there exists an invertible solution S of the Sylvester equation (6.5). Moreover, when we substitute $\tilde{B} = S^{-1}B_\zeta$ (from (6.4)) into the realization (6.1) for $R(z)$ we see that $R(z)$ has the form (6.6) for this solution S of the Sylvester equation (6.5). Similarly, substituting $\tilde{C} = C_\pi S^{-1}$ (from (6.4)) into the realization (6.3) for R^{-1}, we get that then (6.7) gives a correct formula for R^{-1}.

Conversely, suppose that S is any invertible solution of (6.5). We then use (6.6) to define a rational matrix function $R(z)$. By Theorem 4.7 we see that

$$R^{-1}(z) = I - C_\pi(zI - (A_\pi - S^{-1}BC_\pi))^{-1}S^{-1}B_\zeta =$$
$$= I - C_\pi S^{-1}(zI - S(A_\pi - S^{-1}B_\zeta C_\pi)S^{-1})^{-1}B_\zeta$$

But a consequence of (6.5) in

$$S(A_\pi - S^{-1}B_\zeta C_\pi)S^{-1} = A_\zeta. \tag{6.8}$$

Thus we see that formula (6.7) gives a valid formula for R^{-1}. Another consequence of (6.5) is

$$S^{-1}(A_\zeta + B_\zeta C_\zeta S^{-1})S = A_\pi.$$

From this we see that the pair $(C_\pi S^{-1}, A_\zeta + B_\zeta C_\pi S^{-1})$ is similar to the pair (C_π, A_π), and hence is a null kernel pair.

We conclude by arguing as in the proof of Theorem 4.6 that $(C_\pi S^{-1}, A_\zeta)$ is a null kernel pair as well. As by assumption (A_ζ, B_ζ) is a full range pair, we see that (6.6) gives a minimal realization for $R^{-1}(z)$. Then by Theorem 5.2, (A_ζ, B_ζ) is a left pole pair for $R^{-1}(z)$, i.e. a left null pair for $R(z)$.

It remains only to show that distinct invertible solutions S_1 and S_2 of (6.4) lead via (6.5) to distinct rational matrix functions $R_1(z)$ and $R_2(z)$. Thus suppose S_1 and S_2 are two invertible solutions of (6.4) such that

$$R(z) = I + C_\pi(zI - A_\pi)^{-1}S_1^{-1}B_\zeta = I + C_\pi(zI - A_\pi)^{-1}S_2^{-1}B_\zeta.$$

By considering the Laurent series expansion of both expressions at infinity we see that

$$C_\pi A_\pi^j S_1^{-1} B_\zeta = C_\pi A_\pi^j S_2^{-1} B_\zeta$$

for $j = 0, 1, 2, \ldots$. Since (C_π, A_π) is a null kernel pair, we get

$$S_1^{-1} B_\zeta = S_2^{-1} B_\zeta. \tag{6.9}$$

Next, using (6.9) and that both S_1 and S_2 satisfy (6.5) we get

$$S_1^{-1}A_\zeta = (A_\pi - S_1^{-1}B_\zeta C_\pi)S_1^{-1} = (A_\pi - S_2^{-1}B_\zeta C_\pi)S_1^{-1}$$

$$= S_2^{-1}A_\zeta S_2 S_1^{-1}.$$

By induction we conclude that

$$S_1^{-1}A_\zeta^j = S_2^{-1}A_\zeta^j S_2 S_1^{-1}$$

for $j = 0, 1, 2, \ldots$. Combine this with (6.9) to get

$$S_1^{-1}A_\zeta^j B_\zeta = S_2^{-1}A_\zeta^j B_\zeta$$

for $j = 0, 1, 2, \ldots$. Since (A_ζ, B_ζ) is a full range pair, we now conclude that $S_1 = S_2$. ∎

For the scalar case of course the desired scalar function $r(z) := R(z)$ can be found directly (see formula (1.1) in the Introduction). We shall now show that formula (6.6) in Theorem 6.1 is consistent with (1.1). For simplicity we assume that all the prescribed zeros $z_1, \ldots, z_n$ and all the prescribed pols $w_1, \ldots, w_m$ are simple. As is our custom we also prescribe that $r(\infty) = 1$. By Example 2.1.2 and the definitions, prescribing that the zeros of the scalar rational function $r(z)$ occur at $z_1, \ldots, z_n$ and all be simple is the same as to prescribe that (A_ζ, B_ζ) is a global left null pair for $r(z)$, where A_ζ is the $n \times n$ matrix

$$A_\zeta = \begin{bmatrix} z_1 & & 0 \\ & \ddots & \\ 0 & & z_n \end{bmatrix}, \tag{6.10}$$

and B_ζ is the $n \times 1$ matrix

$$B_\zeta = \begin{bmatrix} 1 \\ 1 \\ \vdots \\ 1 \end{bmatrix} \tag{6.11}$$

Similarly, prescribing that the poles occur at $w_1, \ldots, w_m$ and that all these poles are simple is the same as prescribing that (C_π, A_π) is a right pole pair for $r(z)$, where C_π is the $1 \times n$ matrix

$$C_\pi = [1, 1, \ldots, 1] \tag{6.12}$$

and A_π is the $m \times m$ matrix

$$A_\pi = \begin{bmatrix} w_1 & & 0 \\ & \ddots & \\ 0 & & w_m \end{bmatrix} \tag{6.13}$$

By elementary scalar function theory facts, we know that such a function $r(z)$ also satisfying $r(\infty) = 1$ exists if and only if $n = m$ and no zero is also a pole, in which case there is a unique such $r(z)$ given by

$$r(z) = \frac{(z - z_1) \ldots (z - z_n)}{(z - w_1) \ldots (z - w_n)} \tag{6.14}$$

On the other hand, Theorem 6.1 gives a criterion for existence and uniqueness in terms of invertible solutions of a Sylvester equation and an alternative more complicated formula for the solution $r(z)$. The following result gives a direct verification that these two solutions of the interpolation problem are equivalent.

THEOREM 6.2. *Let $z_1, \ldots, z_n$ be n distinct points and $w_1, \ldots, w_m$ another set of m distinct points in $\mathbb{C}$. Define matrices $C_\pi, A_\pi, A_\zeta, B_\zeta$ by (6.10) (6.11), (6.12) and (6.13). Then the Sylvester equation (6.5) has a solution S if and only if the set of $m+n$ points $z_1, \ldots, z_n, w_1, \ldots, w_m$ are all distinct. In this case, the solution is unique, and is invertible if and only if $n = m$. Moreover, if $n = m$ and S is the unique solution, then*

$$1 + C_\pi(zI - A_\pi)^{-1}S^{-1}B_\pi = \frac{(z - z_1)\ldots(z - z_n)}{(z - w_1)\ldots(z - w_n)}. \tag{6.15}$$

The proof will proceed via an elementary lemma.

LEMMA 6.3. *If $C_\pi, A_\pi, A_\zeta, B_\zeta$ are given by (6.10)-(6.13), then the Sylvester equation (6.5) has a solution S if and only if no number z_i coincides with any number w_j, and in this case the solution is unique and given by*

$$S = \left[\frac{1}{w_j - z_i}\right]_{1 \le i \le n, 1 \le j \le n.}$$

Moreover, S is invertible if and only if $n = m$, and in this case the inverse is given by

$$S^{-1} = [t_{\alpha\beta}]_{1 \le \alpha, \beta \le n} \tag{6.16}$$

where

$$t_{\alpha\beta} = \frac{\Pi_{\substack{1 \le j \le n \\ j \ne \alpha}}(w_j - z_\beta) \cdot \Pi_{1 \le k \le n}(w_\alpha - z_k)}{\Pi_{\substack{1 \le j \le n \\ j \ne \beta}}(z_\beta - z_j) \cdot \Pi_{\substack{1 \le k \le n \\ k \ne \alpha}}(w_k - w_\alpha)}. \tag{6.17}$$

PROOF: Using the special form of $C_\pi, A_\pi, A_\zeta, B_\zeta$ we may compute the (i,j)-th entry of both sides of (6.5) to see that $S = [s_{ij}]_{1 \le i \le n; 1 \le j \le m}$ satifies the Sylvester equation if and only if

$$s_{ij}w_j - z_i s_{ij} = 1$$

for all (i,j). Thus a solution exists if and only if $w_j \ne z_i$ for all (i,j) and then $s_{ij} = \frac{1}{w_j - z_i}$ is the unique solution.

Certainly $n = m$ is a necessary condition for S to be invertible. For $n = m$, the determinant of S can be found (after a minor change of notation) in Problem #3(p. 92) of [PS], where it is attributed to Cauchy in the year 1841. The derivation involves elementary row and column operations to obtain a recurrence relation for the determinant with $2n$ points in terms of the determinant with $2(n-1)$ points z_i and w_j. The result is

$$det S = \frac{\Pi_{1 \le k < j \le n}\{(z_k - z_j)(w_j - w_k)\}}{\Pi_{1 \le j,k \le n}(w_j - z_k)} \tag{6.18}$$

From this formula we see that $\det S \neq 0$ (since we are assuming that the z_i's and w_j's are distinct from each other). Next we use the adjoint formula

$$S^{-1} = \frac{1}{det S} adj\, S$$

to compute the inverse of S. The $(n-1) \times (n-1)$ submatrices of S have the same form as S itself and thus all the cofactors of S can be computed from the formula (6.18). The result is the expression (6.17) for the entries of S^{-1}. ∎

PROOF OF THEOREM 6.2: It remains only to verify the identity (6.15). Plugging in the expressions (6.12), (6.13), (6.11) and (6.16) for C_π, A_π, B_ζ and S^{-1}, we see that

$$1 + C_\pi(zI - A_\pi)^{-1} S^{-1} B_\pi = 1 + \sum_{\alpha=1}^{n} \sum_{\beta=1}^{n} (z - w_\alpha)^{-1} t_{\alpha\beta}$$

$$= 1 + \sum_{\alpha=1}^{n} \left\{ \sum_{\beta=1}^{n} t_{\alpha\beta} \right\} (z - w_\alpha)^{-1}.$$

On the other hand, the right side of (6.15) can be expanded in partial fractions

$$\frac{(z - z_1)\cdots(z - z_n)}{(z - w_1)\cdots(z - w_n)} = 1 + \sum_{\alpha=1}^{n} \left\{ \frac{\Pi_{1 \le j \le n}(w_\alpha - z_j)}{\Pi_{1 \le j \le n, j \neq \alpha}(w_\alpha - w_j)} \right\} (z - w_\alpha)^{-1}.$$

Thus (6.15) is verified once we show

$$\sum_{\beta=1}^{n} t_{\alpha\beta} = \frac{\Pi_{1 \le j \le n}(w_\alpha - z_j)}{\Pi_{1 \le j \le n, j \neq \alpha}(w_\alpha - w_j)} \tag{6.19}$$

for $1 \le \alpha \le n$. Plugging in now (6.17) for $t_{\alpha\beta}$, we see that the left side of (6.19) is

$$\sum_{\beta=1}^{n} t_{\alpha\beta} = \frac{\Pi_{1 \le k \le n}(w_\alpha - z_k)}{\Pi_{1 \le k \le n, k \neq \alpha}(w_\alpha - w_k)} \cdot (-1)^{n-1}.$$

$$\cdot \sum_{\beta=1}^{n} \left\{ \frac{\Pi_{1 \le j \le n, j \neq \alpha}(w_j - z_\beta)}{\Pi_{1 \le j \le n, j \neq \beta}(z_\beta - z_j)} \right\}$$

Thus (6.19) is verified once we show

$$1=(-1)^{n-1}\sum_{\beta=1}^{n}\left\{\frac{\Pi_{\substack{1\leq j\leq n\\ j\neq\alpha}}(w_j-z_\beta)}{\Pi_{\substack{1\leq j\leq n\\ j\neq\beta}}(z_\beta-z_j)}\right\}. \tag{6.20}$$

To verify this consider z_n as a complex variable ζ and the other parameters $z_1,...,z_{n-1},w_1,..,w_n$ as fixed. The right side of (6.20) can be expressed as

$$f(\zeta)=(-1)^{n-1}\sum_{\beta=1}^{n-1}\left\{\frac{\Pi_{\substack{1\leq j\leq n\\ j\neq\alpha}}(w_j-z_\beta)}{\Pi_{\substack{1\leq j\neq n-1\\ j\neq\beta}}(z_\beta-z_j)}\right\}(z_\beta-\zeta)^{-1}$$

$$+(-1)^{n-1}\frac{\Pi_{\substack{1\leq j\leq n\\ j\neq\alpha}}(w_j-\zeta)}{\Pi_{1\leq j\leq n-1}(\zeta-z_j)}.$$

Note that $f(\zeta)$ is a rational function in ζ with value 1 at ∞ and with only poles of order at most 1 occurring at the points $\zeta=z_1,z_2,...,z_{n-1}$. Compute the residue R_β of $f(\zeta)$ at the pole $\zeta=z_\beta$ to get

$$R_\beta=\lim_{\zeta\to z_\beta}(\zeta-z_\beta)f(\zeta)=0.$$

Thus $f(\zeta)$ is a rational function with value 1 at infinity and having no poles in the complex plane. Thus f must be identically equal to 1 and (6.20) (and hence also (6.15)) follows as desired. ▪

Theorem 6.1 suggests that the study of invertible solutions S of Sylvester equations

$$SA_\pi-A_\zeta S=\Gamma$$

is of great interest. In general it is known that a unique (possibly not invertible) solution exists if A_π and A_ζ have disjoint spectra, but not much is known concerning when an invertible solution exists except for the case rank $\Gamma=1$. The question of invertibility of a solution also comes up in other applications in control theory; for further discussion, we refer to [DD].

2.7 Notes

The exposition in the first two sections follows mostly [GS]. In that paper the main results are also obtained for some classes of infinite dimensional operator functions. In sections 2.3, 2.4 and 2.5 we follow [BGK] and [GKLR] (see also [GLR2]). Section 2.4 contains the now classical realization theory which was introduced in system theory and is the basis of the state space method. For more details see [K] and [KFA]. The main results of section 2.6 are from [GKLR].

3. INTERPOLATION PROBLEMS FOR J-UNITARY MATRIX FUNCTIONS

It is easy to see that a scalar rational function $u(z)$ which has modulus 1 on the imaginary axis necessarily has the form

$$u(z) = \alpha \frac{z - z_1}{z + \bar{z}_1} \cdot \frac{z - z_2}{z + \bar{z}_2} \cdots \frac{z - z_n}{z + \bar{z}_n}$$

where $|\alpha| = 1$ and the points $z_1, \ldots, z_n$ are its zeros. Indeed, such a $u(z)$ satisfies

$$u(z)\overline{u(-\bar{z})} = 1$$

identically on the imaginary axis, and therefore by analytic continuation, identically at all points of analyticity in the complex plane. Therefore if $u(z)$ has zeros $z_1, \ldots, z_n$, then the poles of $u(z)$ are $-\bar{z}_1, \ldots, -\bar{z}_n$ (including multiplicities). As is clear from the formula the function is uniquely determined by its zeros and the value at one regular point; alternatively the function is uniquely determined by its poles and the value at one regular point. The formula can therefore be considered as a solution to an interpolation problem: namely, construct a rational function $u(z)$ having prescribed zeros (or poles) and assuming values of modulus 1 on the imaginary axis. The main aim of this chapter is to obtain the solution of an analogous interpolation problem for matrix functions. More generally we shall seek a characterization of rational matrix functions $U(z)$ which have J-unitary values on the imaginary axis in terms of null and pole structure. Here J is a fixed signature matrix (i.e. Hermitian unitary) and a matrix X is said to be J-unitary if $X^*JX = J$ (or equivalently if $XJX^* = J$). Note that a J -unitary matrix is necessarily invertible.

3.1 Realization theorems

In this section we study the minimal realizations for rational matrix functions which are J-unitary on the imaginary axis. If $U(z)$ is such a function then the identity

$$U(z)J(U(-\bar{z}))^* = J$$

holds for all regular points on the imaginary axis and then by analytic continuation, for all points of regularity in the complex plane.

The following is the main result of this section. Note that $U(z)$ can have poles on the imaginary axis (unless J is I or $-I$); and any pole of $U(z)$ on the imaginary axis must be also a zero of $U(z)$ with the same partial multiplicities. Also, if $U(z)$ is analytic at infinity, then $U(\infty)$ is necessarily invertible.

THEOREM 1.1. *Let U be a rational matrix function analytic at infinity. Then U is J-unitary on the imaginary axis if and only if any minimal realization $U(z) = D + C(zI - A)^{-1}B$ for U has the properties:*

a) D is J-unitary

b) There exists a unique invertible Hermitian solution H to the Lyapunov equation

$$HA + A^*H = -C^*JC \tag{1.1}$$

for which

$$B = -H^{-1}C^*JD. \tag{1.2}$$

PROOF: If $U(z) = D + C(zI - A)^{-1}B$ is a minimal realization for $U(z)$ and U is J-unitary on the imaginary axis, then certainly $D = U(\infty)$ is J-unitary. Since U is J-unitary on the imaginary axis,

$$U(z)^{-1} = JU(-\bar{z})^*J \tag{1.3}$$

and, since a minimal realization of $U^{-1}(z)$ is

$$U^{-1}(z) = D^{-1} - D^{-1}C(zI - A^\times)^{-1}BD^{-1}$$

with

$$A^\times = A - BD^{-1}C$$

(see Theorem 2.4.7), equation (1.3) may be written as

$$D^{-1} - D^{-1}C(zI - A^\times)^{-1}BD^{-1} = J(D^* - B^*(zI + A^*)^{-1}C^*)J \tag{1.4}$$

Equation (1.4) is an equality between two minimal realizations of the same rational matrix function, hence by corollary 2.4.6, there is a uniquely defined invertible matrix $S = -H$ such that

$$-JB^*H = D^{-1}C, -H^{-1}C^*J = BD^{-1}, -H^{-1}A^*H = A^\times. \tag{1.5}$$

By taking adjoints in each equation and rearranging, we see that H^* also satisfies (1.5); therefore by the uniqueness of $H, H = H^*$. It is straightforward to check that equations (1.5) are equivalent to (1.1) and (1.2), and thus the solution H to (1.1) and (1.2) is unique.

Conversely, suppose $U(z)$ has a minimal realization $D+C(zI-A)^{-1}B$ for which D is J-unitary and $B = -H^{-1}C^*JD$ for some invertible Hermitian solution of (1.1). Then we compute

$$\begin{aligned} U(z)JU(w)^* = \\ &(D + C(zI - A)^{-1}B)J(D^* + B^*(\bar{w}I - A^*)^{-1}C^*) \\ &= DJD^* + C(zI - A)^{-1}BJD^* + DJB^*(\bar{w}I - A^*)^{-1}C^* \\ &+ C(zI - A)^{-1}BJB^*(\bar{w}I - A^*)^{-1}C^*. \end{aligned}$$

Replacing DJD^* by J, BJB^* by $-AH^{-1} - H^{-1}A^*$ and BJD^* by $-H^{-1}C^*$ we obtain

$$U(z)JU(w)^* = J - (z + \bar{w})C(zI - A)^{-1}H^{-1}(\bar{w}I - A^*)^{-1}C^*. \tag{1.6}$$

From (1.6) it follows that U is J-unitary on the imaginary axis. ∎

There are explicit formulas for the similarity matrix $-H$(see Theorem 2.4.3). Thus H is given by the following formula

$$\begin{aligned} H &= -[col(JB^*(A^*)^j)_0^{n-1}]^{-L}[col(D^{-1}C(A^*)^j)_0^{n-1}] \\ &= [C^*J, A^*C^*J, \ldots (A^*)^{n-1}C^*J][BD^{-1}, A^*BD^{-1}, \ldots, (A^*)^{n-1}]^{-R} \end{aligned}$$

We note that property (b) of Theorem 1.1 is equivalent to:

b') *There exists a unique invertible Hermitian solution G to the Lyapunov equation*

$$GA + A^*G = -BJB^* \tag{1.7}$$

for which

$$C = -DJB^*G^{-1}. \tag{1.8}$$

Indeed, given that (b) holds, solve (1.2) for C and substitute in (1.1). The result is that (1.7) and (1.8) hold with $G = H^{-1}$. Analogously one verifies that (b') implies (b).

We conclude this section with some examples of rational functions J-unitary on the imaginary axis.

EXAMPLE 1.1. *Let P be $m \times m$ matrix such that*

$$PJP^* = P$$

and let w be a non imaginary point. Then, the function

$$U(z) = I - P + (z-w)(z+\bar{w})^{-1}P$$

is J-unitary on the imaginary axis.

A typical example of P is given by $P = uu^*J(u^*Ju)^{-1}$ where u is not a J-neutral vector $(u^*Ju \neq 0)$.

EXAMPLE 1.2. *Let u_1 and u_2 be two vectors such that*

$$u_1^*Ju_1 = u_2^*Ju_2 = 0, u_1^*Ju_2 \neq 0$$

and define, for $i \neq j$,

$$W_{ij} = u_i(u_j^*Ju_i)^{-1}u_j^*J.$$

Let w_1 and w_2 be two purely imaginary real points. Then, the function

$$U(z) = I + \left(\frac{z-w_2}{z+\bar{w}_1^*} - 1\right)W_{12} + \left(\frac{z-w_1}{z+\bar{w}_2} - 1\right)W_{21}$$

is J-unitary on the imaginary axis.

EXAMPLE 1.3. *Let α be a real number, let n be a positive integer and u be a J-neutral vector. Then, the function*

$$U(z) = I + i(iz-\alpha)^{-n}uu^*J$$

is J-unitary on the imaginary axis.

3.2 J-unitary functions with given null pair or pole pair.

Let $U(z)$ be a rational matrix function which is J-unitary on the imaginary axis and regular at infinity. Because of the formula

$$U(z) = J[U(-\bar{z})^*]^{-1}J,$$

there is simple connection between global null and pole pairs for $U(z)$ (which can be also seen from equality (1.4) using Theorem 2.5.2). Namely, if (A, B) is a global left pole pair for $U(z)$, then $(-JB^*, -A^*)$ is a (global) right null pair for $U(z)$, and if (C, A) is a right pole pair for $U(z)$, then $(-A^*, -C^*J)$ is a left null pair for $U(z)$. Applying the same result to $U^{-1}(z)$ we see that if (A, B) is a (global) left null pair for $U(z)$, their $(-JB^*, -A^*)$ is a global right pole pair for $U(z)$, and if (C, A) is a (global) right null pair for $U(z)$ then $(-A^*, -C^*J)$ is a (global) left pole pair for $U(z)$.

Solution of the basic interpolation problem for J-unitary rational matrix functions is given by the following theorem.

THEOREM 2.1. *Suppose (C, A) is a null-kernel pair and J is a signature matrix of the same size as A. Then there exists a rational matrix function $U(z)$ invertible at infinity such that (a) $U(z)$ is J-unitary for all z on the imaginary axis apart from the poles of $U(z)$ there (if any) and (b) (C, A) is a global right pole pair for $U(z)$ if and only if there exists an invertible Hermitian solution H to the Lyapunov equation*

$$HA + A^*H = -C^*JC \tag{2.1}$$

In this case any such $U(z)$ has the form

$$U(z) = D - C(zI - A)^{-1}H^{-1}C^*JD \tag{2.2}$$

where H is an invertible Hermitian solution of (2.1) and D is any J-unitary matrix.

PROOF: Let $U(z)$ be a rational matrix function invertible at infinity with the properties a) and b). By Theorem 2.5.2, there is B such that

$$U(z) = D + C(zI - A)^{-1}B$$

is a minimal realization for $U(z)$. Now apply Theorem 1.1.

Conversely, let H be an invertible Hermitian solution of (2.1), and let $U(z)$ be defined by (2.2). We verify first that the realization (2.2) is minimal. By Theorem 2.4.5, we have only to check that the pair $(A, H^{-1}C^*JD)$, or equivalently $(A, H^{-1}C^*)$ is full range.

Observe that $(-A^*, C^*)$ is a full range pair because (C, A) is a null kernel pair. Hence $(-H^{-1}A^*H, H^{-1}C^*)$ is full range as well. Using the equality

$$-H^{-1}A^*H = A + H^{-1}C^*JC,$$

and the fact that the full range property of pairs is preserved under the feedback transformation $(X, Y) \to (X - YF, Y)$ for any matrix F of suitable size (cf. the proof of 2.4.6), we obtain the full range property of $(A, H^{-1}C^*)$. ∎

Using the observation made in the beginning of this section, one can see that an analogous theorem holds for left null pairs:

THEOREM 2.2. *Let (A, B) be a full range pair, and let J be a signature matrix. Then there exists a rational matrix function $U(z)$ invertible at infinity such that $U(z)$ is a J-unitary on the imaginary axis and (A, B) is a global left null pair for $U(z)$ if and only if the Lyapunov equation*

$$GA^* + AG = BJB^* \tag{2.3}$$

has an invertible Hermitian solution G. In this case any such $U(z)$ is given by the formula

$$U(z) = D - JB^*(zI + A^*)^{-1}G^{-1}BD,$$

where G is an invertible Hermitian solution of (2.3) and D is any J-unitary matrix.

Theorems 2.1 and 2.2 have important corollaries which we will state only for the case of left null pair.

THEOREM 2.3. *Let (A, B) be a full range pair with spectrum of A lying in the open right half plane, and let J be a signature matrix. Then there exists a rational matrix function $U(z)$ such that*

(i) $U(z)$ has no poles in the closed right half plane

(ii) $U(z)$ *is* J*-unitary on the imaginary axis*

and

(iii) (A, B) *is a left null pair for* $U(z)$ *with respect to the right half plane*

if and only if the unique solution G *of (2.3) is invertible. In this case such a* $U(z)$ *is unique up to multiplication on the right by a constant* J*-unitary matrix, and the unique* $U(z)$ *with the additional property that* $U(\infty) = I$ *is given by the formula*

$$U(z) = I - JB^*(zI + A^*)^{-1}G^{-1}B, \tag{2.4}$$

where G *satisfies (2.3).*

PROOF: Suppose such a $U(z)$ exists, and let $(\tilde{A}, \tilde{B})$ be a global left null pair for $U(z)$. Then again by the observation at the beginning of this section, $(-J\tilde{B}^*, -\tilde{A}^*)$ is a global right pole pair for $U(z)$. Thus if the spectrum of $\tilde{A}$ contained any point z_0 in the closed left half plane, $-\bar{z}_0$ would be a pole of $U(z)$ in the closed right half plane, contrary to assumption (i). Thus $\tilde{A}$ has no spectrum in the closed left half plane, so by (iii) in fact (A, B) is a global left null pair for $U(z)$. Now Theorem 2.3 implies that the Lyapunov equation (2.3) has an invertible Hermitian solution. Note that since $\sigma(A)$ is in the right half plane, there always exists a unique Hermitian solution.

Conversely, suppose that the unique Hermitian solution of (2.3) is invertible. Then by Theorem 2.3 $U(z)$ as defined by (2.4) is the unique matrix function with $U(\infty) = I$ and with J-unitary values on the imaginary axis which has (A, B) as a global left null pair. In particular, since $\sigma(A)$ is in the right half plane, (A, B) is a left null pair with respect to the right half plane. Then $(-JB^*, -A^*)$ is a global right pole pair for $U(z)$, where $\sigma(-A^*)$ is in the left half plane. Hence $U(z)$ has the additional property (i) as well. ∎

3.3 The associated Hermitian matrix.

Let J be an $m \times m$ signature matrix. An $m \times m$ matrix X is called J-contractive if

$$X^*JX \leq J,$$

i.e. the difference $J - X^*JX$ is positive semidefinite. In this section we study rational matrix functions $U(z)$ that are J-unitary on the imaginary axis and J-contractive in the open right half-plane:

$$U(z)^*JU(z) \leq J \tag{3.1}$$

for all z with $Re z \geq 0$ that are not poles for $U(z)$, and with equality in (3.1) for $Re z = 0$. Note that poles of $U(z)$ in the open right half plane are not ruled out by (3.1); for example, take

$$J = \begin{bmatrix} 1 & 0 \\ 0 & -1 \end{bmatrix}, U(z) = \begin{bmatrix} 1 & 0 \\ 0 & (z+1)(z-1)^{-1} \end{bmatrix}.$$

We start with describing the invariants of the invertible Hermitian solution of (1.1).

THEOREM 3.1. *Let U be a rational matrix function, analytic at infinity and J-unitary on the imaginary axis. Let $U(z) = D + C_i(zI - A_i)^{-1}B_i$, $i = 1, 2$, be two minimal realizations of U with the unique invertible Hermitian solutions H_1 and H_2 of (1.1) and (1.2) associated with the realization of $U(z)$ indexed by 1 and 2, respectively. Then the two minimal realizations are similar*

$$C_1 = C_2S, A_1 = S^{-1}A_2S, B_1 = S^{-1}B_2$$

for a unique invertible matrix S. Moreover

$$H_1 = S^*H_2S.$$

In particular, the matrices H_1 and H_2 have the same number of positive eigenvalues and the same number of negative eigenvalues.

PROOF: The existence and uniqueness of the similarity matrix S follows from Corollary 2.4.6. As $(-A_1^*, -C_1^*, J)$ and $(-A_2^*, -C_2^*J)$ are global left null pairs for $U(z)$, they are similar:

$$-A_1^* = -S_2^{-1}A_2^*S_2, -C_1^*J = -S_2^{-1}C_2^*J \tag{3.2}$$

for some invertible S_2 (which is unique). Comparing (3.2) with the unique similarity S between the global right pole pairs (C_1, A_1) and (C_2, A_2) of $U(z)$, we conclude that $S_2 = (S^*)^{-1}$. Now one verifies that

$$(S^*H_2S)A_1 + A_1^*(S^*H_2S) = -C_1^*JC_1$$

and

$$(S^*H_2S)B_1 = -C_1^*JD$$

By the uniqueness of H_1 it follows that $H_1 = S^*H_2S$. ■.

From Lemma 3.1 we see that the number of positive and negative eigenvalues of the Hermitian matrix H associated with a rational matrix function $U(z)$ analytic and invertible at infinity and J-unitary valued on the imaginary axis is independent of the choice of a minimal realization for $U(z)$, and thus is a characteristic intrinsic to the function $U(z)$ itself. An alternative characterization of the number of negative eigenvalues of H more intrinsic to the funciton $U(z)$ is in terms of the number of negative squares had by a kernel function associated with $U(z)$.

We recall (see [KL]) that a $m \times m$ matrix valued function $K(z,s)$ defined for z and w in some set E and such that $K(z,w)^* = K(w,z)$ is said to have ν negative squares if for any positive integer r, points $w_1, ..., w_r$ in E and any vectors $c_1, ..., c_r$ in C^m the $r \times r$ hermitian matrix with ij entry

$$c_j^* K(w_j, w_i) c_i \tag{3.3}$$

has at most ν strictly negative eigenvalues and exactly ν strictly negative eigenvalues for some choice of $r, w_1, ..., w_r, c_1, ..., c_r$. With this definition at hand, we can now state the following theorem, which gives a characterization of the number of negative eigenvalues of the associated matrix H.

THEOREM 3.2. *Let U be a rational $m \times m$ function J-unitary on the imaginary axis and analytic at infinity, and let $U(z) = D + C(zI - A)^{-1}B$ be a minimal realization of U, with associated Hermitian matrix H. Then,, the number of negative eigenvalues of the matrix H is equal to the number of negative squares of the function*

$$K_U(z,w) = \frac{J - U(z)JU(w)^*}{(z + \bar{w})} \tag{3.4}$$

Finally, let $K(U)$ be the span of the function $z \to K_U(z,w)e$ where w is in the resolvent set of A and where c is in C^m. Then $K(U)$ is a finite dimensional space of rational functions analytic on the resolvent set of A and the dimension of $K(U)$ is equal to the MacMillan degree of U(i.e. the size of A).

PROOF: From the formula (1.6) we have for z and w in the resolvent set of

$$K_U(z,w) = C(zI - A)^{-1}H^{-1}(\bar{w}I - A^*)^{-1}C^* \tag{3.5}$$

Let r be a positive integer and $w_1, ..., w_r$ be in the resolvent set of $A, c_1, ..., c_r$ be in C^m. Then, the matrix equality

$$(c_j^* K_U(w_j, w_i)c_i)_{ij=1,r} = X^* H^{-1} X \tag{3.6}$$

with

$$X = ((\bar{w}_1 - A^*)^{-1}C^*c_1, ..., (\bar{w}_r - A^*)^{-1}C^*c_r)$$

makes it clear that the function K_U has at most ν_H negative squares, where ν_H denotes the number of negative eigenvalues of the hermitian matrix H. The pair (C, A) is null kernel, and thus we can choose a basis of C^n of the form $x_i = (w_i^* - A_i^*)^{-1}C^*c_i \ \ i = 1, ..., n$. In particular, det $X \neq 0$ for $X = (x_1, ..., x_n)$ and the matrix $X^*H^{-1}X$ has exactly v_H negative squares, and thus K_U has ν_H negative squares. ∎

In applications the case where the number of negative squares is zero is of particular interest. For a given signature matrix J let us say that the matrix function $U(z)$ is J-<u>inner</u> if

(i) $U(z)$ is J-unitary at all regular points on the imaginary axis

and

(ii) $U(z)$ is J-contractive in the right half plane.

The following Theorem characterizes when there exists a J-inner function with a preassigned global left null pair.

THEOREM 3.3. *Suppose (A, B) is a full-range pair and J is a signature matrix. Then there exists a rational matrix function $U(z)$ invertible at infinity and such that*

a) $U(z)$ is J-inner

and

b) (A, B) is a global left null pair for $U(z)$

if and only if the Lyapunov equation (2.3) has a positive definite solution G. In this case all such $U(z)$ are given by (2.4) with positive definite solutions G of (2.3).

PROOF: Simply combine Theorems 2.2 and 3.2. Note that the case when $U(z)$ has poles in the closed right half plane is not excluded. ▪

Analogous corollaries can be stated concerning left null pairs of J-contractive rational functions, as well as concerning rational function $U(z)$ that are J-unitary on the imaginary axis and J-expansive (i.e. $(U(z))^*JU(z) \geq J$) in the open right half-plane.

3.4 Notes

This chapter mostly follows the preprint [AG]. A revised version of this preprint is appearing in this volume . Results close to the realization theorems one can find in [G]. Theorem 2.3 is taken from [BR]. For results about inner matrix functions see [GVKDM] (where the functions are called J-lossless). This chapter also has intersections with the paper [Sakh].

4. NEVANLINNA-PICK INTERPOLATION

In this chapter we consider the simplest matrix valued version of the classical Nevanlinna-Pick interpolation problem. The solution will be based on the global interpolation problem for rational matrix functions studied in Section 2.6.

4.1 Classical Nevanlinna-Pick interpolation

The classical Nevanlinna-Pick interpolation problem is as follows. We are given a collection of points $z_1, ..., z_n$ in the right halfplane Π_+ together with a collection of complex number $w_1, ..., w_n$. Let us say that a scalar function f analytic on Π_+ interpolates if

$$f(z_j) = w_j, 1 \leq j \leq n \tag{1.1}$$

The problem is to describe all interpolating functions f for which

$$sup|f(z)| < 1 \tag{1.2}$$

In the spirit of this paper we will be interested only in rational solutions. The data set $w = \{z_j, w_j : 1 \leq j \leq n\}$ we call the data set for the interpolation problem. The set of all rational interpolating functions (i.e. functions satisfying (1.2)) we denote by $I(w)$. Denote by BR all rational functions

f with no poles on Π_+ for which $\sup_{z\in\Pi_+}|f(z)| \leq 1$. With this notation, the problem is to describe the set $I(w) \subset BR$. Define the matrix (called the associated Pick matrix)

$$\Lambda(w) = \left[\frac{1 - w_i\bar{w}_j}{z_i + \bar{z}_j}\right]_{1\leq i,j\leq n} \tag{1.3}$$

Then the result is as follows.

THEOREM 1.1. *There exists a solution $f \in I(w) \cap BR$ of the interpolation problem if and only if the associated Pick matrix $\Lambda(w)$ given by (1.3) is positive definite. In this case there is a rational 2×2 matrix function $[\Theta_{ij}(z)]_{1\leq i,j\leq 2}$ such that the infinitely many solutions $f(z)$ of the classical Nevanlinna-Pick interpolation problem are described by*

$$f(z) = [\Theta_{11}(z)g(z) + \Theta_{12}(z)][\Theta_{21}(z)g(z) + \Theta_{22}(z)]^{-1}$$

where g is any function in BR, and the rational 2×2 matrix function

$$\Theta(z) = \begin{bmatrix} \Theta_{11}(z) & \Theta_{12}(z) \\ \Theta_{21}(z) & \Theta_{22}(z) \end{bmatrix}$$

is given by

$$\Theta(z) = I - C(zI + A^*)^{-1}[\Lambda(w)]^{-1}B, \tag{1.4a}$$

where

$$A = diag.(z_1, z_2, ..., z_n) \tag{1.4b}$$

$$B = \begin{bmatrix} 1 & -w_1 \\ \vdots & \vdots \\ 1 & -w_n \end{bmatrix}, C = \begin{bmatrix} 1 & \cdots & 1 \\ \bar{w}_1 & \cdots & \bar{w}_n \end{bmatrix}. \tag{1.4c}$$

We shall obtain Theorem 1.1 later as a particular (scalar) case of the solution of a more general tangential Nevanlinna-Pick interpolation problem.

4.2 Tangential Nevanlinna-Pick interpolation with simple multiplicities

Let $BR_{M\times N}$ be the set of all rational $M \times N$ matrix functions $F(z)$ with no poles in the open right halfplane Π_+ such that

$$sup\{\|F(z)\| : z \in \Pi_+\} < 1$$

The simplest case of the tangential Nevanlinna-Pick interpolation problem is to find all matrix functions F from $BR_{M\times N}$ which satisfy a set of interpolation conditions of the form

$$x_j^*F(z_j) = y_j^* \text{ for } 1 \leq j \leq n \tag{2.1}$$

Here $z_1, ..., z_n$ are distinct points in $\Pi_+, x_1, ..., x_n$ are nonzero column vectors in $\mathbb{C}^M$ and $y_1, ..., y_n$ are vectors in $\mathbb{C}^N$; by x_j^* and y_j^* we mean the conjugate transposes of x_j and y_j. Experience from the classical case (see [Nev]) suggests that the solutions of the problem (when they exist) should be parametrized by a linear fractional map

$$F = T_\Theta[G] = (\Theta_{11}G + \Theta_{12})(\Theta_{21}G + \Theta_{22})^{-1}$$

The matrix function

$$\Theta(z) = \begin{bmatrix} \Theta_{11}(z) & \Theta_{12}(z) \\ \Theta_{21}(z) & \Theta_{22}(z) \end{bmatrix}$$

is called the symbol of the linear fractional map T_Θ. Note that $F = T_\Theta(G)$ is equivalent to the identity

$$\begin{bmatrix} F \\ I \end{bmatrix} = \Theta \begin{bmatrix} G \\ I \end{bmatrix} (\Theta_{21}G + \Theta_{22})^{-1}. \tag{2.2}$$

Set $X(z) = (\Theta_{21}(z)G(z) + \Theta_{22}(z))^{-1}$. From the identity (where $J = \begin{bmatrix} I & 0 \\ 0 & -I \end{bmatrix}$)

$$F(z)^*F(z) - I = [F(z)^*I]J \begin{bmatrix} F(z) \\ I \end{bmatrix}$$

$$= X(z)^*[G(z)^*I]\Theta(z)^*J\Theta(z) \begin{bmatrix} G(z) \\ I \end{bmatrix} X(z)$$

we see that if

$$\Theta(z) \ is \ J - unitary \ on \ the \ imaginary \ axis \ \partial\Pi_+ \tag{2.3}$$

then $\|F(z)\| < 1$ for all $z \in \partial\Pi_+$ if and only if $\|G(z)\| < 1$ for all $z \in \partial\Pi_+$, and if

$$\Theta(z) \ is \ J - contractive \ for \ z \in \Pi_+ \tag{2.4}$$

then $\|F(z)\| < 1$ for $z \in \Pi_+$ whenever $\|G(z)\| < 1$ for $z \in \Pi_+$. The interpolation conditions (2.1) can be rewritten as

$$[x_j^*, -y_j^*]\begin{bmatrix} F(z_j) \\ I \end{bmatrix} = 0 \text{ for } 1 \le j \le n$$

If we demand in addition that

$$[x_j^*, -y_j^*]\Theta(z_j) = 0 \tag{2.5}$$

then we see from (2.2) that then F also satisfies the interpolation conditions, after we rule out the possibility of a pole of $X(z)$ at z_j cancelling out the zero of $[x_j^*, -y_j^*]\Theta(z)$ at z_j. If the single constant row vector function $[x_j^*, -y_j^*]$ forms a canonical set of left null functions for Θ at z_j and Θ has no other zeros and no poles in Π_+, the formula $F = T_\Theta(G)$ with G chosen arbitrarily in $BR_{M\times N}$ has a chance of producing all $F \in BR_{M\times N}$ which satisfy the interpolating conditions (2.1).

By the results of Chapter 3, it is clear how to build such a matrix function Θ. Set B equal to the $n \times (M+N)$ matrix

$$B = \begin{bmatrix} x_1^* & -y_1^* \\ \vdots & \vdots \\ x_n^* & -y_n^* \end{bmatrix} \tag{2.6}$$

and A equal to the $n \times n$ diagonal matrix

$$A = diag(z_1, , ..., z_n). \tag{2.7}$$

The construction of the rational matrixx function Θ which has $\{[x_j^*, -y_j^*]\}$ as a canonical set of left null functions at z_j and which is regular and J-unitary on the imaginary axis involves solving the Lyapunov equation

$$SA^* + AS = BJB^*. \tag{2.8}$$

For A and B given by (2.7) and (2.6) and letting S_{ij} denote the (i,j)-th entry of the unknown matrix S, we see that equations (2.8) reduce to

$$S_{ij}\bar{z}_j + z_i S_{ij} = x_i^* x_j - y_i^* y_j \tag{2.9}$$

from which we see that

$$S_{ij} = [x_i^* x_j - y_i^* y_j](z_i + \bar{z}_j)^{-1} \text{ for } 1 \le i, j \le n.$$

Let us denote by w the set

$$w = \{z_j, x_j, y_j : 1 \leq j \leq n\} \tag{2.10}$$

of data for the interpolation problem. Denote by $\Lambda(w)$ the matrix

$$\Lambda(w) = [(x_i^* x_j - y_i^* y_j)(z_i + \bar{z}_j)^{-1}]_{1 \leq i,j \leq n} \tag{2.11}$$

which solves the Lyapunov equation (2.8); it turns out that $\Lambda(w)$ plays the role of the classical Pick matrix for the problem. We now give the construction of the required matrix function $\Theta(z)$; the proof is immediate from the following lemma which is a particular case of Corollary 3.2.3 and 3.3.3.

LEMMA 2.1. *Suppose the data set $w = \{z_j; x_j, y_j : 1 \leq j \leq n\}$ for a tangential Nevanlinna-Pick interpolatin problem on Π_+ is given. From this data set construct matrices B, A and $\Lambda = \Lambda(w)$ as in (2.6), (2.7) and (2.11). Then there exists a rational matrix function $\Theta(z)$ which is regular and J-unitary on the imaginary axis and which has (A, B) as a left null pair over Π_+ and no poles in Π_+ if and only if Λ is invertible. In this case one choice of $\Theta(z)$ is given by*

$$\Theta(z) = I - JB^*(zI + A^*)^{-1}\Lambda^{-1}B \tag{2.12}$$

Moreover $\Theta(z)$ is J-contractive on Π_+ if and only if Λ is positive definite.

We are now ready to state the solution of the tangential Nevanlinna-Pick interpolation problem in a precise form.

THEOREM 2.2. *Suppose that the data set $w = \{z_j, x_j, y_j : 1 \leq j \leq n\}$ for a tangential interpolation problem on Π_+ is given. Construct matrices B, A and $\Lambda = \Lambda(w)$ according to (2.6), (2.7) and (2.11). Then there exist functions F in $BR_{M\times N}$ satisfying the interpolation conditions (2.1) if and only if Λ is positive definite. In this case, let*

$$\Theta(z) = \begin{bmatrix} \Theta_{11}(z) & \Theta_{12}(z) \\ \Theta_{21}(z) & \Theta_{22}(z) \end{bmatrix} \tag{2.13}$$

be the function given by Lemma 2.1 (2.12); then $F \in BR_{M\times N}$ satisfies the interpolation conditions (2.1) if and only if

$$F = (\Theta_{11}G + \Theta_{12})(\Theta_{21}G + \Theta_{22})^{-1}$$

for a rational matrix function $G \in BR_{M\times N}$.

Before proving the theorem, we specialize its statement to the scalar case. In this case we may as well take the vectors x_j^* all to be the number 1, and the vectors y_j^* are simply complex numbers $w_j (1 \le j \le n)$. The matrix $\Lambda = \Lambda(w)$ specializes to

$$\Lambda = [(1 - w_i \bar{w}_j)(z_i + \bar{z}_j)^{-1}]_{1 \le i,j \le n}$$

which is the classical Pick matrix. The matrix function $\Theta(z)$ given by (2.12) then specializes to $\Theta(z)$ as given in (1.4). In this way we see that Theorem 1.1 is an immediate consequence of the more general Theorem 2.2.

4.3 Proof of Theorem 2.2

Assume first the $\Lambda = \Lambda(w)$ is invertible. Let $\Theta(z)$ be the function with the properties delineated in Lemma 2.1, with $\Theta(\infty) = I$ partitioned as is (2.13). We prove first the following lemma:

LEMMA 3.1. *The matrix* Λ *is positive definite if and only if* $\Theta_{22}(z)^{-1}$ *has no poles in* Π_+.

PROOF: We have

$$\Theta(-\bar{z})^* J \Theta(z) = J,$$

where $J = \begin{bmatrix} I_M & 0 \\ 0 & -I_N \end{bmatrix}$. So

$$\Theta_{11}(-\bar{z})^*\Theta_{11}(z) - \Theta_{21}(-\bar{z})^*\Theta_{21}(z) = I; \tag{3.1}$$

$$\Theta_{12}(-\bar{z})^*\Theta_{12}(z) - \Theta_{22}(-\bar{z})^*\Theta_{22}(z) = -I; \tag{3.2}$$

$$\Theta_{11}(-\bar{z})^*\Theta_{12}(z) - \Theta_{21}(-\bar{z})^*\Theta_{22}(z) = 0 \tag{3.3}$$

for all $z \in \mathbb{C}$ which are not poles of $\Theta(z)$ or of $\Theta(-\bar{z})^*$. Equality (3.2) implies that for $Re z = 0$.

$$\Theta_{22}(z)^*\Theta_{22}(z) = I + \Theta_{12}(z)^*\Theta_{12}(z) \ge I$$

and hence $\Theta_{22}(z)$ is invertible on the imaginary axis. In particular, $\det\Theta_{22}(z)$ is not identically zero.

For an $(M+N)\times(M+N)$ matrix X partitioned as $X=\begin{bmatrix} X_{11} & X_{12} \\ X_{21} & X_{22}\end{bmatrix}$ and such that X_{22} is invertible, define the Redheffer transformation (see [R])

$$R(X)=\begin{bmatrix} U_{11} & U_{12} \\ U_{21} & U_{22}\end{bmatrix}$$

$$U_{11}=X_{12}X_{22}^{-1}, U_{12}=X_{11}-X_{12}X_{22}^{-1}X_{21}; U_{21}=X_{22}^{-1}; U_{22}=-X_{22}^{-1}X_{21} \tag{3.4}$$

It is easy to verify, by using the equalities

$$X_{11}^{*}X_{11}-X_{21}^{*}X_{21}=I; X_{12}^{*}X_{12}-X_{22}^{*}X_{22}=-I; X_{11}^{*}X_{12}=X_{21}^{*}X_{22}$$

that X is J-unitary if and only if $R(X)$ is unitary. Moreover, X is J-contractive if and only if $R(X)$ is contractive: $\|R(X)\|\leq 1$.

Applying the Redheffer transformation to $\Theta(z)$, we see that $W(z):=R(\Theta(z))$ is a rational matrix function which is unitary on the imaginary axis and takes the value $\begin{bmatrix} 0 & I \\ I & 0\end{bmatrix}$ at infinity.

Assume now that Λ is positive definite. Then by Corollary 3.3.3 $\Theta(z)$ is J-contractive in Π_+ and hence $W(z)$ is contractive in Π_+. In particular the third formula in (3.4) shows that $\Theta_{22}(z)^{-1}$ has no poles in Π_+.

Suppose now that $\Theta_{22}(z)^{-1}$ has no poles in Π_+. Then $W(z)$ is analytic in Π_+. An application of the maximum modulus principle shows that actually $W(z)$ is contractive in Π_+. (Indeed, fix $\epsilon>0$ and let $r>0$ be so big that $\|W(z)\|\leq 1+\epsilon$ for $|z|=r$. Apply the maximum modulus principle for the analytic function $W(z)$ on the closed set

$$I_r=\{z\in\mathbb{C}: Re z\geq 0, |z|\leq r\}$$

The result is $\|W(z_0)\|\leq 1+\epsilon$ for every $z_0\in I_r$. Taking $\epsilon\to 0$ and consequently $r\to\infty$ proves that $W(z)$ is contractive in Π_+). But then $\Theta(z)$ is J-contractive on Π_+, and by Corollary 3.3.3 Λ is positive definite. ∎

Proof of Theorem 2.2 Let us suppose first that solution $F\in BR_{M\times N}$ of the interpolation conditions exist. Then the block column matrix function $\begin{bmatrix} F(z) \\ I_N\end{bmatrix}$ has no poles in Π_+ and satisfies

$$[x_j^{*}, -y_j^{*}]\begin{bmatrix} F(z_j) \\ I_N\end{bmatrix}=0 \tag{3.5}$$

Let us assume for the moment that $\Lambda = \Lambda(w)$ is invertible and let $\Theta(z)$ be the matrix function with $\Theta(\infty) = I$ delineated in Lemma 2.1. Write

$$\begin{bmatrix} F(z) \\ I \end{bmatrix} = \Theta(z) \begin{bmatrix} G_1(z) \\ G_2(z) \end{bmatrix} \tag{3.6}$$

for some rational matrix function $G_1(z)$, and $G_2(z)$. We claim that $G_1(z)$ and $G_2(z)$ have no poles in Π_+. Indeed, using formula (2.12), the equation

$$\Lambda A^* + A\Lambda = BJB^*$$

and a general formula (2.4.14) for a realization of the inverse function, we see that

$$\Theta(z)^{-1} = I + JB^*\Lambda^{-1}(zI - A)^{-1}B.$$

So to prove that $G_1(z)$ and $G_2(z)$ are analytic in Π_+ we have only to verify that $K(z) := (zI - A)^{-1}B \begin{bmatrix} F(z) \\ I \end{bmatrix}$ is such. From the formulas (2.6), (2.7) and the fact that $F(z)$ is analytic in Π_+ it follows that the singular part of the Laurent series of $K(z)$ is a neighborhood of every z_j is

$$\begin{bmatrix} 0 & \dots & 0\ x_j & 0 & \dots & 0 \\ 0 & \dots & 0 - y_j & 0 & \dots & 0 \end{bmatrix}^* \begin{bmatrix} F(z_j) \\ I \end{bmatrix},$$

which is zero in view of (3.5). Thus indeed $G_1(z)$ and $G_2(z)$ are analytic in Π_+.

Since $\|F(z)\| < 1$ and $\Theta(z)$ is J-unitary on the imaginary axis, we have for $Rez = 0$:

$$\begin{aligned} 0 &> F(z)^*F(z) - I \\ &= [G_1(z)^*, G_2(z)^*]\Theta(z)^*J\Theta(z) \begin{bmatrix} G-1(z) \\ G_2(z) \end{bmatrix} \\ &= [G_1(z)^*, G_2(z)^*]J \begin{bmatrix} G_1(z) \\ G_2(z) \end{bmatrix} \\ &= G_1(z)^*G_1(z) - G_2(z)^*G_2(z). \end{aligned} \tag{3.3}$$

This inequality has several consequences. First of all as $\begin{bmatrix} F(z) \\ I \end{bmatrix}$ clearly has full rank N on the imaginary axis and $\Theta(z)$ is by construction regular there from (3.6) we see that $\begin{bmatrix} G_1(z) \\ G_2(z) \end{bmatrix}$ has full rank N on the axis. Moreover, if $G_2(z_0)x = 0$ for some z_0 with $Rez_0 = 0$ and vector $x \in \mathbb{C}^N$, then (3.7) would imply that $G_1(z_0)x = 0$ as well, a contradiction to the fact that $\begin{bmatrix} G_2(z_0) \\ G_1(z_0) \end{bmatrix}$ has full rank N. Thus $G_2(z)$ is necessarily invertible for each z on the imaginary axis and (3.7) gives that $G(z) := G_1(z)G_2(z)^{-1}$ has the property

$$\sup_{Rez=0} \|G(z)\| < 1. \tag{3.8}$$

Next, from the bottom block row of (3.6) we deduce

$$I = \Theta_{21}G_1 + \Theta_{22}G_2 \tag{3.9}$$

As $\Theta(z)$ is J-unitary and regular on the imaginary axis, we know that $\Theta_{22}(z)$ is regular and invertible on the imaginary axis (see the first paragraph of the proof of Lemma 3.1).

Moreover,

$$\sup_{Rez=0} \|\Theta_{22}^{-1}(z)\Theta_{21}(z)\| < 1. \tag{3.10}$$

Indeed, for pure imaginary z we have

$$\Theta(z)J\Theta(z)^* = J,$$

which implies

$$\Theta_{21}(z)\Theta_{21}(z)^* - \Theta_{22}(z)\Theta_{22}(z)^* = -I.$$

So

$$\Theta_{22}(z)^{-1}\Theta_{21}(z)\Theta_{21}(z)^*\Theta_{22}(z)^{*-1} = I - \Theta_{22}(z)^{-1}\Theta_{22}(z)^{*-1},$$

and since $\Theta_{22}(z)$ is invertible (including at infinity where $\Theta_{22}(\infty) = I$), (3.10) follows.

Thus (3.9) may be rewritten as

$$I = \Theta_{22}(\Theta_{22}^{-1}\Theta_{21}G + I)G_2 \tag{3.11}$$

Since by (3.8) and (3.10) $\|\Theta_{22}^{-1}\Theta_{21}G(z)\| < 1$ for $Rez = 0$, we see that $t[\Theta_{22}^{-1}\Theta_{21}G](z)+I$ is invertible for $Rez = 0$ and for $0 \le t \le 1$. Hence

$$wno\ det(\Theta_{22}^{-1}\Theta_{21}G + I) \tag{3.12}$$

$$= wno\ detI = 0$$

where *wno f* denotes the *winding number* of f, i.e. the net change in the argument $f(z)$ along the imaginary axis for a rational scalar function f having no poles or zeros on that axis. Since Θ_{22} and G_2 by construction have no poles in Π_+, by the argument principle

$$wno\ det\Theta_{22} \geq 0, wno\ detG_2 \geq 0 \tag{3.13}$$

On the other hand from (3.11) we have

$$\begin{aligned} 0 &= wno\ detI \\ &= wno\ det\Theta_{22} + wno\ det(\Theta_{22}^{-1}\Theta_{21}G + I) + wno\ detG_2 \end{aligned}$$

This combined with (3.12) and (3.13) forces (3.13) to hold with equality. Thus both Θ_{22}^{-1} and G_2^{-1} have no poles in Π_+, and by Lemma 3.1 Λ is positive definite.

Suppose now that we do not know a priori that Λ is invertible, but nevertheless a solution F in $BR_{M\times N}$ of the inteprolation conditions exists. If P is the matrix

$$P = [x_i^* x_j (z_i + \bar{z}_j)^{-1}]_{1\leq i,j\leq n}$$

then P is positive definite; one way to see this is by the result of the first part of the proof, since the interpolation problem with each $y_j = 0$ certainly has a solution. Thus for all $\delta > 0$ sufficiently small, the matrix

$$\Lambda - \delta P = P^{\frac{1}{2}}[P^{-\frac{1}{2}}\Lambda P^{-\frac{1}{2}} - \delta I]P^{\frac{1}{2}}$$

is invertible. On the other hand, if $F \in BR_{M\times N}$ satisfies the interpolation conditions, then for all $r > 1$ sufficiently close to 1, $F_r = rF$ is still in $BR_{M\times N}$ and satisfies interpolation conditions

$$x_i^* F_r(z_i) = r y_i^*, 1 \leq i \leq n.$$

The Pick matrix corresponding to these interpolation conditions is

$$\begin{aligned} \Lambda_r &= [(x_i^* x_j - r^2 y_i^* y_j)(z_i + \bar{z}_j)^{-1}] \\ &= r^2\{\Lambda - (1 - \frac{1}{r^2})P\}. \end{aligned}$$

By the observation above, Λ_r is invertible for all $r > 1$ sufficietly close to 1. By the first part of the proof, Λ_r must be positive definite. But then, since

$$\Lambda = r^{-2}\Lambda_r + (1 - r^{-2})P$$

where P is positive definite, it follows that Λ itself is positive definite (and in particular invertible).

Conversely, suppose that Λ is positive definite. Then by Lemma 3.1 we see that Θ_{22}^{-1} has no poles in Π_+; in particular,

$$wno\ det\Theta_{22} = 0 \tag{3.14}$$

Let now G be an arbitrary function in $BR_{M\times N}$ and define

$$\begin{bmatrix} F_1 \\ F_2 \end{bmatrix} = \Theta \begin{bmatrix} G \\ I \end{bmatrix} = \begin{bmatrix} \Theta_{11}G + \Theta_{12} \\ \Theta_{21}G + \Theta_{22}. \end{bmatrix} \tag{3.15}$$

From the bottom block row

$$\begin{aligned} F_2 &= \Theta_{21}G + \Theta_{22} \\ &= \Theta_{22}(\Theta_{22}^{-1}\Theta_{21}G + I) \end{aligned} \tag{3.16}$$

Using the inequality (3.10) and the fact that $\|G(z)\| < 1$ for $Rez = 0$, we have

$$wno\ det(\Theta_{22}^{-1}\Theta_{21}G + I) = 0$$

Taking into account (3.14) and (3.15) we conclude that $wno\ detF_2 = 0$. Thus F_2^{-1} has no poles in Π_+. We rewrite (3.15) as

$$\begin{bmatrix} F \\ I \end{bmatrix} = \Theta \begin{bmatrix} G \\ I \end{bmatrix} F_2^{-1}. \tag{3.17}$$

where $F = F_1F_2^{-1}$. As Θ, $\begin{bmatrix} G \\ I \end{bmatrix}$ and F_2^{-1} each have no poles in Π_+ and (A, B) is a null pair for $\Theta(z)$ over Π_+ we conclude from (3.17) that

$$[x_j^*, -y_j^*] \begin{bmatrix} F(z_j) \\ I \end{bmatrix} = 0 \text{ for } 1 \le j \le n,$$

i.e. F satisfies the interpolation conditions (2.1). Finally, we have (for $Rez = 0$)

$$F(z)^*F(z) - I = F_2(z)^{*-1}[G(z)^*I]\Theta(z)^*J\Theta(z) \begin{bmatrix} G(z) \\ I \end{bmatrix} F_2(z)$$

$$= F_2(z)^{*-1}[G(z)^*G(z) - I]F_2(z)^{-1}$$

Combining this with invertibility of $F_2(z)$ everywhere on the imaginary axis and at infinity (which follows from (3.16)), together with

$$\sup_{Rez=0} \|G(z)\| < 1,$$

we conclude that $F \in BR_{M\times N}$. Thus $F = (\Theta_{11}G + \Theta_{12})(\Theta_{21}G + \Theta_{22})^{-1}$ is a solution of (2.1) in $BR_{M\times N}$ for any $G \in BR_{M\times N}$. Theorem 2.2 is proved completely. ∎

The reader will note that we used only the simplest cases of the theory presented in Chapters 2 and 3 for our solution of the matrix Nevanlinna-Pick interpolation problem in Chapter 4. This is because we considered only the simplest type of matrix interpolation problem to simplify the exposition in Chapter 4. The forthcoming monograph [BGR2], will present solutions of more complicated interpolation problems in the same spirit as for the simple case presented here. Also essentially the same procedure applies when the underlying domain is taken to be the unit disk rather than a half plane.

4.4 Notes

Classical work on scalar Nevanlinna-Pick interpolation begins with the original work of Nevanlinna [Nev] and of Pick [P]; Pick obtained the Pick matrix test for existence of solutions while Nevanlinna derived a linear fractional parametrization of all solutions. Tangential interpolation was introduced first in [Fe 1, Fe2], then in a more abstract form in [N1], and in the West in [H1]. State space formulas very close to ours for the linear fractional parametrizer of the set of all solutions have been given by Soviet authors [Go, KP, Kt, N1] using techniques different from ours. The identification of the parametrizer Θ as itself the solution of an interpolatin problem comes from [BR]; these latter authors based their analysis on the Grassmannian Krein space approach to interpolation in [BH]. Here we have kept the derivation at an elementary level by adapting the winding number argument in [H2].

5. NEVANLINNA-PICK-TAKAGI INTERPOLATION PROBLEMS

5.1 The Scalar Case

We have see that, in order that there exist a (scalar) rational function f having modulus < 1 at all points in the closed right half plane and satisfying interpolation conditions

$$f(z_i) = w_i, 1 \leq i \leq n$$

at preassigned points $z_1, ..., z_n$ in the right half plane Π_+ a necessary and sufficient condition is that the Pick matrix

$$\Lambda = [(1 - w_i \bar{w}_j)(z_i + \bar{z}_j)^{-1}] \tag{1.1}$$

be positive definite. In this case a rational 2×2 matrix function $\Theta(z) = [\Theta_{ij}(z)]_{1 \leq i,j \leq 2}$ can be constructed from the data of the problem such that all interpolating rational functions $f \in BR$ are described as

$$f = (\Theta_{11} g + \Theta_{12})(\Theta_{21} g + \Theta_{22})^{-1}$$

where g is an arbitrary element of BR (scalar rational functions with modulus < 1 in the closed right half plane). Nevertheless, the formula for Θ continues to make sense under the assumption that Λ is merely invertible. In this section we describe what can be said when Λ is invertible but possibly has some negative eigenvalues. Quite simply, we have the following result.

THEOREM 1.1. *Let the points $z_1, , ..., z_n$ in the right half plane Π_+ and the complex numbers $w_1, ..., w_n$ be given. Let Λ be the Pick matrix given by (1.1), and suppose that Λ is invertible. Then the number κ of negative eigenvalues of Λ is the smallest integer for which one can find a rational function f such that*

(i) $|f(z)| < 1$ for z on the imaginary axis

(ii) f is analytic at the points $z_1, ..., z_n$ and satisfies the interpolation conditions

$$f(z_i) = w_i, \quad 1 \leq i \leq n$$

and

(iii) the total number of poles of $f(z)$ (counting multiplicities) in the right half plane is κ.

Moreover, if $\Theta(z) = [\Theta_{ij}(z)]_{1 \le i,j \le n}$ *is given by (4.1.4), any such* f *has the form*

$$f = (\Theta_{11}g + \Theta_{12})(\Theta_{21}g + \Theta_{22})^{-1} \tag{1.2}$$

for a $g \in BR$. *Conversely, if* $g \in BR$ *and* f *defined by (1.2) is analytic at the points* $z_1, ..., z_n$ *(a generic class of* $g's$ *in* BR*), then* f *satisfies (i), (ii) and (iii).*

We shall see that this result is just the specialization to the scalar case of the results on matrix interpolation problems which we discuss in the next section.

The conclusion of Theorem 1.1 concerning the range of the linear fractional map can be made more precise if we enlarge the class of solutions of the inteprolation problem. Namely, we consider rational functions having modulus < 1 on the imaginary axis and k poles in Π_+, some of which may occur at one of the interpolation nodes $z_1, ..., z_n$. We then demand that f satisfy the interpolation condition

$$f(z_i) = w_i$$

at all those ponts z_i at which it happens to be analytic. Then it turns out that the image of the linear fractional map $g \to f = (\Theta_{11}g + \Theta_{12})(\Theta_{21}g + \Theta_{22})^{-1}$ $(g \in BR)$ coincides with all solutions f of the interpolation problem in this broader sense.

5.2 The Matrix Case

Let us now return to the tangential Nevanlinna-Pick interpolation problem discussed in Section 4.2. For κ a nonnegative integer, let $BR_{M \times N}(\kappa)$ be the set of rational $M \times N$ matrix functions $F(z)$ such that

(i) $sup\{\|F(z)\| : Rez = 0\} < 1$

and

(ii) F has κ poles (counting multiplicities) in Π_+.

For the matrix case, by the multiplicity of a pole z_0 we mean the sum of the absolute values of all the negative partial multiplicities appearing in the Smith form of $F(z)$ at z_0; the total pole multiplicity over Π_+ is then the sum of all these partial pole multiplicities over all points z_0 in Π_+.

Equivalently, the total pole multiplicity over Π_+ for $F(z)$ is the size of the matrix A in a left pole pair (C, A) for $F(z)$ over Π_+. When $\kappa = 0$ the class $BR_{M\times N}(\kappa)$ collapses to the class $BR_{M\times N}$ considered in Section 4.2. We again consider interpolation conditions of the form

$$x_j^* F(z_j) = y_j^* \text{ for } 1 \le j \le n \tag{2.1}$$

where $x_1, ..., x_n$ are given nonzero column vectors in $\mathbb{C}^M, y_1, ..., y_n$ are given vectors in $\mathbb{C}^N$ and $z_1, ..., z_n$ are given points in Π_+. From this set w of interpolation data build the Pick matrix

$$\Lambda(w) = [(x_i^* x_j - y_i^* y_j)(z_i + \bar{z}_j)^{-1}]_{1\le i,j\le n} \tag{2.2}$$

If $\Lambda = \Lambda(w)$ is invertible we can also form the rational $(M+N)\times(M+N)$ matrix function $\Theta(z)$ as in formula (4.2.12) (with A and B given by (2.6) and (2.7) and $J = I_M \oplus -I_N$). Then we have the following result.

THEOREM 2.1. *Let the data set $w = \{z_j, x_j, y_m : 1 \le j \le n\}$ for a tangential Nevanlinna-Pick interpolation problem be given. Assume $\Lambda = \Lambda(w)$ given by (2.2) is invertible and let $\Theta(z) = [\Theta_{ij}(z)]_{1\le i,j\le 2}$ be given by (4.2.12). Then the number κ of negative eigenvalues of Λ is the smallest κ for which there exists an F in $BR_{M\times N}(\kappa)$ which is analytic at $z_1, ..., z_n$ and satisfies the interpolation conditions (2.1). Moreover all such F's are given by*

$$F = (\Theta_{11}G + \Theta_{12})(\Theta_{21}G + \Theta_{22})^{-1}$$

with G an arbitrary matrix funciton in $BR_{M\times N}$ such that the poles of $(\Theta_{21}G + \Theta_{22})^{-1}$ do not include any of the points $z_1, ..., z_n$.

For the scalar case this theorem specializes to Theorem 1.1 in the same way that Theorem 4.2.2 specializes to Theorem 4.1.1. As for the scalar case it is possible to describe the set of all functions of the form $F = (\Theta_{11}G + \Theta_{12})(\Theta_{21}G + \Theta_{22})^{-1}$ with G an arbitrary function in $BR_{M\times N}$, but for the matrix case this is trickier than for the scalar case, so we omit the details.

5.3 Sketch of the Proof of Theorem 2.1

The proof of Theorem 2.1 depends on the following lemma.

LEMMA 3.1. *Let an interpolation data set* $w = \{z_j, x_j, y_j : 1 \le j \le n\}$ *be given and assume that* $\Lambda = \Lambda(w)$ *given by (4.2.11) is invertible. Let* $\Theta(z) = [\Theta_{ij}(z)]_{1\le i,j\le 2}$ *be given by (4.2.12). Then the number of negative eigenvalues* κ *of* Λ *coincides with the total pole multiplicity of* Θ_{22}^{-1} *over* Π_+.

Sketch of Proof. The starting point for the proof is the characterization of the number k of negative eigenvalues of Λ in terms of the number of negative squares in the kernel

$$K_\Theta(z,w) = [J - \Theta(z)J\Theta(w)^*](z+\bar{w})^{-1}$$

over Π_+, together with a finer analysis of the Redheffer transform $U = R[\Theta]$ of Θ (see Section 4.4.3). Indeed one can prove that the number of negative squares of K_Θ over Π_+ is the same as the total pole multiplicity of $U(z)$ over Π_+; a finer analysis using the special form of $\Theta(z)$ then shows that this in turn actually equals the total pole multiplicity of $U_{21} = \Theta_{22}^{-1}$ over Π_+. We omit the details. ∎

Once Lemma 3.1 is established, the proof of Theorem 2.1 parallels rather closely the proof of Theorem 4.2.2 (i.e. the case $\kappa = 0$). If F is a matrix function in $BR_{M\times N}(\tilde{\kappa})$ we may choose a regular $N \times N$ matrix function $W(z)$ analytic on Π_+ having total null multiplicity over Π_+ equal to $\tilde{\kappa}$ such that $F(z)W(z)$ has no poles in Π_+. Then,, if F also satisfies the interpolation conditions (2.1), as in the proof of Theorem 4.2.2 we will have the factorization

$$\begin{bmatrix} F(z) \\ I \end{bmatrix} W(z) = \Theta(z) \begin{bmatrix} G_1(z) \\ G_2(z) \end{bmatrix} \tag{3.1}$$

with rational matrix functions G_1 and G_2 having no poles in Π_+. Since $\|F(z)\| < 1$ and $\Theta(z)$ is J-unitary on the imaginary axis, we get that

$$0 > G_1(z)G_1(z)^* - G_2(z)G_2(z)^*$$

on the imaginary axis. As in the proof of Theorem 4.2.2, this then implies that $G_2(z)$ is full rank on the imaginary axis, and that $G := G_1G_2^{-1}$ has the property

$$sup\{\|G(z)\| : Re z = 0\} < 1 \tag{3.2}$$

Next, from the bottom block row of (3.1) we deduce

$$W = \Theta_{21}G_1 + \Theta_{22}G_2$$

$$= \Theta_{22}(\Theta_{22}^{-1}\Theta_{21}G + I)G_2 \tag{3.3}$$

By the same argument as in the proof of Theorem 4.2.2,

$$wno\ det(\Theta_{22}^{-1}\Theta_{21}G + I) = 0$$

Since G_2 is regular on the imaginary axis and has no poles in Π_+,

$$wno\ detG_2 \geq 0.$$

Finally, from (3.3) we get

$$wno\ detW = wno\ det\Theta_{22} + wno\ det(\Theta_{22}^{-1}\Theta_{21}G + 1) + wno\ detG_2$$

We conclude

$$\tilde{\kappa} = wno\ detW \geq wno\ det\Theta_{22}.$$

Thus Θ_{22}^{-1} can have total pole multiplicity over Π_+ of at most $\tilde{\kappa}$ and thus by Lemma 3.1, Λ has at most $\tilde{\kappa}$ negative eiganvalues. We have thus established $\kappa \leq \tilde{\kappa}$.

Conversely, suppose Λ has κ negative eigenvalues. Choose $G \in BR_{M\times N}$ so that $\Theta_{21}G + \Theta_{22}$ is invertible at each of the points $z_1, ..., z_n$ (this is true for a generic subset of such G's). Define matrix functions F_1 and F_2 by

$$\begin{bmatrix} F_1 \\ F_2 \end{bmatrix} = \Theta \begin{bmatrix} G \\ I \end{bmatrix} = \begin{bmatrix} \Theta_{11}G + \Theta_{12} \\ \Theta_{21}G + \Theta_{22} \end{bmatrix} \tag{3.4}$$

As before, $F_2(z)$ necessarily has full rank and $F(z) := F_1(z)F_2(z)^{-1}$ is a strict contraction for z on the imaginary axis. From the bottom block row of (3.4) we have

$$F_2 = \Theta_{21}G + \Theta_{22}$$

$$= \Theta_{22}(\Theta_{22}^{-1}\Theta_{21}G + I).$$

As before, $wno\ det(\Theta_{22}^{-1}\Theta_{21}G + I) = 0$, so

$$wno\ det F_2 = wno\ det\Theta_{22}$$

Now by Lemma 3.1, Θ_{22}^{-1} has total pole multiplicity over Π_+ equal to κ, so

$$wno\ det\Theta_{22} = \kappa$$

Thus F_2^{-1} has total pole multiplicity κ so $F = F_1F_2^{-1}$ has at most κ poles in Π_+. Since by assumption F_2 has fall rank at each point $z_1, ..., z_n$, we see from (3.4) and the construction of Θ by the same argument as in the proof of Theorem 4.2.2 that F satisfies the interpolation conditions (2.1), and that F has the form $F = (\Theta_{11}G + \Theta_{12})(\Theta_{21}G + \Theta_{22})^{-1}$. Thus the smallest value $\tilde{\kappa}$ of κ for which there exist solutions of the interpolation conditions (2.1) in $BR_{M\times N}(\kappa)$ is no more than the number κ of negative eigenvalues of Λ, i.e. $\tilde{\kappa} \leq \kappa$. ∎.

5.4 Notes

Historically, the variant of Nevanlinna-Pick interpolation discussed in this chapter was first considered by Takagi [T] in connection with the Caratheodory-Toeplitz problem of prescribing the first few Taylor coefficients at the origin for a function on the unit disk. More recent elaborations are [AAK] (in connection with the Nehari problem), [BH] and [N2]. In engineering this type of interpolation has connections with model reduction when the approximation is done in the Hankel norm [G].

REFERENCES

[AAK] V. M. Adamjan, D. Z. Arov and M. G. Krein Analytic properties of Schmidt pairs for a Hankel operator and the generalized Schur-Takagi problem, Math U.S.S.R. Sbornik 15 (1971), 31-73.

[AG] D. Alpay and I. Gohberg, Unitary rational matrix functions and orthogonal matrix polynomials, to appear.

[BGK] H. Bart, I. Gohberg and M. A. Kaashoek. Minimal Factorization of Matrix and Operator Functions, Birkhauser, Basel, 1979.

[BGR1] J. A. Ball, I. Gohberg and L. Rodman, Minimal factorization of meromorphic matrix functions in terms of local data, Integral Equations and Operator Theory 10(1987), 309-348.

[BGR2] J. A. Ball, I. Gohberg and L. Rodman, Interpolation Problems for Matrix Valued Functions. Part I: Rational Functions, monograph in preparation.

[BH] J. A. Ball and J W. Helton, A Beurling-Lax theorem for the Lie group $U(m,n)$ which contains most classical interpolation, J. Operator Theory 9 (1983), 107-142.

[BR] J. A. Ball and A. C. M. Ran, Local inverse spectral problems for rational matrix functions, Integral Equations and Operator Theory 10 (1987), 349-415.

[DD] B. N. Datta and K. Datta, Theoretical and computational aspects of some linear algebra problems in control theory, in Computational and Combinatorial Aspects in Control Theory (ed C. I. Byrnes and A. Lindquist), Elsevier (Amsterdam) (1986), pp. 201-212.

[DGK] P. Delsarte, Y. Genin and Y. Kamp, The Nevanlinna-Pick problem for matrix valued functions, SIAM J. Appl. Math 36 (1979), 47-61.

[F] B. A. Francis, A course in H_∞control theory, Lecture Notes in Control and Information Science, Vol 88, Springer-Verlag, 1987.

[F1] I. I. Fedchina, Tangential Nevanlinna-Pick problem with multiple points, Akad. Nauk Armjan. SSR Dokl 61 (1975), 214-218 [in Russian].

[F2] I. I. Fedchina, Description of solutions of the tangential Nevanlinna-Pick problem, Akad. Nauk Armjan. SSR Dokl. 60 (1975), 37-42 [in Russian].

[G] K. Glover, All optimal Hankel-norm approximations of linear multivariable systems and their L^∞-error bounds. Inter. J. Control 39(1984), 1115-1193.

[GKLR] I. Gohberg, M. A. Kaashoek, L. Lerer and L. Rodman, Minimal divisors of rational matrix functions with prescribed zero and pole structure, in: Topics in Operator Theory, Systems and Networks (eds. H. Dym, I. Gohberg), Operator Theory Advances and Applications, 12 (1984), Birkhauser Verlag, Basel, pp. 241-275.

[GKvS] I. Gohberg, M. A. Kaashoek and F. van Schagen, Rational matrix and operator functions with prescribed singularities, Integral Equations and Operator Theory 5 (1982),673-717.

[GLR1] I. Gohberg, P. Lancaster and L. Rodman, Matrix Polynomials, Academic Press (New York), 1982.

[GLR2] I. Gohberg, P. Lancaster, L. Rodman, Invariant Subspaces of Matrices with Applications, J. Wiley, (New York), 1986.

[Go] L. B. Golinsky, On a generalization of the matrix Nevanlinna-Pick problem, Proc. Armenian Academy of Science 18 (1983), 187-205 [in Russian].

[GS] I. C. Gohberg and E. I. Sigal, On operator generalizations of the logarithmic residue theorem and the theorem of Rouche, Math. USSR - Sb 13 (1971), 603-625.

[GVKDM] Y. Genin, P. Van Dooren, T. Kailath, J-M. Delosme, M. Morf, On $\sum$-lossless transfer functions and related questions, Linear Algebra and its Applications, 50, 1983, 251-275.

[H1] J. W. Helton, Orbit structure of the Mobius transformation semigroup acting on H^∞(broadband matching), in Topics in Functional Analysis, Advances in Math. Supplemental Studies Vol 3,

Academic Press, 1978, pp. 129-157.

[H2] J. W. Helton, The distance of a function to H^∞ in the Poincare metric, electrical power transfer, J. Functional Anal. 38 (1980), 273-314.

[K] T. Kailath, Linear Systems, Prentice Hall, Inc., Englewood Cliffs, N. J., 1980.

[KFA] R. E. Kalman, P. L. Falb and M. A. Arbib, Topics in Mathematical System Theory, McGraw Hill, New York, 1969.

[KL] M. G. Krein and H. Langer, Uber die Verallgemeinerten Resolvanten und die characteristische Funktion eines isometrisches Operator in Raume π_k. Colloquia Mathematica Societatis Janos Bolyai 5. Hilbert Space Operators (1970), 353-399.

[KP] I. V. Kovalischina and V. P. Potapov, Integral representations of Hermitian positive functions, Private translation by T. Ando (1982), Sapporo, Japan.

[Kt] V. E. Katsnelson, Methods of J-theory in continuous interpolation problems of analysis, Part I, Private translation by T. Ando (1982), Sapporo, Japan.

[Nev] R. Nevanlinna, Ueber beschraenkte analytische Funktionen, Ann. Acad. Sci. Fenn. 32 (1929), No. 7.

[N1] A. A. Nudelman, On a new problem of moment problem type, Soviet Math. Doklady 18(1977), 507-510 [Doklady Akademii Nauk SSSR (1977)].

[N2] A. A. Nudelman, A generalization of classical interpolation problems, Soviet Math Doklady 23(1981), 125-128 [Doklady Akademii Nauk SSSR (1981)].

[P] G. Pick, Ueber beschraenkungen analytischer Funktionen durch vorgegebene Funktionswerte, Math. Ann. 78 (1918), 270-275.

[PS] G. Polya and G. Szego, Problems and Theorems in Analysis, Volume II, Springer-Verlag (New York), 1976.

[R] R. Redheffer, On a certain linear fractional transformation, J. Math. Physics 39 (1960), 269-286.

[RR] M. Rosenblum and J. Rovnyak, An operator-theoretic approach to theorems of Pick-Nevanlinna and Loewner types I, Integral Equations and Operator Theory 3 (1980), 408-436.

[Sakh] L. A. Sakhnovich, Factorization problems and operator identities, Russian Mathematical Surveys 41:1 (1986), 1-64.

[T] T. Takagi, On an algebraic problem related to an analytic theorem of Caratheodory and Fejer, Japan J. Math. 1 (1924), 83-93.

J. A. Ball
Department of Mathematics
Virginia Polytechnic Institute and State University
Blacksburg, Virginia 24061
USA

I. Gohberg
School of Mathematical Sciences
The Raymond and Beverly Sackler Faculty of Exact Sciences
Tel-Aviv University
Tel-Aviv, Ramat-Aviv 69978
ISRAEL

L. Rodman
Department of Mathematics
Arizona State University
Tempe, Arizona 85287
USA
and
School of Mathematical Sciences
Tel-Aviv University
Tel-Aviv, Ramat-Aviv 69978
ISRAEL

Operator Theory:
Advances and Applications, Vol. 33

INTERPOLATION PROBLEMS FOR RATIONAL MATRIX FUNCTIONS WITH INCOMPLETE DATA AND WIENER-HOPF FACTORIZATION

I. Gohberg, M.A. Kaashoek and A.C.M. Ran

This paper concerns the problem to construct a rational matrix function with a prescribed zero and pole structure and which has the value I at infinity. In general, even in the scalar case the problem is only solvable after the given data have been completed with a complementary zero and pole structure. A description of all solutions of minimal possible McMillan degree is given in realized form. The results are applied to obtain Wiener-Hopf factorizations from a new point of view.

0. INTRODUCTION

This paper concerns the problem of constructing a regular $m \times m$ rational matrix function W with a prescribed zero and pole structure. We assume that the function W, which has to be built, does not have a zero or a pole at infinity; more precisely, we assume that W is analytic and has the value I (the $m \times m$ identity matrix) at infinity. In the scalar case this interpolation problem is trivial and looks as follows. Find a rational (scalar) function ω with $\omega(\infty) = 1$ and with given zeros and poles (multiplicities taken into account). If the number of the given zeros is equal to the number of given poles, then the solution is obvious, and if not, then a solution can be obtained only after some poles and zeros have been added. For matrix functions the situation is much more complicated. In the matrix case the description of the given zeros and poles will contain geometric information involving eigenvectors and generalized eigenvectors, and these geometric data have to be completed in a special way.

To make the latter more precise, let us first describe the form in which the zero and pole data will be given. A rational $m \times m$ matrix function W which is analytic and has the value I at infinity may be represented (see [4]) in the following way:

$$(0.1) \qquad W(\lambda) = I + C(\lambda - A)^{-1}B.$$

Here A is a square matrix of which the order n may be much larger than m, B and C are matrices of sizes $n\times m$ and $m\times m$, respectively, and

(0.2) $\quad W(\lambda)^{-1} = I - C(\lambda - A^{\times})^{-1}B,$

where $A^{\times} = A - BC$. In what follows we assume that the number n is chosen as small as possible. In that case the pair (C,A) describes the right pole data and the pair $(A^{\times},B)$ the left zero data (see [4] for more details). A pair of matrices (C_p,A_p) will be called a *right pole pair* for W if there exists an invertible matrix E such that

(0.3) $\quad C_p = CE, \quad A_p = E^{-1}AE.$

Similarly, (A_z,B_z) is called a *left zero pair* for W if there exists an invertible matrix F such that

(0.4) $\quad B_z = F^{-1}B, \quad A_z = F^{-1}A^{\times}F.$

Note that in that case

(0.5) $\quad \bigcap_{j=0}^{n-1} \operatorname{Ker} C_pA_p^{j-1} = (0), \qquad \operatorname{Im}[B_z A_zB_z \dots A_z^{n-1}B_z] = \mathbb{C}^n.$

It is clear that in order to rebuild the rational matrix function W from the data (C_p,A_p) and (A_z,B_z) one needs additional information connected with the similarities in (0.3) and (0.4). In fact (see [9, 2]) one has to take into account the matrix $\Gamma := F^{-1}E$, which is an invertible matrix satisfying the equation

(0.6) $\quad \Gamma A_p - A_z\Gamma = B_zC_p.$

Here the invertibility of Γ replaces the condition appearing in the scalar problem which requires that the number of zeros is equal to the number of poles

Now, assume that we have incomplete data, i.e., a triple

(0.7) $\quad \tau = \{(C_1,A_1),(A_2,B_2);\Gamma\}$

with the following properties:

(P.1) A_1 and A_2 are square matrices of orders n_1 and n_2, respectively, and with eigenvalues in some set σ;

(P.2) C_1 and B_2 are matrices of sizes $m\times n_1$ and $n_2\times m$, respectively, and

$$\bigcap_{j=0}^{n_1-1} \operatorname{Ker} C_1A_1^j = 0, \quad \operatorname{Im}[B_2A_2B_2 \dots A_2^{n_2-1}B_2] = \mathbb{C}^{n_2};$$

(P.3) Γ is an $n_2\times n_1$ matrix satisfying the equation $\Gamma A_1 - A_2\Gamma = B_2C_1$.

Such a triple τ will be called a *σ-admissible triple*. If $n_1 = n_2$ and Γ is invertible, then we have an interpolation problem with a complete set of data and (see [9]) there exists a unique rational matrix function W, with $W(\infty) = I$, such that (C_1,A_1) is a right pole pair and (A_2,B_2) is a left zero pair for W, namely

$$W(\lambda) = I + C_1(\lambda - A_1)^{-1}\Gamma^{-1}B_2.$$

The problem we deal with in this paper concerns the case when Γ is not invertible. Note that the non-invertibility of Γ is the matrix analogue of the case appearing in the scalar problem when the number of given zeros is not equal to the number of given poles.

To deal with singular coupling matrices Γ we introduce the notion of a complementary admissible triple. Let τ be the σ-admissible triple given by (0.7), and let

$$\tau_0 = \{(C_1^{(0)},A_1^{(0)}),(A_2^{(0)},B_2^{(0)});\Gamma_0\}$$

be another σ_0-admissible triple. We call τ_0 a *complement* of τ if $\sigma \cap \sigma_0 = \emptyset$ and

$$\det\begin{pmatrix} \Gamma & \Gamma_{10} \\ \Gamma_{01} & \Gamma_0 \end{pmatrix} \neq 0,$$

where Γ_{10} and Γ_{01} are the unique solutions of

$$\Gamma_{10}A_1^{(0)} - A_2\Gamma_{10} = B_2C_1^{(0)},$$

$$\Gamma_{01}A_1 - A_2^{(0)}\Gamma_{01} = B_2^{(0)}C_1.$$

The complement τ_0 of τ will be called *minimal* if among all complements of τ

$$\operatorname{rank}\begin{pmatrix} \Gamma & \Gamma_{10} \\ \Gamma_{01} & \Gamma_0 \end{pmatrix}$$

is as small as possible. If τ_0 is such a minimal complement of τ, then

$$W(\lambda) = I + \begin{pmatrix} C_1 & C_1^{(0)} \end{pmatrix} \begin{pmatrix} \lambda - A_1 & 0 \\ 0 & \lambda - A_1^{(0)} \end{pmatrix}^{-1} \begin{pmatrix} \Gamma & \Gamma_{10} \\ \Gamma_{01} & \Gamma_0 \end{pmatrix}^{-1} \begin{pmatrix} B_2 \\ B_2^{(0)} \end{pmatrix}$$

is a rational matrix function with the desired spectral data and a minimal set of additional zeros and poles (see Section 2 below for the precise statement).

The existence of such a minimal complementary triple, with σ_0 consisting of a single point, has been proved in [8]. The main problem solved in the present paper is the problem to describe all minimal complementary triples of a given admissible triple τ. The full solution is given in Section 2. In our construction an essential role is played by a concept from [5], based on incoming and outgoing data, which we review in the first section. In the third section we apply the theory developed in Section 2 and solve an analogous interpolation problem connected with standard pairs for regular matrix polynomials. The results here extend an earlier theorem in [13] (see also [11], Theorem 6.7).

Section 4 concerns Wiener-Hopf factorization of a rational matrix function W, with $W(\infty) = I$, which has no zeros and poles on a given closed contour γ. We derive formulas for the factors in such a factorization and the corresponding indices using the point of view developed in Sections 1 and 2. The starting point is a minimal realization $W(\lambda) = I + C(\lambda - A)^{-1}B$. Since W has no poles and zeros on γ, the matrices A and $A^{\times} := A - BC$ have no eigenvalue on γ. Let P and $P^{\times}$ be the spectral projections of A and $A^{\times}$, respectively, corresponding to the inner domain γ_+ of γ. Then

$$\tau_+ = \{(C|_{\operatorname{Im} P}, A|_{\operatorname{Im} P}), (A^{\times}|_{\operatorname{Im} P^{\times}}, P^{\times}B), P^{\times}|_{\operatorname{Im} P} : \operatorname{Im} P \to \operatorname{Im} P^{\times}\}$$

is a γ_+-admissible triple. We derive the factors in a Wiener-Hopf factorization of W by constructing a suitable complement of τ_+. This approach makes more transparent the construction of the Wiener-Hopf factorization given in [5]. The last section contains analogous results for matrix polynomials (cf. [11]).

1. (C,A)-INVARIANT AND (A,B)-INVARIANT SUBSPACES

This section is of a preliminary character. In it we collect some facts about (C,A)-invariant subspaces and (A,B)-invariant subspaces.

Let (C,A) be a pair of matrices of sizes $m \times n$ and $n \times n$, respectively, and let (C,A) be observable, i.e. $\bigcap_{j\geq 0} \operatorname{Ker} CA^j = 0$. A subspace M is called (C,A)-*invariant* if

$$A(M \cap \operatorname{Ker} C) \subset M.$$

In case M is (C,A)-invariant we introduce the *outgoing spaces* for (C,A;M) as follows:

$$K_0 := M,$$

$$K_j := M \cap \operatorname{Ker} C \cap \dots \cap \operatorname{Ker} CA^{j-1}, \qquad j = 1,2,\dots .$$

Note that $K_0 \supset K_1 \supset K_2 \ldots$. We also define *outgoing indices* $\alpha_1 \geq \ldots \geq \alpha_t$ by $t = \dim K_0/K_1$ and

$$\alpha_j = \#\{m \mid \dim K_{m-1}/K_m \geq j\}, \qquad j = 1,\ldots,t.$$

Let ε_1 be an arbitrary complex number. A basis $\{e_{jk}\}_{k=1}^{\alpha_j}{}_{j=1}^{t}$ of M can be constructed such that

$$(A-\varepsilon_1)e_{jk} = e_{j,k+1}, \quad k = 1,\ldots,\alpha_j-1, \quad j = 1,\ldots,t,$$

$$\{e_{jk} \mid 1 \leq k \leq \alpha_j-1,\ j = 1,\ldots,t\} \text{ is a basis for } M \cap \operatorname{Ker} C.$$

(See e.g. [5].) Such a basis will be called an *outgoing basis* for M with respect to $A - \varepsilon_1$. Connected with an outgoing basis we define vectors $\{z_j\}_{j=1}^{t}$ in $\mathbb{C}^m$ by $z_j = Ce_{j\alpha_j}$. It is easily seen that these vectors are linearly independent.

Choose a projection ρ on M. The numbers $\alpha_1,\ldots,\alpha_t$ are the observability indices of the pair $(C|_M, A|_M)$. Indeed, extend the set of vectors $\{z_j\}_{j=1}^{t}$ to a basis $\{z_j\}_{j=1}^{m}$ of $\mathbb{C}^m$. With respect to the basis $\{e_{jk}\}_{k=1}^{\alpha_j}{}_{j=1}^{t}$ in M and $\{z_j\}_{j=1}^{m}$ in $\mathbb{C}^m$ the pair $(C|_M, \rho A|_M)$ is in Brunovsky canonical form (see e.g. [7, 10] for this canonical form). Hence the numbers $\alpha_1,\ldots,\alpha_t$ are indeed the observability indices of $(C|_M, \rho A|_M)$.

Note also that the pair $(C|_M, \rho A|_M)$ is observable. By Rosenbrock's theorem (see [15]), there exist $S : M \to M$ and $G : \mathbb{C}^m \to M$ such that

$$(1.1) \qquad \rho(A - GC)|_M = S$$

if and only if for the degrees of the invariant factors $\beta_1 \geq \ldots \geq \beta_r$ of λ-S we have

$$(1.2) \qquad r \leq t, \quad \sum_{j=1}^{\nu} \beta_j - \alpha_j \geq 0 \quad (\nu = 1,\ldots,r).$$

Any such pair (S,G) will be called a *right correction pair* for (C,A) relative to M. Let (S,G) be a right correction pair for (C,A) and assume S has only one eigenvalue, ε say. Then the inequalities (1.2) give a restriction on the sizes $\beta_1,\ldots,\beta_r$ of the Jordan blocks of S. In particular we can take the sizes of the Jordan blocks equal to $\alpha_1,\ldots,\alpha_t$. In the other extreme we can take $r = 1$, in which case S has a single Jordan block with eigenvalue ε.

Next, we discuss the dual concepts. Let (A,B) be a controllable pair of matrices, of sizes $n \times n$ and $n \times m$, respectively. A subspace M is called (A,B)-*invariant* if

$$AM \subset M + \operatorname{Im} B.$$

In that case we introduce the *incoming subspaces*

$$H_0 := M$$

$$H_j := M + \operatorname{Im} B + \dots + \operatorname{Im} A^{j-1}B, \qquad j = 1,2,\dots .$$

Note: $H_0 \subset H_1 \subset H_2 \dots$. We define the *incoming indices* $\omega_1 \geq \dots \geq \omega_s$ by $s = \dim H_1/H_0$ and

$$\omega_j = \#\{m \mid \dim H_m/H_{m-1} \geq j\}, \qquad j = 1,\dots,s.$$

A basis $\{f_{jk}\}_{k=1}^{\omega_j}{}_{j=1}^{s}$ can be constructed for a complement K of M such that

$$(A - \varepsilon_1)f_{jk} - f_{jk+1} \in M + \operatorname{Im} B \quad (f_{j,\omega_j+1} := 0), \quad k = 1,\dots,\omega_j,$$

$$\{f_{j1}, j = 1,\dots,s\} \text{ is a basis for } (M + \operatorname{Im} B)/M.$$

(See e.g. [5].) Such a basis will be called an *incoming basis* for K with respect to $A - \varepsilon_1$. Connected with an incoming basis we have vectors $\{y_j\}_{j=1}^{s}$ in $\mathbb{C}^m$ such that

$$f_{j1} - By_j \in M.$$

These vectors form a basis for $\mathbb{C}^m$ modulo $B^{-1}[M]$. Finally we choose a projection ρ on K along M.

The numbers $\omega_1,\dots,\omega_s$ are the controllability indices of the pair $(\rho A|_K, \rho B)$. Indeed extend $\{y_j\}_{j=1}^{s}$ to a basis in $\mathbb{C}^m$. With respect to this basis and $\{f_{jk}\}_{k=1}^{\omega_j}{}_{j=1}^{s}$ in K the pair $(\rho A|_K, \rho B)$ is in Brunovsky canonical form.

The pair $(\rho A|_K, \rho B)$ is controllable. By Rosenbrock's theorem there exist $T : K \to K$ and $F : K \to \mathbb{C}^m$ such that

$$(1.3) \qquad \rho(A - BF)|_K = T$$

if and only if for the degrees of the invariant factors $\nu_1 \geq \dots \geq \nu_p$ of λ-T we have

$$(1.4) \qquad p \leq s, \quad \sum_{j=1}^{\nu} \nu_j - \omega_j \geq 0, \qquad \nu = 1,\dots,p.$$

Any such pair (F,T) will be called a *left correction pair* for (A,B) relative to a complement K of M. Let (F,T) be a left correction pair, and suppose T has only one eigenvalue ε. Then (1.4) gives restrictions on the sizes of the Jordan blocks of T. As for a right correction pair, we can take in particular the sizes of the Jordan blocks of T to be equal to $\omega_1,\dots,\omega_s$.

2. MINIMAL COMPLEMENTS OF ADMISSIBLE TRIPLES

In this section $\tau = \{(C_p, A_p), (A_z, B_z), \Gamma\}$ is a σ-*admissible triple*, i.e. the pair (C_p, A_p) is observable, the pair (A_z, B_z) is controllable, both $\sigma(A_p)$ and $\sigma(A_z)$ are in σ and the coupling operator Γ satisfies

$$\Gamma A_p - A_z \Gamma = B_z C_p. \tag{2.1}$$

Finally A_p acts on the space X_p, A_z acts on X_z.

The problem we shall consider in this section is the following: for which ε-admissible triples τ_0, where $\varepsilon \subset \mathbb{C}$ is a set with $\varepsilon \cap \sigma = \emptyset$, is $\tau \oplus \tau_0$ a global spectral triple for a rational $m \times m$ matrix function $W(\lambda)$ such that $W(\infty) = I_m$ and its McMillan degree is as small as possible?

Recall from [8] that if $\tau_0 = \{(C_{p0}, A_{p0}), (A_{z0}, B_{z0}), \Gamma_0\}$ is an ε-admissible triple, then $\tau \oplus \tau_0$ is defined by

$$\tau \oplus \tau_0 = \left\{([C_p, C_{p0}], A_p \oplus A_{p0}), (A_z \oplus A_{z0}, \begin{pmatrix} B_z \\ B_{z0} \end{pmatrix}), \begin{pmatrix} \Gamma & \Gamma_{12} \\ \Gamma_{21} & \Gamma_0 \end{pmatrix}\right\}$$

where Γ_{12} and Γ_{21} are the unique solutions of

$$\Gamma_{12} A_{p0} - A_z \Gamma_{12} = B_z C_{p0},$$

$$\Gamma_{21} A_p - A_{z0} \Gamma_{21} = B_{z0} C_p.$$

Also recall that an admissible triple $\tau = \{(C_p, A_p), (A_z, B_z), \Gamma\}$ is called a *global spectral triple* if Γ is invertible, and in that case it is a global spectral triple for the rational matrix valued function $W(\lambda)$ given by

$$W(\lambda) = I + C_p(\lambda - A_p)^{-1}\Gamma^{-1}B_z.$$

Note that the McMillan degree of W equals rank Γ.

An ε-admissible triple τ_0 will be called a *complement* of τ if the coupling operator $\Gamma_{\tau\oplus\tau_0}$ of $\tau \oplus \tau_0$ is invertible. We say that τ_0 is a *minimal complement* of τ if for any other complement τ_0' we have

$$\operatorname{rank} \Gamma_{\tau\oplus\tau_0} \leq \operatorname{rank} \Gamma_{\tau\oplus\tau_0'} .$$

The problem as stated in the beginning of this section can now be reformulated as follows: describe all minimal complements of τ. In [8], Lemma 4.4, a minimal complement was constructed for an arbitrary admissible triple τ, at present our interest lies in the description of all minimal complements.

First we introduce some spaces and projections. Let N be a complement in X_p to $\operatorname{Ker} \Gamma$, and let K be a complement in X_z to $\operatorname{Im} \Gamma$. So

$$X_p = N \dot{+} \operatorname{Ker}\Gamma, \quad X_z = \operatorname{Im}\Gamma \dot{+} K.$$

Let ρ_p be the projection along N onto Ker Γ and ρ_z the projection along Im Γ onto K. Further, let η_p be the imbedding of Ker Γ onto X_p and η_z the imbedding of K into X_z.

Note that Ker Γ is (C_p,A_p)-invariant. Indeed, by (2.1), we have for $x \in \operatorname{Ker}\Gamma \cap \operatorname{Ker} C_p$ that $A_p x \in \operatorname{Ker}\Gamma$. Likewise Im Γ is (A_z,B_z)-invariant. Let (S,G) be a right correction pair for (C_p,A_p) with respect to Ker Γ, i.e. $S : \operatorname{Ker}\Gamma \to \operatorname{Ker}\Gamma$, $G : \mathbb{C}^m \to \operatorname{Ker}\Gamma$ and

(2.2) $$\rho_p(A_p - GC_p)\big|_{\operatorname{Ker}\Gamma} = S.$$

Such a pair will be called a *zero correction pair* for τ. Also, let (F,T) be a left correction pair for (A_z,B_z) relative to a complement K of Im Γ, i.e. $T : K \to K$, $F : K \to \mathbb{C}^m$ and

(2.3) $$\rho_z(A_z - B_z F)\big|_K = T.$$

Such a pair will be called a *pole correction pair* for τ.

We select a generalized inverse $\Gamma^\dagger$ of Γ such that $\Gamma\Gamma^\dagger = I - \rho_z$, $\Gamma^\dagger\Gamma = I - \rho_p$.

The next theorem describes all minimal complements of τ.

THEOREM 2.1. *Let* $\tau = \{(C_p,A_p),(A_z,B_z),\Gamma\}$ *be an admissible triple, let* (S,G) *be a zero correction pair for* τ, *and let* (F,T) *be a pole correction pair for* τ, *such that* $\sigma(T) \subset \varepsilon$, $\sigma(S) \subset \varepsilon$. *Here* ε *is an arbitrary set with* $\sigma \cap \varepsilon = \emptyset$. *Then*

$$\tau_0 = \{(-C_pX - F,T),(S,-YB_z + G),\Gamma_0\}$$

is a minimal complement for τ, *where* $X : K \to X_p$ *and* $Y : X_z \to \operatorname{Ker}\Gamma$ *are the unique solutions of the Lyapunov equations*

(2.4) $$A_pX - XT = A_{12},$$

(2.5) $$YA_z - SY = A_{21}.$$

Here $A_{12} : K \to X_p$ *and* $A_{21} : X_z \to \operatorname{Ker}\Gamma$ *are defined as follows: let* ρ_pA_{12} *be arbitrary and put*

(2.6) $$(I - \rho_p)A_{12} = \Gamma^\dagger(A_z - \eta_zT - B_zF)\eta_z.$$

Furthermore, let

(2.7) $$A_{21}(I - \rho_z) = \rho_p(A_p - S\rho_p - GC_p)\Gamma^\dagger,$$

(2.8) $$A_{21}\rho_z = \rho_pA_{12} + GF + \Gamma_1T - S\Gamma_1,$$

where $\Gamma_1 : K \to \operatorname{Ker}\Gamma$ *is an arbitrary operator. Finally,* $\Gamma_0 = Y\Gamma X - Y\eta_z - \rho_p X + \Gamma_1$. *Conversely, every minimal complement*

$$\tau_0 = \{(C_{p0},A_{p0}),(A_{z0},B_{z0}),\Gamma_0\}$$

with $A_{p0} : K \to K$ *and* $A_{z0} : \operatorname{Ker}\Gamma \to \operatorname{Ker}\Gamma$ *is obtained as in the first part of the theorem, up to a change of basis in* K *and* Ker Γ. *Consequently, every minimal complement is similar to a minimal complement obtained as in the first part of the theorem.*

PROOF. First let us show that (2.2) and (2.7) are consistent and likewise (2.3) and (2.6). From $\Gamma\Gamma^{\dagger} = (I - \rho_z)$ one sees that $\operatorname{Im}\Gamma^{\dagger}\rho_z \subset \operatorname{Ker}\Gamma$, which implies that (2.7) and (2.2) are not contradictory. From $\rho_p\Gamma^{\dagger}\Gamma = 0$ one sees that $\rho_p\Gamma^{\dagger} = \rho_p\Gamma^{\dagger}\rho_z$. This implies that (2.6) and (2.3) do not contradict each other.

Observe that (2.4) and (2.5) indeed have unique solutions X and Y, since $\varepsilon \cap \sigma = \emptyset$.

To show that τ_0 is an ε-admissible triple we first show that the pair $(C_pX + F, T)$ is observable. Assume that $0 \neq x \in K$ is an unobservable eigenvector corresponding to λ. Then $-C_pXx = Fx$ and $Tx = \lambda x$. In particular $\lambda \in \varepsilon$. By (2.6) we have

$$B_zFx = -\Gamma A_{12}x - Tx + A_zx = -\Gamma A_{12}x - (\lambda - A_z)x.$$

On the other hand, by (2.1) and (2.4) we have

$$-B_zC_pXx = A_z\Gamma Xx - \Gamma A_pXx = A_z\Gamma Xx - \Gamma A_{12}x - \lambda\Gamma Xx.$$

So, by equating these two, we obtain

$$(\lambda - A_z)(x - \Gamma Xx) = 0.$$

Since $\lambda \in \varepsilon$, we have $\lambda \notin \sigma(A_z)$. Hence $x = \Gamma Xx$. But $x \in K$, which leads to $x = 0$, a contradiction.

Likewise, by duality one shows that the pair $(S, -YB_z + G)$ is controllable.

Finally we shall show that Γ_0 satisfies

$$\Gamma_0T - S\Gamma_0 = (-YB_z + G)(-C_pX - F). \tag{2.9}$$

Write out the right hand side to obtain

$$YB_zC_pX - GC_pX + YB_zF - GF.$$

By (2.1), (2.6), (2.7) and (2.8) this equals

$$Y(\Gamma A_p - A_z\Gamma)X + A_{21}\Gamma X + S\rho_p X - \rho_p A_p X + YA_z\eta_z - Y\eta_z T - Y\Gamma A_{12} +$$

$$+ \rho_p A_{12} - A_{21}\eta_z + \Gamma_1 T - S\Gamma_1,$$

which by (2.4) and (2.5) is equal to

$$Y\Gamma(XT + A_{12}) - (A_{21} + SY)\Gamma X + A_{21}\Gamma X - Y\Gamma A_{12} + S\rho_p X - \rho_p XT - Y\eta_z T + SY\eta_z$$

$$+ \Gamma_1 T - S\Gamma_1 = \Gamma_0 T - S\Gamma_0.$$

So τ_0 is well defined and an ε-admissible triple.

Let us compute $\Gamma_{\tau\oplus\tau_0}$, the coupling operator of $\tau \oplus \tau_0$. It is given by

$$\Gamma_{\tau\oplus\tau_0} = \begin{pmatrix} \Gamma & \Gamma_{12} \\ \Gamma_{21} & \Gamma_0 \end{pmatrix}$$

where Γ_{12} and Γ_{21} are the solution of

(2.10) $\quad \Gamma_{21}A_p - S\Gamma_{21} = (-YB_z + G)C_p$

(2.11) $\quad \Gamma_{12}T - A_z\Gamma_{12} = -B_z(C_pX + F).$

In fact we shall prove that $\Gamma_{12} = -\Gamma X + \eta_z$ and $\Gamma_{21} = -Y\Gamma + \rho_p$. To do this it suffices to show that $-Y\Gamma + \rho_p$ satisfies (2.10) and $-\Gamma X + \eta_z$ satisfies (2.11), since (2.10) and (2.11) are uniquely solvable. Compute

$$(-Y\Gamma + \rho_p)A_p - S(-Y\Gamma + \rho_p) = -Y(A_z\Gamma + B_zC_p) + \rho_pA_p + (YA_z - A_{21})\Gamma - S\rho_p =$$

$$= -YB_zC_p + \rho_pA_p - A_{21}\Gamma - S\rho_p = (-YB_z + G)C_p.$$

Likewise

$$(-\Gamma X + \eta_z)T - A_z(-\Gamma X + \eta_z) = \Gamma(A_{12} - A_pX) + \eta_zT + (\Gamma A_p - B_zC_p)X - A_z\eta_z$$

$$= \Gamma A_{12} + \eta_zT - A_z\eta_z - B_zC_pX = -B_z(C_pX + F).$$

So

$$\Gamma_{\tau\oplus\tau_0} = \begin{pmatrix} \Gamma & -\Gamma X+\eta_z \\ -Y\Gamma+\rho_p & \Gamma_0 \end{pmatrix} : X_p \oplus K \to X_z \oplus \operatorname{Ker}\Gamma.$$

Clearly $\dim X_p \oplus K = \dim X_z \oplus \operatorname{Ker}\Gamma$. To prove that $\Gamma_{\tau\oplus\tau_0}$ is invertible it suffices to show $\operatorname{Ker}\Gamma_{\tau\oplus\tau_0} = (0)$. Suppose $\Gamma_{\tau\oplus\tau_0}\begin{pmatrix} x_p \\ x_0 \end{pmatrix} = 0$. Then:

$$\Gamma x_p - \Gamma X x_0 + x_0 = 0\,,$$

$$-Y\Gamma x_p + \rho_p x_p + \Gamma_0 x_0 = 0.$$

From the first equation we see $x_0 \in K \cap \operatorname{Im}\Gamma = (0)$. But then $x_p \in \operatorname{Ker}\Gamma$, so the second equation implies $\rho_p x_p = x_p = 0$. Hence $\operatorname{Ker}\Gamma_{\tau\oplus\tau_0} = (0)$.

We have shown that τ_0 is a complement to τ. To see that τ_0 is a minimal complement note that if τ_0' is a minimal complement then

$$\operatorname{rank} \Gamma_{\tau\oplus\tau_0'} = \dim X_p \oplus K = \dim X_z \oplus \operatorname{Ker} \Gamma.$$

Indeed, if

$$\hat{\Gamma} = \begin{pmatrix} \Gamma & \Gamma_{12} \\ \Gamma_{21} & \Gamma_{22} \end{pmatrix} : X_p \oplus K_2 \to X_z \oplus Z$$

is invertible then $\operatorname{Im} \Gamma_{12} \supset K$. So $\dim K_2$ = number of columns of $\Gamma_{12} \geq \dim K$, hence $\dim Z \geq \dim \operatorname{Ker} \Gamma$.

For the converse, let

$$\tau_0' = \{(C_1,A_1),(A_2,B_2),\Gamma_{22}\}$$

be a minimal complement of τ, where $\Gamma_{22} : K_2 \to Z$. As we have seen above $\dim K_2 = \dim K$ and $\dim Z = \dim \operatorname{Ker} \Gamma$. By applying a similarity we may assume $K_2 = K$ and $Z = \operatorname{Ker} \Gamma$ without loss of generality. Let

$$\hat{\Gamma} = \Gamma_{\tau\oplus\tau_0'} = \begin{pmatrix} \Gamma & \Gamma_{12} \\ \Gamma_{21} & \Gamma_{22} \end{pmatrix}$$

be invertible, with inverse

$$\hat{\Gamma}^{-1} = \begin{pmatrix} \Gamma_{11}^{\times} & \Gamma_{12}^{\times} \\ \Gamma_{21}^{\times} & \Gamma_{22}^{\times} \end{pmatrix} : X_z \oplus \operatorname{Ker} \Gamma \to X_p \oplus K$$

Since Γ and $\Gamma_{22}^{\times}$ are coupled we have ([6])

$$\dim \operatorname{Ker} \Gamma = \dim \operatorname{Ker} \Gamma_{22}^{\times}.$$

Hence $\Gamma_{22}^{\times} = 0$.

Since $\hat{\Gamma}^{-1}$ is injective we have that $\Gamma_{12}^{\times}$ is injective. Moreover, since $\Gamma\Gamma_{12}^{\times} = 0$ one sees that $\Gamma_{12}^{\times} = \rho_p\Gamma_{12}^{\times} : \operatorname{Ker} \Gamma \to \operatorname{Ker} \Gamma$ in a 1 - 1 and onto way. Then from $I = \Gamma_{21}\Gamma_{12}^{\times}$ we see that $\Gamma_{21}\eta_p : \operatorname{Ker} \Gamma \to \operatorname{Ker} \Gamma$ is the inverse of $\rho_p\Gamma_{12}^{\times}$.

Likewise, since $\hat{\Gamma}^{-1}$ is onto $\Gamma_{21}^{\times} : X_z \to K$ is onto. From $\Gamma_{21}^{\times}\Gamma = 0$ we have $\Gamma_{21}^{\times} = \Gamma_{21}^{\times}\rho_z$. Hence $\Gamma_{21}^{\times}\eta_z : K \to K$ is invertible and from $\Gamma_{21}^{\times}\Gamma_{12} = I_K$ we see that its inverse is $\rho_z\Gamma_{12}$.

Define $\Phi : X_p \oplus K \to X_p \oplus K$ and $\Psi : X_z \oplus \operatorname{Ker} \Gamma \to X_z \oplus \operatorname{Ker} \Gamma$ by

$$\Phi = \begin{pmatrix} I & 0 \\ 0 & \rho_z\Gamma_{12} \end{pmatrix}, \qquad \Psi = \begin{pmatrix} I & 0 \\ 0 & \Gamma_{21}\eta_p \end{pmatrix}.$$

Then Φ and Ψ are invertible, with inverses

$$\Phi^{-1} = \begin{pmatrix} I & 0 \\ 0 & \Gamma_{21}^{\times}\eta_z \end{pmatrix}, \qquad \Psi^{-1} = \begin{pmatrix} I & 0 \\ 0 & \rho_p\Gamma_{12}^{\times} \end{pmatrix}.$$

We claim that $\tau \oplus \tau_0'$ is (Φ,Ψ)-similar to $\tau \oplus \tau_0$ where τ_0 is of the form described in the theorem. In other words, up to a change of basis in K and Ker Γ $\tau \oplus \tau_0'$ is obtained as in the first part of the theorem. We have

$$\Phi\begin{pmatrix} A_p & 0 \\ 0 & A_1 \end{pmatrix}\Phi^{-1} = \begin{pmatrix} A_p & 0 \\ 0 & T \end{pmatrix},$$

$$\begin{pmatrix} C_p & C_1 \end{pmatrix}\Phi^{-1} = \begin{pmatrix} C_p & C_1\Gamma_{21}^{\times}\eta_z \end{pmatrix},$$

$$\Psi^{-1}\begin{pmatrix} A_z & 0 \\ 0 & A_2 \end{pmatrix}\Psi = \begin{pmatrix} A_z & 0 \\ 0 & S \end{pmatrix},$$

$$\Psi^{-1}\begin{pmatrix} B_z \\ B_2 \end{pmatrix} = \begin{pmatrix} B_z \\ \rho_p\Gamma_{12}^{\times}B_2 \end{pmatrix},$$

$$\Psi^{-1}\Gamma_{\tau\oplus\tau_0'}\Phi^{-1} = \begin{pmatrix} \Gamma & \Gamma_{12}\Gamma_{21}^{\times}\eta_z \\ \rho_p\Gamma_{12}^{\times}\Gamma_{21} & \rho_p\Gamma_{12}^{\times}\Gamma_{22}\Gamma_{21}^{\times}\eta_z \end{pmatrix} \tag{2.12}$$

where $T = \rho_z\Gamma_{12}A_1\Gamma_{21}^{\times}\eta_z$, $S = \rho_p\Gamma_{12}^{\times}A_2\Gamma_{21}\eta_p$. Now $\Gamma_{12}\Gamma_{21}^{\times} + \Gamma\Gamma_{11}^{\times} = I$ so for $x \in K$ we have

$$x = \eta_z x = \Gamma_{12}\Gamma_{21}^{\times}\eta_z x + \Gamma\Gamma_{11}^{\times}\eta_z x.$$

Likewise $\Gamma_{12}^{\times}\Gamma_{21} + \Gamma_{11}^{\times}\Gamma = I$ so for $x \in X_p$ we have

$$\rho_p x = \rho_p\Gamma_{12}^{\times}\Gamma_{21}x + \rho_p\Gamma_{11}^{\times}\Gamma x.$$

So (2.12) is equal to

$$\Psi^{-1}\Gamma_{\tau\oplus\tau_0'}\Phi^{-1} = \begin{pmatrix} \Gamma & \eta_z - \Gamma\Gamma_{11}^{\times}\eta_z \\ \rho_p - \rho_p\Gamma_{11}^{\times}\Gamma & \rho_p\Gamma_{12}^{\times}\Gamma_{22}\Gamma_{21}^{\times}\eta_z \end{pmatrix}.$$

Define

$$G = \rho_p\Gamma_{11}^{\times}(I-\rho_z)B_z + \rho_p\Gamma_{12}^{\times}B_2, \tag{2.13}$$

$$F = -C_p(I-\rho_p)\Gamma_{11}^{\times}\eta_z - C_1\Gamma_{21}^{\times}\eta_z, \tag{2.14}$$

$$A_{12} = A_p(I-\rho_p)\Gamma_{11}^{\times}\eta_z - (I-\rho_p)\Gamma_{11}^{\times}\eta_z T, \tag{2.15}$$

$$A_{21} = \rho_p\Gamma_{11}^{\times}(I-\rho_z)A_z - S\rho_p\Gamma_{11}^{\times}(I-\rho_z), \tag{2.16}$$

$$(2.17)\qquad X=(I-\rho_p)\Gamma_{11}^{\times}\eta_z,\qquad Y=\rho_p\Gamma_{11}^{\times}(I-\rho_z),$$

$$(2.18)\qquad \Gamma_1=-\rho_p\Gamma_{11}^{\times}\eta_z.$$

We shall show that the relations (2.2) - (2.8) in the statement of the theorem hold, and that

$$(2.19)\qquad C_1\Gamma_{21}^{\times}\eta=-C_pX-F$$

$$(2.20)\qquad \rho_p\Gamma_{12}^{\times}B_2=-YB_z+G$$

$$(2.21)\qquad \rho_p\Gamma_{12}^{\times}\Gamma_{22}\Gamma_{21}^{\times}\eta_z=Y\Gamma X-Y\eta_z-\rho_pX+\Gamma_1.$$

Once this is established we have shown that τ_0' is similar to a minimal complement obtained as in the first part of the theorem.

Note that (2.19) and (2.20) are immediate from (2.13), (2.14) and (2.17). Also (2.4) and (2.5) are immediate from (2.15), (2.16) and (2.17).

From

$$\hat{\Gamma}(A_p\oplus A_1)-(A_z\oplus A_2)\hat{\Gamma}=\begin{pmatrix}B_z\\ B_2\end{pmatrix}\begin{pmatrix}C_p & C_1\end{pmatrix}$$

we obtain

$$(2.22)\qquad \begin{pmatrix}\Gamma & \eta_z-\Gamma\Gamma_{11}^{\times}\eta_z\\ \rho_p-\rho_p\Gamma_{11}^{\times}\Gamma & \rho_p\Gamma_{12}^{\times}\Gamma_{22}\Gamma_{21}^{\times}\eta_z\end{pmatrix}\begin{pmatrix}A_p & 0\\ 0 & T\end{pmatrix}-$$

$$-\begin{pmatrix}A_z & 0\\ 0 & S\end{pmatrix}\begin{pmatrix}\Gamma & \eta_z-\Gamma\Gamma_{11}^{\times}\eta_z\\ \rho_p-\rho_p\Gamma_{11}^{\times}\Gamma & \rho_p\Gamma_{12}^{\times}\Gamma_{22}\Gamma_{21}^{\times}\eta_z\end{pmatrix}=$$

$$=\begin{pmatrix}B_z\\ \rho_p\Gamma_{12}^{\times}B_2\end{pmatrix}\begin{pmatrix}C_p & C_1\Gamma_{21}^{\times}\eta_z\end{pmatrix}$$

To show that (2.2) holds, take $x\in\operatorname{Ker}\Gamma$:

$$A_px-Sx-GC_px=A_px-Sx-\rho_p\Gamma_{11}^{\times}(I-\rho_z)B_zC_px-\rho_p\Gamma_{12}^{\times}B_2C_px.$$

By (2.1) and (2.22) this is equal to

$$A_px-Sx-\rho_p\Gamma_{11}^{\times}(I-\rho_z)(\Gamma A_p-A_z\Gamma)x+S(\rho_p-\rho_p\Gamma_{11}^{\times}\Gamma)x-\rho_pA_px+\rho_p\Gamma_{11}^{\times}\Gamma A_px$$

$$=(I-\rho_p)A_px\in N.$$

So (2.2) holds. Next, to show (2.3), take $x\in K$:

$$A_zx-Tx-B_zFz=A_zx-Tx+B_zC_p(I-\rho_p)\Gamma_{11}^{\times}x+B_zC_1\Gamma_{21}^{\times}x.$$

By (2.1) and (2.22) this equals

$$A_z x - Tx + (\Gamma A_p - A_z\Gamma)(I-\rho_p)\Gamma_{11}^{\times}x + (\eta_z - \Gamma\Gamma_{11}^{\times}\eta_z)Tx - A_z(\eta_z - \Gamma\Gamma_{11}^{\times}\eta_z)x =$$

$$= \Gamma(A_p(I-\rho_p)\Gamma_{11}^{\times}x - \Gamma_{11}^{\times}\eta_z Tx) \in \operatorname{Im}\Gamma.$$

So (2.3) holds.

Let us check (2.6) and (2.7) next. Again using (2.1) and (2.22) we have

$$\rho_p A_p - \rho_p S\rho_p - \rho_p GC_p = \rho_p A_p - \rho_p S\rho_p - \rho_p\Gamma_{11}^{\times}(I-\rho_z)B_z C_p - \rho_p\Gamma_{12}^{\times}B_2 C_p$$

$$= \rho_p A_p - \rho_p S\rho_p - \rho_p\Gamma_{11}^{\times}(I-\rho_z)(\Gamma A_p - A_z\Gamma) - (\rho_p - \rho_p\Gamma_{11}^{\times}\Gamma)A_p + S(\rho_p - \rho_p\Gamma_{11}^{\times}\Gamma) =$$

$$= \{\rho_p\Gamma_{11}^{\times}(I-\rho_z)A_z - S\rho_p\Gamma_{11}^{\times}(I-\rho_z)\}\Gamma = A_{21}\Gamma,$$

which proves (2.7) upon postmultiplying with $\Gamma^{\dagger}$. Further

$$A_z\eta_z - \eta_z T\eta_z - B_z F\eta_z = A_z\eta_z - \eta_z T\eta_z + B_z C_p(I-\rho_p)\Gamma_{11}^{\times}\eta_z + B_z C_1\Gamma_{21}^{\times}\eta_z =$$

$$= A_z\eta_z - \eta_z T\eta_z + (\Gamma A_p - A_z\Gamma)(I-\rho_p)\Gamma_{11}^{\times}\eta_z + (\eta_z - \Gamma\Gamma_{11}^{\times}\eta_z)T\eta_z - A_z(\eta_z - \Gamma\Gamma_{11}^{\times}\eta_z)$$

$$= \Gamma(A_p(I-\rho_p)\Gamma_{11}^{\times}\eta_z - (I-\rho_p)\Gamma_{11}^{\times}\eta_z T) = \Gamma A_{12},$$

which proves (2.6) upon premultiplying with $\Gamma^{\dagger}$.

Before proving (2.8) and (2.21) we give an alternative formula for Γ_1:

$$\Gamma_1 = \rho_p\Gamma_{12}^{\times}\Gamma_{22}\Gamma_{21}^{\times}\eta_z - \rho_p\Gamma_{11}^{\times}\Gamma\Gamma_{11}^{\times}\eta_z. \tag{2.23}$$

Indeed, from $\hat{\Gamma}^{-1} = \hat{\Gamma}^{-1}\hat{\Gamma}\,\hat{\Gamma}^{-1}$ we obtain

$$\Gamma_{11}^{\times} = \Gamma_{11}^{\times}\Gamma\Gamma_{11}^{\times} + \Gamma_{11}^{\times}\Gamma_{12}\Gamma_{21}^{\times} + \Gamma_{12}^{\times}\Gamma_{21}\Gamma_{11}^{\times} + \Gamma_{12}^{\times}\Gamma_{22}\Gamma_{21}^{\times}$$

So

$$\rho_p\Gamma_{11}^{\times}\eta_z = \rho_p\Gamma_{11}^{\times}\Gamma\Gamma_{11}^{\times}\eta_z + \rho_p\Gamma_{11}^{\times}\Gamma_{12}\Gamma_{21}^{\times}\eta_z + \rho_p\Gamma_{12}^{\times}\Gamma_{21}\Gamma_{11}^{\times}\eta_z + \rho_p\Gamma_{12}^{\times}\Gamma_{22}\Gamma_{21}^{\times}\eta_z.$$

Using $\Gamma_{12}\Gamma_{21}^{\times}\eta_z = \eta_z - \Gamma\Gamma_{11}^{\times}\eta_z$ and $\rho_p\Gamma_{12}^{\times}\Gamma_{21} = \rho_p - \rho_p\Gamma_{11}^{\times}\Gamma$ we obtain:

$$\rho_p\Gamma_{11}^{\times}\eta_z = -\rho_p\Gamma_{11}^{\times}\Gamma\Gamma_{11}^{\times}\eta_z + 2\rho_p\Gamma_{11}^{\times}\eta_z + \rho_p\Gamma_{12}^{\times}\Gamma_{22}\Gamma_{21}^{\times}\eta_z$$

which proves (2.23).

To prove (2.21) note that $Y\eta_z = \rho_p X = 0$. So by (2.23)

$$Y\Gamma X - Y\eta_z - \rho_p X + \Gamma_1 = \rho_p\Gamma_{11}^{\times}\Gamma\Gamma_{11}^{\times}\eta_z + \Gamma_1 = \rho_p\Gamma_{12}^{\times}\Gamma_{22}\Gamma_{21}^{\times}\eta_z.$$

Finally we prove (2.8):

$$\rho_p A_{12} - A_{21}\eta_z = \rho_p A_p(I-\rho_p)\Gamma_{11}^{\times}\eta_z - \rho_p\Gamma_{11}^{\times}(I-\rho_z)A_z\eta_z.$$

On the other hand by (2.13), (2.14) and (2.22)

$$-GF = (\rho_p\Gamma_{11}^{\times}(I-\rho_z)B_z + \rho_p\Gamma_{12}^{\times}B_2)(C_p(I-\rho_p)\Gamma_{11}^{\times}\eta_z + C_1\Gamma_{21}^{\times}\eta_z) =$$

$$= \rho_p\Gamma_{11}^{\times}(I-\rho_z)(\Gamma A_p - A_z\Gamma)(I-\rho_p)\Gamma_{11}^{\times}\eta_z +$$

$$+ \rho_p\Gamma_{11}^{\times}(I-\rho_z)\{(\eta_z - \Gamma\Gamma_{11}^{\times}\eta_z)T - A_z(\eta_z - \Gamma\Gamma_{11}^{\times}\eta_z)\}$$

$$+ \{(\rho_p - \rho_p\Gamma_{11}^{\times}\Gamma)A_p - S(\rho_p - \rho_p\Gamma_{11}^{\times}\Gamma)\}(I-\rho_p)\Gamma_{11}^{\times}\eta_z$$

$$+ \rho_p\Gamma_{12}^{\times}\Gamma_{22}\Gamma_{21}^{\times}\eta_z T - S\rho_p\Gamma_{12}^{\times}\Gamma_{22}\Gamma_{21}^{\times}\eta_z =$$

$$= \rho_p A_p(I-\rho_p)\Gamma_{11}^{\times}\eta_z - \rho_p\Gamma_{11}^{\times}(I-\rho_z)A_z\eta_z + \{\rho_p\Gamma_{12}^{\times}\Gamma_{22}\Gamma_{21}^{\times}\eta_z - \rho_p\Gamma_{11}^{\times}\Gamma\Gamma_{11}^{\times}\eta_z\}T +$$

$$- S\{\rho_p\Gamma_{12}^{\times}\Gamma_{22}\Gamma_{21}^{\times}\eta_z - \rho_p\Gamma_{11}^{\times}\Gamma\Gamma_{11}^{\times}\eta_z\} =$$

$$= \rho_p A_{12} - A_{21}\eta_z + \Gamma_1 T - S\Gamma_1,$$

where the last equality uses (2.23). □

The next theorem describes all rational matrix functions W with $W(\infty) = I$ and of minimal degree such that τ is a σ-spectral triple for W.

THEOREM 2.2. *Let $\tau = \{(C_p,A_p),(A_z,B_z),\Gamma\}$ be a σ-spectral triple, and let τ_0 be a minimal complement to τ, as constructed in the first part of Theorem 2.1. The function W which is proper and has value I_m at infinity for which $\tau \oplus \tau_0$ is a global spectral triple is given by*

$$W(\lambda) = I_m + C_p(\lambda - A_p)^{-1}\{(\Gamma_{11}^{\times} - \eta_p Y)B_z + \eta_p G\} - (C_p X + F)(\lambda - T)^{-1}\rho_z B_z,$$

its inverse is given by

$$W(\lambda)^{-1} = I_m - \{(C_p(\Gamma_{11}^{\times} - X\rho_z) - F\rho_z\}(\lambda - A_z)^{-1}B_z - C_p(\lambda - S)^{-1}(-YB_z + G).$$

Here $\Gamma_{11}^{\times}$ is fixed by the identities

$$\Gamma\Gamma_{11}^{\times} = I - \eta_z\rho_z + \Gamma X\rho_z,$$

$$\rho_p\Gamma_{11}^{\times} = Y + \rho_p X\rho_z - \Gamma_1\rho_z.$$

PROOF. The theorem follows immediately from Theorem 2.1 and [8], Theorem 4.1 (see also [2,3,9]), noting that

$$\Gamma_{\tau\oplus\tau_0}^{-1} = \begin{pmatrix} \Gamma_{11}^{\times} & \eta_p \\ \rho_z & 0 \end{pmatrix}.$$

Indeed, the formulas above fixing Γ_{11} are precisely those which force $\begin{pmatrix} \Gamma_{11}^{\times} & \eta_p \\ \rho_z & 0 \end{pmatrix}$ to be $\Gamma_{\tau\oplus\tau_0}^{-1}$. □

Note by the way that $\Gamma_{11}^{\times}$ is a generalized inverse of Γ in the sense that $\Gamma = \Gamma\Gamma_{11}^{\times}\Gamma$.

Let us consider a special choice of a minimal complement for τ. Let (F,T) be a pole correction pair for τ and (S,G) a zero correction pair such that both S and T have only one eigenvalue $\lambda_0 \notin \sigma$. Further we take the sizes of the Jordan blocks of S to be equal to the outgoing indices $\alpha_1,\dots,\alpha_t$, and the sizes of the Jordan blocks of T to be equal to the incoming indices $\omega_1,\dots,\omega_s$. With this choice of zero and pole correction pair construct a minimal complement τ_0 of τ as in Theorem 2.1. For the function $W(\lambda)$ corresponding to $\tau \oplus \tau_0$ we have the following remark concerning its local Smith form.

PROPOSITION 2.3. *Let* $W(\lambda)$ *be the rational matrix function with global spectral triple* $\tau \oplus \tau_0$ *and let*

$$W(\lambda) = E_1(\lambda)\,\mathrm{diag}\,((\lambda-\lambda_0)^{\mu_i})_{i=1}^{m} E_2(\lambda)$$

be the local Smith form at λ_0 *of the function corresponding to* $\tau\oplus\tau_0$. *Assume* $\mu_1 \le \mu_2 \le \dots \le \mu_m$. *Then* $\mu_{m-j+1} = \alpha_j$ *for* $j = 1,\dots,t$, $\mu_j = -\omega_j$ *for* $j = 1,\dots,s$, *and* $\mu_j = 0$ *for all other* j.

PROOF. By Theorem 2.2 and the special choice of S and T the partial pole multiplicities of $W(\lambda)$ at λ_0 are $\omega_1,\dots,\omega_s$ and the partial zero multiplicities of $W(\lambda)$ at λ_0 are $\alpha_1,\dots,\alpha_t$. This proves the proposition. □

We shall see in Section 4 that the function $W(\lambda)$ corresponding to $\tau \oplus \tau_0$ with this specific choice of τ_0 actually admits the following factorization:

$$W(\lambda) = W_-(\lambda)\,\mathrm{diag}\left(\left(\frac{\lambda-\varepsilon_1}{\lambda-\varepsilon_2}\right)^{\kappa_j}\right)_{j=1}^{m}$$

where $\varepsilon_1 \notin \sigma \cup \{\varepsilon_2\}$, $\kappa_1 \le \dots \le \kappa_m$ are integers, and $W_-(\lambda)$ is analytic and invertible outside $\sigma \cup \{\varepsilon_1\}$. In fact, the numbers κ_j are given by $\kappa_j = -\alpha_j$, $(j = 1,\dots,t)$, $\kappa_j = \omega_{m-j+1}$, $(j = m-s+1,\dots,m)$ and $\kappa_j = 0$ otherwise, as one can see from Proposition 2.3.

3. COMPLETION OF STANDARD PAIRS FOR POLYNOMIALS

In this section we study completion problems for regular polynomials, using an approach via admissible triples. Using a transformation in the argument of the polynomial, and premultiplying if necessary by a constant matrix, we may assume without loss of generality that the polynomial is comonic, i.e. it takes the value I at zero. We shall also assume throughout this section that the leading coefficient of the polynomial under consideration is invertible.

Let $L(\lambda) = I + \lambda A_1 + \dots + \lambda^\ell A_\ell$ be an $n \times n$ matrix polynomial with A_ℓ invertible. Recall that the pair (X,T), where X is $n \times n\ell$ matrix and T is an $n\ell \times n\ell$ matrix, is called a *standard pair* for $L(\lambda)$ if $\mathrm{col}\,(XT^i)_{i=0}^{\ell-1}$ is invertible and

(3.1) $$L(\lambda) = I - XT^{-1}(\lambda U_1 + \dots + \lambda^\ell U_\ell),$$

where

(3.2) $$\begin{pmatrix} XT^{\ell-1} \\ \vdots \\ X \end{pmatrix}^{-1} = \begin{pmatrix} U_\ell & \cdots & U_1 \end{pmatrix}.$$

In that case

(3.3) $$L(\lambda)^{-1} = X(\lambda - T)^{-1}Y,$$

where Y is given by

(3.4) $$Y = -T\begin{pmatrix} XT^{\ell-1} \\ \vdots \\ X \end{pmatrix}^{-1}\begin{pmatrix} 0 \\ \vdots \\ 0 \\ I_n \end{pmatrix} = -TU_1$$

(See, e.g., [11], Chapter 7.) Note also that by (3.3), since $L(0) = I_n$, we obtain that T is invertible.

The problem we are concerned with in this section is the following: given a pair (X,T) of matrices of sizes $n \times r$ and $r \times r$, respectively, with T invertible, describe all pairs $(\tilde{X},\tilde{T})$ with $\sigma(T) \cap \sigma(\tilde{T}) = \emptyset$ such that $\left([X,\tilde{X}], \begin{pmatrix} T & 0 \\ 0 & \tilde{T} \end{pmatrix}\right)$ is a standard pair for a comonic $n \times n$ matrix polynomial $L(\lambda)$ with invertible leading coefficient and of degree as small as possible. Such a pair $(\tilde{X},\tilde{T})$ will be called a *completion* of (X,T). In [13] (see also [11], Theorem 6.7) already all possible choices for $\tilde{T}$ such that $\sigma(\tilde{T})$ consists of only one point are described. A necessary condition was already derived in [11], Proposition 6.1;

(3.5) $$\operatorname{rank} \mathrm{col}\,(XT^i)_{i=0}^{\ell-1} = r$$

is necessary for the existence of a completion $(\tilde{X},\tilde{T})$ such that $L(\lambda)$ has degree ℓ, in fact (3.5) is also sufficient for the existence of a completion([11], Theorem 6.7; [13]).

Our approach to the problem will be based on Theorem 2.1, which we will apply to obtain $L(\frac{1}{\lambda})^{-1}$. If $L(\lambda)$ is a comonic matrix polynomial, then $L(\frac{1}{\lambda})^{-1}$ is a rational matrix function with the value I_n at infinity. The next lemma describes a global spectral triple for this function

LEMMA 3.1. *Let* $L(\lambda)$ *be a comonic matrix polynomial of degree* ℓ *with invertible leading coefficient and standard pair* (X,T). *Then a global spectral triple for* $L(\frac{1}{\lambda})^{-1}$ *is given by*

$$(3.6)\qquad \left\{(XT^{-1},T^{-1}),\left(\begin{pmatrix} 0 & I_n & & \\ & \ddots & \ddots & \\ & & \ddots & I_n \\ & & & 0 \end{pmatrix},\begin{pmatrix} 0 \\ \vdots \\ 0 \\ I_n \end{pmatrix}\right),\begin{pmatrix} XT^{-1} \\ \vdots \\ X \end{pmatrix}\right\}.$$

PROOF. Let $L(\lambda)$ be given by

$$L(\lambda) = \lambda^{\ell}A_{\ell} + \dots + \lambda A_1 + I,$$

then $L(\frac{1}{\lambda})$ has the following realization:

$$(3.7)\qquad L(\tfrac{1}{\lambda}) = I + \begin{bmatrix} A_{\ell} & \dots & A_1 \end{bmatrix}\left(\lambda - \begin{pmatrix} 0 & I_n & & \\ & \ddots & \ddots & \\ & & \ddots & I_n \\ & & & 0 \end{pmatrix}\right)^{-1}\begin{pmatrix} 0 \\ \vdots \\ 0 \\ I_n \end{pmatrix}$$

since A_{ℓ} is invertible this realization is minimal. Further, since (X,T) is a standard pair for $L(\lambda)$ we can use (3.3) to get a minimal realization for $L(\frac{1}{\lambda})^{-1}$:

$$(3.8)\qquad L(\tfrac{1}{\lambda})^{-1} = I - XT^{-1}(\lambda - T^{-1})^{-1}T^{-1}Y.$$

From (3.7) one sees that

$$\left(\begin{pmatrix} 0 & I & & \\ & \ddots & \ddots & \\ & & \ddots & I \\ & & & 0 \end{pmatrix},\begin{pmatrix} 0 \\ \vdots \\ 0 \\ I \end{pmatrix}\right)$$

is a zero pair for $L(\frac{1}{\lambda})^{-1}$, whereas from (3.8) one sees that (XT^{-1},T^{-1}) is a pole pair for $L(\frac{1}{\lambda})^{-1}$. A straightforward calculation shows that $\mathrm{col}\,(XT^{i})_{i=\ell-1}^{0}$ is the coupling operator corresponding to these pairs. □

Now, let (X,T) be a pair of matrices of sizes $n\times r$ and $r\times r$, respectively, such that (3.5) holds, and let ℓ be the smallest number for which (3.5) holds. Introduce $\sigma = \sigma(T^{-1}) \cup \{0\}$. If $L(\lambda)$ is a comonic polynomial of degree ℓ with invertible leading coefficient such that $\left(\begin{bmatrix} X, \tilde{X} \end{bmatrix}, \begin{bmatrix} T & 0 \\ 0 & \tilde{T} \end{bmatrix}\right)$ is a standard pair for $L(\lambda)$, and $\sigma(T) \cap \sigma(\widetilde{T}) = \emptyset$, then a σ-spectral triple for $L(\frac{1}{\lambda})^{-1}$ is given by

$$(3.9) \qquad \tau = \left\{(XT^{-1},T^{-1}), \left(\begin{pmatrix} 0 & I_n & & \\ & \ddots & \ddots & \\ & & \ddots & I_n \\ & & & 0 \end{pmatrix}, \begin{pmatrix} 0 \\ \vdots \\ 0 \\ I_n \end{pmatrix} \right), \begin{pmatrix} XT^{\ell-1} \\ \vdots \\ X \end{pmatrix} \right\}$$

Note that by (3.5) we have

$$\operatorname{Ker} \operatorname{col} (XT^i)_{i=\ell-1}^{0} = (0).$$

THEOREM 3.2. *Let* K *be a complement to* $\operatorname{Im} \operatorname{col} (XT^i)_{i=\ell-1}^{0}$ *in* $\mathbb{C}^{n\ell}$, *and let* ε *be an arbitrary set in* $\mathbb{C}$ *with* $\sigma \cap \varepsilon = \emptyset$. *Let*

$$(3.10) \qquad \tau_0 = \{(X_0,T_0),(0,0),0\}$$

be an ε*-minimal complement of* τ, *where the zero operators in the second pair act on the zero dimensional space and* $T_0 : K \to K$. *Put*

$$(3.11) \qquad \widetilde{X} = X_0T_0^{-1}, \quad \widetilde{T} = T_0^{-1}$$

Then $(\widetilde{X},\widetilde{T})$ *is a completion of* (X,T).

Conversely, any completion $(\widetilde{X},\widetilde{T})$ *of* (X,T) *is similar to a pair obtained in the above manner.*

The polynomial for which $\left(\begin{bmatrix} X, \widetilde{X} \end{bmatrix}, \begin{bmatrix} T & 0 \\ 0 & \widetilde{T} \end{bmatrix} \right)$ *is a standard pair is given by*

$$(3.12) \qquad L(\lambda) = I - \sum_{j=1}^{\ell} (XT^{-1}U_{1j} + \widetilde{X}\widetilde{T}^{-1}U_{2j})\lambda^j,$$

where

$$\begin{bmatrix} U_{1\ell} & \cdots & U_{11} \\ U_{2\ell} & \cdots & U_{21} \end{bmatrix} = \begin{bmatrix} XT^{\ell-1} & \widetilde{X}\widetilde{T}^{\ell-1} \\ \vdots & \vdots \\ X & \widetilde{X} \end{bmatrix}^{-1},$$

and

$$(3.13) \qquad L(\lambda)^{-1} = \begin{bmatrix} X & \widetilde{X} \end{bmatrix} \left(\lambda - \begin{bmatrix} T & 0 \\ 0 & \widetilde{T} \end{bmatrix} \right)^{-1} \begin{bmatrix} -TU_{11} \\ -\widetilde{T}U_{21} \end{bmatrix}.$$

PROOF. Since $\operatorname{Ker} \operatorname{col} (XT^i)_{i=\ell-1}^{0} = (0)$ an ε-minimal complement of τ is indeed of the form (3.10). By the choice of σ, T_0 is invertible, so $\widetilde{X}$ and $\widetilde{T}$ are well-defined. Using (3.11) a little computation gives that $\tau \oplus \tau_0$ is given by

$$\left\{\left(\begin{pmatrix} XT^{-1}, \tilde{X}\tilde{T}^{-1}\end{pmatrix}, \begin{pmatrix} T^{-1} & 0 \\ 0 & \tilde{T}^{-1}\end{pmatrix}\right), \left(\begin{pmatrix} 0 & I_n & & \\ & \ddots & \ddots & \\ & & \ddots & I_n \\ & & & 0 \end{pmatrix}, \begin{pmatrix} 0 \\ \vdots \\ 0 \\ I_n \end{pmatrix}\right), \begin{pmatrix} XT^{\ell-1} & \tilde{X}\tilde{T}^{\ell-1} \\ \vdots & \vdots \\ X & \tilde{X}\end{pmatrix}\right\}$$

This triple is the global spectral triple of $L(\frac{1}{\lambda})^{-1}$, where $L(\lambda)^{-1}$ is given by (3.13). Indeed:

$$L(\tfrac{1}{\lambda})^{-1} = I + \begin{pmatrix} XT^{-1}, \tilde{X}\tilde{T}^{-1}\end{pmatrix}\begin{pmatrix} (\lambda - T^{-1})^{-1} & 0 \\ 0 & (\lambda - \tilde{T}^{-1})^{-1}\end{pmatrix}\begin{pmatrix} U_{11} \\ U_{21}\end{pmatrix}$$

$$= (I - XU_{11} - \tilde{X}U_{21}) + X(T - \tfrac{1}{\lambda})^{-1}TU_{11} + \tilde{X}(\tilde{T} - \tfrac{1}{\lambda})^{-1}\tilde{T}U_{21}$$

$$= X(\tfrac{1}{\lambda} - T)^{-1}(-TU_{11}) + \tilde{X}(\tfrac{1}{\lambda} - \tilde{T})^{-1}(-\tilde{T}U_{21}),$$

which shows (3.13). Next, we check (3.12)

$$L(\tfrac{1}{\lambda}) = I - \begin{pmatrix} XT^{-1}, \tilde{X}\tilde{T}^{-1}\end{pmatrix}\begin{pmatrix} XT^{\ell-1} & \tilde{X}\tilde{T}^{\ell-1} \\ \vdots & \vdots \\ X & \tilde{X}\end{pmatrix}^{-1}\begin{pmatrix} \frac{1}{\lambda^\ell} I_n \\ \vdots \\ \frac{1}{\lambda} I_n \end{pmatrix}$$

$$= I - \sum_{j=1}^{\ell} (XT^{-1}U_{1j} + \tilde{X}\tilde{T}^{-1}U_{2j})\frac{1}{\lambda^j} .$$

This shows (3.12), as well as the fact that $L(\lambda)$ is a polynomial of degree ℓ indeed. Hence $(\begin{pmatrix} X & \tilde{X}\end{pmatrix}, \begin{pmatrix} T & 0 \\ 0 & \tilde{T}\end{pmatrix})$ is indeed a standard pair for a comonic $n \times n$ matrix polynomial of degree ℓ. The fact that $L(\lambda)$ has an invertible leading coefficient follows from the minimality of the realization for $L(\frac{1}{\lambda})$ corresponding to the global spectral triple $\tau \oplus \tau_0$.

For the converse, suppose $(\tilde{X},\tilde{T})$ is a pair as desired, and let $L(\lambda)$ be the comonic $n \times n$ matrix polynomial with standard pair $(\begin{pmatrix} X, \tilde{X}\end{pmatrix}, \begin{pmatrix} T & 0 \\ 0 & \tilde{T}\end{pmatrix})$. Construct a global spectral triple for $L(\frac{1}{\lambda})^{-1}$ as in Lemma 3.1. From that one sees that

$$\tau_0 = \{(\tilde{X}\tilde{T}^{-1}, \tilde{T}^{-1}), (0,0), 0\}$$

is a complement of τ. It is a minimal complement as $\tilde{T}$ is of size $(n\ell-r) \times (n\ell-r)$. Applying Theorem 2.1 now yields the desired result. □

As a corollary we obtain a generalization of the result of [13] (see also [11], Theorem 6.7).

THEOREM 3.3. *Let* (X,T) *be a pair of matrices of sizes* $n \times r$ *and* $r \times r$, *respectively, such that* (3.5) *holds, and with* T *invertible. Let* $\tilde{T}$ *be an*

$(n\ell-r)\times(n\ell-r)$ *matrix with* $\tilde{T}$ *invertible and* $\sigma(T)\cap\sigma(\tilde{T})=\emptyset$. *Then for some matrix* $\tilde{X}$ *the pair* $(\tilde{X},\tilde{T})$ *is a completion of* (X,T) *if and only if for the degrees of the invariant factors* $\nu_1\geq\ldots\geq\nu_p$ *of* $\lambda-\tilde{T}$ *we have* $p\leq n-r+q_{\ell-2}$ *and*

$$(3.14)\qquad \sum_{j=1}^{\mu}\nu_j-\omega_j\geq 0 \quad \textit{for } \mu=1,\ldots,p,$$

where for $k=1,2,\ldots,n-r+q_{\ell-2}$,

$$(3.15)\qquad \omega_k=\#\{j\mid n+q_{\ell-j-1}-q_{\ell-j}\geq k,\ j=0,\ldots,\ell-1\}$$

and $q_j := \operatorname{rank}\operatorname{col}(XT^i)_{i=0}^{j}$.

PROOF. Suppose the pair $(\tilde{X},\tilde{T})$ is a completion of (X,T). Then the triple τ_0 given by (3.10), (3.11) is a minimal complement of the triple τ given by (3.9). In particular (X_0,T_0) is a pole correction pair for τ. Note that the degrees of the invariant factors of $\lambda-\tilde{T}$ are the same as the degrees of the invariant factors of $\lambda-T_0$. Then (3.14) follows from (1.4) if we can show that the incoming indices are given by (3.15). The incoming subspaces for τ are given by

$$H_j=\operatorname{Im}\operatorname{col}(XT^i)_{i=\ell-1}^{0}+\operatorname{Im}\begin{pmatrix}0\\ \vdots\\ 0\\ I_n\end{pmatrix}+\ldots+\operatorname{Im}\begin{pmatrix}0 & I_n & & \\ & \ddots & \ddots & \\ & & \ddots & I_n\\ & & & 0\end{pmatrix}^{j-1}\begin{pmatrix}0\\ \vdots\\ 0\\ I_n\end{pmatrix}$$

$$=\operatorname{Im}\begin{pmatrix}XT^{\ell-1} & & \\ \cdot & & 0\\ \cdot & & I_n\\ \cdot & \cdot^{\cdot^{\cdot}} & \\ X & \underbrace{I_n}_{jn} & \end{pmatrix}=\operatorname{Im}\begin{pmatrix}XT^{\ell-j-1} & & \\ \vdots & & 0\\ X & & \\ 0 & & I_n\\ \vdots & \cdot^{\cdot^{\cdot}} & \\ 0 & I_n & \end{pmatrix}$$

So $\dim H_1/H_0=\dim H_1-\dim\operatorname{Im}\operatorname{col}(XT^i)_{i=0}^{\ell-1}=n+\dim\operatorname{Im}\operatorname{col}(XT^i)_{i=0}^{\ell-2}-r=n+q_{\ell-2}-r$, which is the number of incoming indices; and

$$\dim H_j/H_{j-1}=n_j+q_{\ell-j-1}-n(j-1)-q_{\ell-j}=n+q_{\ell-j-1}-q_{\ell-j}.$$

Then (3.15) follows from the definition of the incoming indices.

Conversely, suppose (3.14) holds for $\tilde{T}$. Define $T_0=\tilde{T}^{-1}$. By Rosenbrock's theorem there exists an X_0 such that (X_0,T_0) is a pole correction pair for τ. Hence $\tilde{X}=X_0T_0^{-1}$ is a matrix such that $(\tilde{X},\tilde{T})$ is a completion of (X,T). □

In [13] this theorem was proved for the case when $\sigma(\tilde{T})$ consists of only one point. Note that the assumption that T is invertible is really not a loss of generality, since this reflects our assumption that $L(\lambda)$ is comonic.

As a corollary of Theorem 3.3 we have the following remark on the local Smith form of the polynomial corresponding to a completion of a specific type. Let $(\tilde{X},\tilde{T})$ be a completion of (X,T) such that $\tilde{T}$ has only one eigenvalue, λ_0 say. Let $L(\lambda)$ be the polynomial given by (3.12), and let

$$L(\lambda) = E(\lambda)\,\mathrm{diag}\,((\lambda-\lambda_0)^{\kappa_j})_{j=1}^{m}F(\lambda)$$

be its local Smith form with $\kappa_1 \geq \ldots \geq \kappa_m$. Then with ω_j given by (3.15) inequality (3.14) holds for κ_j instead of ν_j. Indeed, the κ_j's are precisely the ν_j's for $j=1,\ldots,p$, and zero for $j > p$.

4. WIENER-HOPF FACTORIZATION

4.1. Motivation and canonical factorization.

In this section we study Wiener-Hopf factorization, both canonical and non-canonical, via an inverse spectral approach. Let W be an $m\times m$ rational matrix function with $W(\infty) = I_m$ such that W has no zeros and poles on a given simple closed contour γ. The region inside γ will be denoted by γ_+, the region outside γ by γ_-. Note that $\infty \in \gamma_-$. Fix two points $\varepsilon_1 \in \gamma_+$ and $\varepsilon_2 \in \gamma_-$. A *(right) Wiener-Hopf factorization* of W with respect to γ is defined as

$$(4.1) \qquad W(\lambda) = W_-(\lambda)D(\lambda)W_+(\lambda), \qquad \lambda \in \gamma,$$

where W_+ has no zeros and poles in $\gamma_+ \cup \gamma$, the function W_- has no zeros and poles in $\gamma_- \cup \gamma$, and

$$D(\lambda) = \mathrm{diag}\left(\left(\frac{\lambda-\varepsilon_1}{\lambda-\varepsilon_2}\right)^{\kappa_j}\right)_{j=1}^{m}.$$

Here $\kappa_1 \leq \ldots \leq \kappa_m$ are integers, which are called the *(right) factorization indices*. In case all $\kappa_j = 0$ the factorization is said to be *canonical*. Since $W(\infty) = I_m$, we may assume in what follows that also the factors W_- and W_+ in (4.1) are normalized to I_m at ∞.

Formulas for the indices κ_j and the factors W_- and W_+ in terms of a realization of W have been derived in [5] (see also [14]). Here we shall derive formulas for W_-, W_+ and the indices using the point of view developed in Sections 1 and 2.

As in [5] the starting point is a minimal realization of W:

$$W(\lambda) = I_m + C(\lambda I_n - A)^{-1}B.$$

Since W has no poles and zeros on γ, both A and $A^\times$ (= A - BC) have no eigenvalues on γ. Let P be the spectral projection of A corresponding to the region γ_+, and $P^\times$ the spectral projection of $A^\times$ corresponding to the same region. The image of P will be denoted by M and we write $M^\times$ for the kernel of $P^\times$. Then a γ_+-spectral triple for W is given by

$$(4.2)\qquad \tau_+ = \{(C|_M, A|_M), (A^\times|_{\operatorname{Im} P^\times}, P^\times B), P^\times|_M : M \to \operatorname{Im} P^\times\}.$$

Suppose we have a factorization (4.1) for W. Since $W_+ = (W_-D)^{-1}W$ has no poles and zeros on γ_+, it follows from [8], Theorem 5.1 (see, also [1], Lemma 1.4) that the γ_+-spectral triples of W_-D and W coincide; in other words τ_+ is also a γ_+-spectral triple of W_-D. By our results from Section 2 it follows that there is a specific minimal complement τ_0 of τ_+ such that the function corresponding to $\tau_+ \oplus \tau_0$ is W_-D. In this section we describe precisely which minimal complement τ_0 one has to take. Once this is done we have formulas for the function $\widetilde{W} = W_-D$ corresponding to $\tau_+ \oplus \tau_0$, and using the proof of Theorem 5.1 in [8] we can derive formulas for W_+ and W_+^{-1}. It then remains to derive the formulas for W_- and W_-^{-1}.

The above plan will be carried out in the next two subsections. We conclude the present subsection with the construction of a canonical factorization which corresponds to the case when τ_+ does not need an extension, i.e. when τ_+ itself is already a global spectral triple. The latter means that $P^\times|_M : M \to \operatorname{Im} P^\times$ is invertible, which in turn is equivalent to the statement that $\mathbb{C}^n = M \oplus M^\times$, where $M^\times = \operatorname{Ker} P^\times$. Now, let W_- be the rational matrix function corresponding to τ_+, i.e.,

$$W_-(\lambda) = I + C(\lambda - A|_M)^{-1}(P^\times|_M)^{-1}P^\times B,$$

$$W_-(\lambda)^{-1} = I - C(P^\times|_M)^{-1}(\lambda - A^\times|_{\operatorname{Im} P^\times})^{-1}P^\times B,$$

and put $W_+(\lambda) = W_-(\lambda)^{-1}W(\lambda)$. Then the discussion in the previous paragraph yields that $W(\lambda) = W_-(\lambda)W_+(\lambda)$ is a canonical Wiener-Hopf factorization relative to γ, and we can employ the proof of Theorem 5.1 in [8] to show that

$$W_+(\lambda) = I + C((P - I)|_{M^\times})^{-1}(\lambda - A|_{\operatorname{Ker} P})^{-1}(I - P)B,$$

$$W_+(\lambda)^{-1} = I - C(\lambda - A^\times|_{M^\times})^{-1}((P - I)|_{M^\times})^{-1}(I - P)B.$$

The equivalence between canonical factorization and the invertibility of $P^\times|_M : M \to \operatorname{Im} P^\times$ is already contained in [6], and the formulas for the factors as given above are a variation of those in [6] (see also [2] where they appear precisely in the form given above).

4.2. Preliminaries.

In this subsection we present some elementary material necessary to describe the factors in a non-canonical Wiener-Hopf factorization. We start again from the γ_+-spectral triple τ_+ for W given by (4.2). Note that $\mathrm{Ker}\,(P^\times|_M) = M \cap M^\times$. Furthermore

$$\mathrm{codim}\,\mathrm{Im}\,P^\times|_M = \dim \mathrm{Im}\,P^\times - \dim \mathrm{Im}\,P^\times|_M = n - \dim M^\times - (\dim M - \dim M \cap M^\times) =$$
$$= \mathrm{codim}\; M + M^\times.$$

Choose a complement N in M of $M \cap M^\times$, and a complement K in $\mathrm{Im}\,P^\times$ of $\mathrm{Im}\,P^\times|_M$. Introduce the projections ρ_p in M along N onto $M \cap M^\times$ and ρ_z in $\mathrm{Im}\,P^\times$ along $\mathrm{Im}\,P^\times|_M$ onto K. Also, let η_p be the embedding of $M \cap M^\times$ into M and η_z the embedding of K in $\mathrm{Im}\,P^\times$.

Choose a point ε_2 in γ_-, and let (S,G) be a zero correction pair for τ_+ such that $\sigma(S) = \{\varepsilon_2\}$, $S : M \cap M^\times \to M \cap M^\times$, and such that S has Jordan canonical form with respect to an outgoing basis $\{e_{jk}\}_{k=1}^{\alpha_j}{}_{j=1}^{t}$ in $M \cap M^\times$. In particular the sizes of the Jordan blocks of S are $\alpha_1,\dots,\alpha_t$; i.e. (1.2) holds with equality for every ν. Further, let (F,T) be a pole correction pair for τ_+ such that $\sigma(T) = \{\varepsilon_2\}$, $T : K \to K$, and such that T has Jordan canonical form with respect to an incoming basis $\{g_{jk}\}_{k=1}^{\omega_j}{}_{j=1}^{s}$ in K, i.e. (1.4) holds with equality for every ν.

Connected to this zero correction pair and pole correction pair we shall define also operators $\widetilde{S} : M \cap M^\times \to M \cap M^\times$, $\widetilde{T} : K \to K$, $\widetilde{F} : K \to \mathbb{C}^m$ and $\widetilde{G} : \mathbb{C}^m \to M \cap M^\times$ as follows. Choose ε_1 in γ_+. Define an outgoing basis $\{d_{jk}\}_{k=1}^{\alpha_j}{}_{j=1}^{t}$ from $\{e_{jk}\}_{k=1}^{\alpha_j}{}_{j=1}^{t}$ by

$$d_{jk} = \sum_{\nu=0}^{k-1} \binom{k-1}{\nu} (\varepsilon_2 - \varepsilon_1)^\nu e_{j\,k-\nu},$$

and define $\widetilde{S}$ by

$$(4.3) \qquad (\widetilde{S} - \varepsilon_1) d_{jk} = d_{j\,k+1} \qquad (d_{j\alpha_j+1} := 0).$$

Also, define an incoming basis $\{f_{jk}\}_{k=1}^{\omega_j}{}_{j=1}^{s}$ from the basis $\{g_{jk}\}_{k=1}^{\omega_j}{}_{j=1}^{s}$ by

$$f_{jk} = \sum_{\nu=0}^{k-1} \binom{\nu+\omega_j-k}{\nu} (\varepsilon_1 - \varepsilon_2)\; g_{j\,k-\nu},$$

and define $\widetilde{T}$ by

$$(4.4) \qquad (\widetilde{T} - \varepsilon_1) f_{jk} = f_{j\,k+1} \qquad (f_{j,\alpha_j+1} := 0).$$

Next, put

$$z_j = Ce_{j\alpha_j} = Cd_{j\alpha_j} \qquad (j=1,\dots,t),$$

and let y_j, $j=1,\dots,s$, be such that

$$f_{j1} - P^\times By_j = g_{j1} - P^\times By_j \in \operatorname{Im} P^\times|_M$$

(this can be done as $\{g_{j1}\}_{j=1}^s$ forms a basis for $\operatorname{Im} P^\times|_M + \operatorname{Im} P^\times B$ modulo $\operatorname{Im} P^\times|_M$.) Finally, choose a space $Y_0 \subset (P^\times B)^{-1} \operatorname{Im} P^\times|_M$ such that

$$\mathbb{C}^m = \operatorname{span}\{y_j\}_{j=1}^s \dot{+} Y_0 \dot{+} \operatorname{span}\{z_j\}_{j=1}^t .$$

Note that indeed $\operatorname{span}\{y_j\}_{j=1}^s$ is a complement of $(P^\times B)^{-1} \operatorname{Im} P^\times|_M$, and that $z_j \in (P^\times B)^{-1} \operatorname{Im} P^\times|_M$. Indeed, for the latter one uses

$$P^\times Bz_j = P^\times BCPe_{j\alpha_j} = (P^\times|_M A|_M - A^\times|_{\operatorname{Im} P^\times} P^\times|_M)e_{j\alpha_j} =$$

$$= P^\times|_M A|_M e_{j\alpha_j} \in \operatorname{Im} P^\times|_M .$$

Now introduce $\widetilde{F} : K \to \mathbb{C}^m$ and $\widetilde{G} : \mathbb{C}^m \to M \cap M^\times$ by

$$\widetilde{F}f_{jk} = \binom{\omega_j}{k}(\varepsilon_1 - \varepsilon_2)^k y_j, \tag{4.5}$$

$$\begin{cases} \widetilde{G}y = 0 \text{ for } y \in \operatorname{span}\{y_j\} \dot{+} Y_0 \\ \widetilde{G}z_j = (\widetilde{S} - S)d_{j\alpha_j} = (\widetilde{S} - S)e_{j\alpha_j} \end{cases} \tag{4.6}$$

4.3. Non-canonical Wiener-Hopf factorization

Start by choosing operators X, Y and Γ_1 such that

$$\tau_0 = \{(-CPX - F,T),(S,-YP^\times B + G),\Gamma_0\},$$

where $\Gamma_0 = YP^\times|_M X - Y\eta_z - \rho_p X + \Gamma_1$, is an $\{\varepsilon_2\}$-minimal complement of τ_+ (cf. Theorem 2.1). Let $\widetilde{W}(\lambda)$ be the function corresponding to $\tau_+ \oplus \tau_0$. As observed before we have $W(\lambda) = \widetilde{W}(\lambda)W_+(\lambda)$, where W_+ has no poles and zeros on γ_+ (Theorem 5.1 in [8]). In the next theorem we shall provide formulas for W_+ and W_+^{-1}. Note that formulas for $\widetilde{W}$ and its inverse are given by

$$\widetilde{W}(\lambda) = I + C(\lambda - A|_M)^{-1}\{(\Gamma_{11}^\times - \eta_p Y)P^\times B + \eta_p G\} + (-CPX - F)(\lambda - T)^{-1}\rho_z P^\times B, \tag{4.7}$$

$$\widetilde{W}(\lambda)^{-1} = I - \{C(\Gamma_{11}^\times - PX\rho_z) + F\rho_z\}(\lambda - A^\times)^{-1}P^\times B - C\eta_p(\lambda - S)^{-1}(-YP^\times B + G), \tag{4.8}$$

where $\begin{pmatrix} \Gamma_{11}^\times & \eta_p \\ \rho_z & 0 \end{pmatrix}$ is the inverse of the coupling operator of $\tau_+ \oplus \tau_0$ (see Theorem 2.2).

THEOREM 4.1. *The function* $W(\lambda) = I + C(\lambda - A)^{-1}B$ *factorizes as*

$$W(\lambda) = \widetilde{W}(\lambda)W_+(\lambda)$$

where

$$(4.9)\qquad W_+(\lambda) = I + \\ + [C + \{C(PX\rho_z - \Gamma_{11}^\times) - F\rho_z\}P^\times, C\eta_p]\left(\lambda - \begin{pmatrix} A|_{\operatorname{Ker} P} & 0 \\ V_+ & S \end{pmatrix}\right)^{-1} \begin{pmatrix} (I-P)B \\ \{\rho_p P + YP^\times(I-P)\}B - G \end{pmatrix}$$

$$(4.10)\qquad W_+(\lambda)^{-1} = I + \\ - [C, C(P-P^\times)X + F + C\eta_z]\left(\lambda - \begin{pmatrix} A^\times|_{M^\times} & V_+^\times \\ 0 & T \end{pmatrix}\right)^{-1} \begin{pmatrix} B - (\Gamma_{11}^\times - \eta_p Y)P^\times B - \eta_p G \\ \rho_z P B \end{pmatrix}$$

with

$$V_+ = -YP^\times BC + GC$$

$$V_+^\times = -BCPX - BF.$$

PROOF. From [8], Theorem 5.1 we know that if $W_1(\lambda)$ and $W_2(\lambda)$ are given by the minimal realizations $W_i(\lambda) = I + C_i(\lambda - A_i)^{-1}B_i$ and have the same γ_+-spectral triple, then

$$W_+(\lambda) := W_2(\lambda)^{-1}W_1(\lambda) =$$

$$(4.11)\qquad = (I - C_2(I-Q_2^\times)(\lambda - A_2^\times)^{-1}B_2)(I + C_1(I-Q_1)(\lambda - A_1)^{-1}B_1) \\ + C_2(\lambda - A_2^\times)^{-1}(I-Q_2^\times)EQ_1B_1 - C_2FQ_1^\times(I-Q_1)(\lambda - A_1)^{-1}B_1$$

and

$$W_+(\lambda)^{-1} := W_1(\lambda)^{-1}W_2(\lambda) =$$

$$(4.12)\qquad = (I - C_1(I-Q_1^\times)(\lambda - A_1^\times)^{-1}B_1)(I + C_2(I-Q_2)(\lambda - A_2)^{-1}B_2) \\ + C_1(\lambda - A_1^\times)^{-1}(I-Q_1^\times)E^{-1}Q_2B_2 - C_1F^{-1}Q_2^\times(I-Q_2)(\lambda - A_2)^{-1}B_2$$

where Q_i, resp. $Q_i^\times$, is the Riesz projection of A_i, resp. $A_i^\times = A_i - B_iC_i$, corresponding to γ_+, and $E : \operatorname{Im} Q_1 \to \operatorname{Im} Q_2$, $F : \operatorname{Im} Q_1^\times \to \operatorname{Im} Q_2^\times$ are invertible operators such that

$$A_2|_{\operatorname{Im} Q_2}E = EA_1|_{\operatorname{Im} Q_1}, \qquad C_2|_{\operatorname{Im} Q_2}E = C_1|_{\operatorname{Im} Q_1}$$

$$A_2^\times|_{\operatorname{Im} Q_2^\times}F = FA_1^\times|_{\operatorname{Im} Q_1^\times}, \qquad Q_2^\times B_2 = FQ_1^\times B_1$$

$$Q_2^\times|_{\operatorname{Im} Q_2}E = FQ_1^\times|_{\operatorname{Im} Q_1}.$$

Applying these formulas to the realization $W(\lambda) = I + C(\lambda - A)^{-1}B$ for $W = W_1$ and (4.8) for $\widetilde{W}^{-1} = W_2^{-1}$ we obtain the formula for W_+. Indeed we have for this realization of $\widetilde{W}^{-1}$ that

$$A_2^\times = \begin{pmatrix} A^\times|_{\operatorname{Im} P^\times} & 0 \\ 0 & S \end{pmatrix}, \qquad B_2 = \begin{pmatrix} P^\times B \\ -YP^\times B + G \end{pmatrix}$$

$$C_2 = \begin{pmatrix} C(\Gamma_{11}^\times - PX\rho_z) + F\rho_z, & C\eta_p \end{pmatrix},$$

$$A_2 = \hat{\Gamma}_{\tau_+\oplus\tau_0} \begin{pmatrix} A_M & 0 \\ 0 & T \end{pmatrix} \hat{\Gamma}_{\tau_+\oplus\tau_0}^{-1}.$$

So

$$E = \begin{pmatrix} P^\times|_M \\ -YP^\times|_M + \rho_p \end{pmatrix}, \quad F = \begin{pmatrix} I_{\operatorname{Im} P^\times} \\ 0 \end{pmatrix}, \quad Q_1^\times = P^\times, \quad Q_1 = P.$$

Inserting these into (4.11) yields (4.9) easily. Indeed, it gives

$$W_+(\lambda) = (I - C\eta_p(\lambda - S)^{-1}(-YP^\times B + G))(I + C(\lambda - A|_{\operatorname{Ker} P})^{-1}(I-P)B) +$$

$$+ C\eta_p(\lambda - S)^{-1}(-YP^\times + \rho_p)PB - (C(\Gamma_{11}^\times - PX\rho_z) + F\rho_z)(\lambda - A|_{\operatorname{Ker} P})^{-1}(I-P)B$$

which is (4.9) after a little rewriting.

Likewise one proves (4.10) from (4.12) using the realization (4.7) for $\widetilde{W}$. □

Next, we provide formulas for the factors in a Wiener-Hopf factorization of $W(\lambda)$, and describe also the indices.

THEOREM 4.2. *Let* $W(\lambda) = I_m + C(\lambda I_n - A)^{-1}B$ *be a minimal realization of* W, *and let*

$$\tau_0 = \{(-CPX - F, T), (S, -YP^\times B + G), \Gamma_0\}$$

be an $\{\varepsilon_2\}$*-minimal complement of* τ_+ *as above. Further, let* $\widetilde{S}$, $\widetilde{T}$, $\widetilde{F}$ *and* $\widetilde{G}$ *be given by* (4.3), (4.4), (4.5) *and* (4.6), *respectively; and let* $\Gamma_{11}^\times$ *come from the inverse of the coupling operator of* $\tau_+ \oplus \tau_0$ *as before. Then*

(4.13) $$W(\lambda) = W_-(\lambda)D(\lambda)W_+(\lambda),$$

where W_+ *and* W_+^{-1} *are given by* (4.9) *and* (4.10) *and*

(4.14) $$W_-(\lambda) = I + \begin{pmatrix} C, & -CPX - F + \widetilde{F} \end{pmatrix}\left(\lambda - \begin{pmatrix} A|_M & V_- \\ 0 & \widetilde{T} \end{pmatrix}\right)^{-1}\begin{pmatrix} \Gamma_{11}^\times & \eta_p \\ \rho_z & 0 \end{pmatrix}\begin{pmatrix} P^\times B \\ -YP^\times B + G - \widetilde{G} \end{pmatrix}$$

(4.15) $$W_-(\lambda)^{-1} = I - \begin{pmatrix} C, & -CPX - F + \widetilde{F} \end{pmatrix}\begin{pmatrix} \Gamma_{11}^\times & \eta_p \\ \rho_z & 0 \end{pmatrix}\left(\lambda - \begin{pmatrix} A^\times|_{\operatorname{Im} P} & 0 \\ V_-^\times & \widetilde{S} \end{pmatrix}\right)^{-1}\begin{pmatrix} P^\times B \\ -YP^\times B + G - \widetilde{G} \end{pmatrix}$$

with

$$V_- = \{(\Gamma_{11}^\times - \eta_p Y)P^\times B + \eta_p G\}\widetilde{F}$$

$$V_-^\times = \eta_p \widetilde{G}\{C(\Gamma_{11}^\times - PX\rho_z) - F\rho_z\}.$$

Finally,

$$(4.16) \qquad D(\lambda) = E\,\mathrm{diag}\left(\left(\frac{\lambda-\varepsilon_1}{\lambda-\varepsilon_2}\right)^{\kappa_j}\right)_{j=1}^m E^{-1},$$

where $\kappa_j = -\alpha_j$ $(j = 1,\dots,t)$, $\kappa_j = 0$ $(j = t+1,\dots,m-s)$, $\kappa_j = \omega_{m-j+1}$ $(j = m-s+1,\dots,m)$, *and* E *is the matrix*

$$E = \left(z_1 \dots z_t\ z_{t+1} \dots z_{m-s} y_s \dots y_1\right),$$

where $\{z_{t+1},\dots,z_{m-s}\}$ *is a basis for* Y_0.

The proof will be given in several steps, which will be presented as separate lemmas.

First we will show that $\widetilde{W}(\lambda) = W_-(\lambda)D(\lambda)$, thereby establishing also (4.13), in view of Theorem 4.1. Note that this will also prove the remark at the end of Section 2. Next, we shall show that the function given by (4.15) is indeed the inverse of the function $W_-(\lambda)$ given by (4.14).

The first step is in fact a repetition of the argument which we used to write $W = \widetilde{W}W_+$. In fact, consider the γ_--spectral triple for $\widetilde{W}(\lambda)^{-1}$. By (4.7), (4.8) and the fact that the inverse of the coupling matrix of $\tau_+ \oplus \tau_0$ has the form $\begin{pmatrix} \Gamma_{11}^\times & \eta_p \\ \rho_z & 0 \end{pmatrix}$, this triple is given by

$$(4.17) \qquad \widetilde{\tau}_0 = \{(C\eta_p, S), (T, \rho_z P^\times B), 0\}.$$

We claim that an $\{\varepsilon_1\}$- complement to $\widetilde{\tau}_0$ is given by

$$(4.18) \qquad \tau_0' = \{(\widetilde{F}, \widetilde{T}), (\widetilde{S}, \widetilde{G}), 0\}.$$

Indeed, with $\Gamma_{\widetilde{\tau}_0 \oplus \tau_0'} = \begin{pmatrix} 0 & I_K \\ -I_{M\cap M^\times} & 0 \end{pmatrix} : M \cap M^\times \oplus K \to K \oplus M \cap M^\times$ we have

$$\Gamma_{\widetilde{\tau}_0 \oplus \tau_0'} \begin{pmatrix} S & 0 \\ 0 & \widetilde{T} \end{pmatrix} - \begin{pmatrix} T & 0 \\ 0 & \widetilde{S} \end{pmatrix} \Gamma_{\widetilde{\tau}_0 \oplus \tau_0'} = \begin{pmatrix} \rho_z P^\times B \\ \widetilde{G} \end{pmatrix} \left(C\eta_p, \widetilde{F}\right).$$

Here one uses the fact that $\widetilde{G}\widetilde{F} = 0$ (see (4.5) and (4.6)), and the following useful formulas:

$$(4.19) \qquad (\widetilde{T} - T)f_{jk} = (\varepsilon_1 - \varepsilon_2)^k \binom{\omega_j}{k} f_{j1}$$

$$(4.20)\qquad (\tilde{S}-S)d_{j\alpha_j} = -\sum_{\nu=0}^{\alpha_j-1}\binom{\alpha_j}{\nu+1}(\varepsilon_2-\varepsilon_1)^{\nu+1}e_{j\alpha_j-\nu}.$$

These formulas can be checked by direct computation (cf. [5], formulas II (1.11), (1.13) and III (1.10), (1.13)). Note that (4.19) and (4.5) imply

$$\tilde{T}-T = \rho_Z P^{\times}B\tilde{F},$$

and (4.6) gives

$$\tilde{S}-S = \tilde{G}C\eta_p.$$

Denote the function corresponding to $\tilde{\tau}_0 \oplus \tau_0'$ by $D(\lambda)^{-1}$. We shall show that in fact $D(\lambda)$ is given by (4.16). Then we have that $\tilde{W}(\lambda)^{-1}$ and $D(\lambda)^{-1}$ have the same γ_--spectral triple, so by [8], Theorem 5.1 again one has

$$\tilde{W}(\lambda)^{-1} = D(\lambda)^{-1}W_-(\lambda)^{-1}$$

for a function W_- which has no poles and zeros outside λ. We shall show that W_- is in fact given by (4.14) and (4.15).

A realization for $D(\lambda)^{-1}$ is given by

$$(4.21)\qquad D(\lambda)^{-1} = I + \begin{bmatrix} C\eta_p, \tilde{F}\end{bmatrix}\left(\lambda - \begin{bmatrix} S & 0\\ 0 & \tilde{T}\end{bmatrix}\right)^{-1}\begin{bmatrix} -\tilde{G}\\ \rho_Z P^{\times}B\end{bmatrix},$$

$$(4.22)\qquad D(\lambda) = I - \begin{bmatrix}\tilde{F}, -C\eta_p\end{bmatrix}\left(\lambda - \begin{bmatrix} T & 0\\ 0 & \tilde{S}\end{bmatrix}\right)^{-1}\begin{bmatrix}\rho_Z P^{\times}B\\ \tilde{G}\end{bmatrix}.$$

LEMMA 4.3. *With* $D(\lambda)$ *given by* (4.22) *we have*

$$D(\lambda)y = \begin{cases} \left(\frac{\lambda-\varepsilon_2}{\lambda-\varepsilon_1}\right)^{\alpha_j} z_j & y = z_j,\quad j=1,\dots,t\\ y & y\in Y_0\\ \left(\frac{\lambda-\varepsilon_1}{\lambda-\varepsilon_2}\right)^{\omega_j} y_j & y = y_j,\quad j=1,\dots,s.\end{cases}$$

In other words $D(\lambda)$ *is given by* (4.16).

PROOF. First take $y \in Y_0$. Since $P^{\times}By \in \operatorname{Im} P^{\times}|_M$ we have $\rho_Z P^{\times}By = 0$. Also $\tilde{G}y = 0$, so $D(\lambda)y = y$ as claimed. Next take $y = z_j$. Then $\rho_Z P^{\times}Bz_j = 0$, so by (4.22)

$$D(\lambda)z_j = z_j + C\eta_p(\lambda-\tilde{S})^{-1}\tilde{G}z_j = z_j + C\eta_p(\lambda-\tilde{S})^{-1}(\tilde{S}-S)e_{j\alpha_j}.$$

Analogous to (4.20) we also have

$$(\tilde{S}-S)e_{j\alpha_j} = \sum_{\nu=0}^{\alpha_j-1}\binom{\alpha_j}{\nu+1}(\varepsilon_1-\varepsilon_2)^{\nu+1}d_{j\alpha_j-\nu}.$$

So:

$$D(\lambda)z_j = z_j + C\eta_p\sum_{\nu=0}^{\alpha_j-1}\binom{\alpha_j}{\nu+1}(\varepsilon_1-\varepsilon_2)^{\nu+1}(\lambda-S)^{-1}d_{j\alpha_j-\nu}.$$

With respect to the basis $\{d_{jk}\}_{k=1}^{\alpha_j}{}_{j=1}^{t}$ $(\lambda - \widetilde{S})^{-1}$ has the form $\oplus_{j=1}^{t} J_j$ where

$$J_j = \begin{pmatrix} (\lambda-\varepsilon_1)^{-1} & & & 0 \\ \vdots & \ddots & & \\ \vdots & & \ddots & \\ (\lambda-\varepsilon_1)^{-\alpha_j} & \cdots & & (\lambda-\varepsilon_1)^{-1} \end{pmatrix}.$$

So, using also $z_j = Cd_{j\alpha_j}$ we have

$$D(\lambda)z_j = z_j + \sum_{\nu=0}^{\alpha_j-1} \binom{\alpha_j}{\nu+1}(\varepsilon_1-\varepsilon_2)^{\nu+1}(\lambda-\varepsilon_1)^{-\nu-1}z_j =$$

$$= \sum_{\nu=-1}^{\alpha_j-1} \binom{\alpha_j}{\nu+1}\left(\frac{\varepsilon_1-\varepsilon_2}{\lambda-\varepsilon_1}\right)^{\nu+1} z_j = \left(1+\frac{\varepsilon_1-\varepsilon_2}{\lambda-\varepsilon_1}\right)^{\alpha_j} z_j = \left(\frac{\lambda-\varepsilon_2}{\lambda-\varepsilon_1}\right)^{\alpha_j} z_j,$$

as claimed.

Finally for $y = y_j$ it is more convenient to show that $D(\lambda)^{-1}y_j = \left(\frac{\lambda-\varepsilon_2}{\lambda-\varepsilon_1}\right)^{\omega_j} y_j$. Since $\widetilde{G}y_j = 0$, we have by (4.21) that

$$D(\lambda)^{-1}y_j = y_j - \widetilde{F}(\lambda-\widetilde{T})^{-1}\rho_Z P^{\times}By_j = y_j + \widetilde{F}(\lambda-\widetilde{T})^{-1}f_{j1}.$$

With respect to the basis $\{f_{jk}\}_{k=1}^{\omega_j}{}_{j=1}^{s}$ $(\lambda-\widetilde{T})^{-1}$ has the form $\oplus_{j=1}^{s} \widetilde{J}_j$, where

$$\widetilde{J}_j = \begin{pmatrix} (\lambda-\varepsilon_1)^{-1} & & & \\ \vdots & \ddots & & \\ \vdots & & \ddots & \\ (\lambda-\varepsilon_1)^{-\omega_j} & \cdots & & (\lambda-\varepsilon_1)^{-1} \end{pmatrix}.$$

So by (4.5):

$$D(\lambda)^{-1}y_j = y_j + \sum_{k=1}^{\omega_j} \widetilde{F}\frac{1}{(\lambda-\varepsilon_1)^k} f_{jk} = y_j + \sum_{k=1}^{\omega_j} \binom{\omega_j}{k}\left(\frac{\varepsilon_1-\varepsilon_2}{\lambda-\varepsilon_1}\right)^k y_j =$$

$$= \sum_{k=0}^{\omega_j} \binom{\omega_j}{k}\left(\frac{\varepsilon_1-\varepsilon_2}{\lambda-\varepsilon_1}\right)^k y_j = \left(1+\frac{\varepsilon_1-\varepsilon_2}{\lambda-\varepsilon_1}\right)^{\omega_j} y_j = \left(\frac{\lambda-\varepsilon_2}{\lambda-\varepsilon_1}\right)^{\omega_j} y_j. \qquad \square$$

We shall use (4.21) to show that $\widetilde{W}D^{-1}$ is the function W_- given by (4.14). We could use the formulas in the proof of [8], Theorem 5.1 again, however, we prefer to give a direct proof here.

LEMMA 4.4. *With* $\widetilde{W}$, D *and* W_- *given by* (4.7), (4.16) *and* (4.14), *respectively, we have*

(4.23) $\widetilde{W}(\lambda)D(\lambda)^{-1} = W_-(\lambda).$

PROOF. First take $y \in Y_0$. Then $D(\lambda)^{-1}y \equiv y$. Since $P^{\times}By \in \operatorname{Im} P^{\times}|_M$ we have $\rho_Z P^{\times}By = 0$. So

$$\widetilde{W}(\lambda)D(\lambda)^{-1}y = \widetilde{W}(\lambda)y = y + C(\lambda - A|_M)^{-1}\{(\Gamma_{11}^{\times} - \eta_p Y)P^{\times}B + \eta_p G\}y,$$

and since also $\widetilde{G}y = 0$ we obtain

$$\begin{aligned} W_-(\lambda)y &= \\ &= y + \begin{pmatrix} C, -CPX - F + \widetilde{F} \end{pmatrix}\left(\lambda - \begin{pmatrix} A|_M & V_- \\ 0 & \widetilde{T} \end{pmatrix}\right)^{-1}\begin{pmatrix} (\Gamma_{11}^{\times} - \eta_p Y)P^{\times}B + \eta_p(G - \widetilde{G}) \\ \rho_z P^{\times}B \end{pmatrix} y \\ &= y + C(\lambda - A|_M)^{-1}\{(\Gamma_{11}^{\times} - \eta_p Y)P^{\times}B + \eta_p G\}y. \end{aligned}$$

So for $y \in Y_0$ (4.23) holds.

Next, consider z_j. Again we have $\rho_z P^{\times}Bz_j = 0$. So, by (4.21)

$$\begin{aligned} \widetilde{W}(\lambda)D(\lambda)^{-1}z_j &= \widetilde{W}(\lambda)\left(\frac{\lambda - \varepsilon_1}{\lambda - \varepsilon_2}\right)^{\alpha_j} z_j = \\ &= (I + C(\lambda - A|_M)^{-1}\{(\Gamma_{11}^{\times} - \eta_p Y)P^{\times}B + \eta_p G\})D(\lambda)^{-1}z_j = \\ &= (I + C(\lambda - A|_M)^{-1}\{(\Gamma_{11}^{\times} - \eta_p Y)P^{\times}B + \eta_p G\})(I - C(\lambda - S)^{-1}\widetilde{G})z_j = \\ &= (I + C(\lambda - A|_M)^{-1}\{(\Gamma_{11}^{\times} - \eta_p Y)P^{\times}B + \eta_p G\})z_j - C(\lambda - S)^{-1}\widetilde{G}z_j + \\ &\quad - C(\lambda - A|_M)^{-1}\{(\Gamma_{11}^{\times} - \eta_p Y)P^{\times}B + \eta_p G\}C(\lambda - S)^{-1}\widetilde{G}z_j. \end{aligned}$$

Since $\tau_+ \oplus \tau_0$ is an admissible triple we have

$$\begin{aligned} &\begin{pmatrix} A|_M & 0 \\ 0 & T \end{pmatrix}\begin{pmatrix} \Gamma_{11}^{\times} & \eta_p \\ \eta_z & 0 \end{pmatrix} - \begin{pmatrix} \Gamma_{11}^{\times} & \eta_p \\ \eta_z & 0 \end{pmatrix}\begin{pmatrix} A^{\times}|_{\operatorname{Im} P^{\times}} & 0 \\ 0 & S \end{pmatrix} = \\ &= \begin{pmatrix} \Gamma_{11}^{\times} & \eta_p \\ \eta_z & 0 \end{pmatrix}\begin{pmatrix} P^{\times}B \\ -YP^{\times}B + G \end{pmatrix}\begin{pmatrix} CP, -CPX - F \end{pmatrix}\begin{pmatrix} \Gamma_{11}^{\times} & \eta_p \\ \rho_z & 0 \end{pmatrix}. \end{aligned} \tag{4.24}$$

Using the (1,2)-entry of this identity we have

$$(\lambda - A|_M)^{-1}\{(\Gamma_{11}^{\times} - \eta_p Y)P^{\times}B + \eta_p G\}C(\lambda - S)^{-1}\eta_p = (\lambda - A|_M)^{-1}\eta_p - \eta_p(\lambda - S)^{-1}\eta_p.$$

Hence

$$\begin{aligned} \widetilde{W}(\lambda)D(\lambda)^{-1}z_j &= \\ &= (I + C(\lambda - A|_M)^{-1}\{(\Gamma_{11}^{\times} - \eta_p Y)P^{\times}B + \eta_p G\})z_j - C(\lambda - S)^{-1}\widetilde{G}z_j + \\ &\quad - C\{(\lambda - A|_M)^{-1}\eta_p - (\lambda - S)^{-1}\}\widetilde{G}z_j = \\ &= (I + C(\lambda - A|_M)^{-1}\{(\Gamma_{11}^{\times} - \eta_p Y)P^{\times}B + \eta_p(G - \widetilde{G})\})z_j = W_-(\lambda)z_j. \end{aligned}$$

Finally, consider y_j. We have, by (4.21)

$$\widetilde{W}(\lambda)D(\lambda)^{-1}y_j =$$

$$= [I + C(\lambda - A|_M)^{-1}\{(\Gamma_{11}^{\times} - \eta_p Y)P^{\times}B + \eta_p G\} + (-CPX - F)(\lambda - T)^{-1}\rho_z P^{\times}B] \cdot$$

$$\cdot [I + \widetilde{F}(\lambda - \widetilde{T})^{-1}\rho_z P^{\times}B]y_j$$

$$= (I + C(\lambda - A|_M)^{-1}\{(\Gamma_{11}^{\times} - \eta_p Y)P^{\times}B + \eta_p G\})y_j + (-CPX - F)(\lambda - T)^{-1}\rho_z P^{\times}By_j +$$

$$+ \widetilde{F}(\lambda - \widetilde{T})^{-1}\rho_z P^{\times}By_j + C(\lambda - A|_M)^{-1}V_-(\lambda - \widetilde{T})^{-1}\rho_z P^{\times}By_j +$$

$$+ (-CPX - F)(\lambda - T)^{-1}\rho_z P^{\times}B\widetilde{F}(\lambda - \widetilde{T})^{-1}\rho_z P^{\times}By_j,$$

where we also used (4.11). By (4.5) and (4.19) we have $\rho_z P^{\times}BF = \widetilde{T} - T = (\lambda - T) - (\lambda - \widetilde{T})$. Inserting this into the last term we obtain

$$\widetilde{W}(\lambda)D(\lambda)^{-1}y_j =$$

$$= (I + C(\lambda - A|_M)^{-1}\{(\Gamma_{11}^{\times} - \eta_p Y)P^{\times}B + \eta_p G\})y_j +$$

$$+ (-CPX - F + \widetilde{F})(\lambda - \widetilde{T})^{-1}\rho_z P^{\times}By_j + C(\lambda - A|_M)^{-1}V_-(\lambda - \widetilde{T})^{-1}\rho_z PBy_j.$$

Noting that $\widetilde{G}y_j = 0$ this equals $W_-(\lambda)y_j$. □

Our next and final step is to show that the function given by (4.15) is indeed the inverse of W_-, thereby also establishing that W_- and its inverse are analytic in γ_-.

LEMMA 4.5. *Let* $W_-(\lambda)$ *be given by* (4.14). *Then its inverse is given by* (4.15).

PROOF. The inverse of $W_-(\lambda)$ is given by

$$W_-(\lambda)^{-1} = I - \begin{pmatrix} C, -CPX - F + \widetilde{F} \end{pmatrix}(\lambda - \widetilde{A}^{\times})^{-1}\begin{pmatrix} \Gamma_{11}^{\times} & \eta_p \\ \rho_z & 0 \end{pmatrix}\begin{pmatrix} P^{\times}B \\ -YP^{\times}B + G \end{pmatrix}$$

where

$$\widetilde{A}^{\times} = \begin{pmatrix} A|_M & V_- \\ 0 & \widetilde{T} \end{pmatrix} - \begin{pmatrix} \Gamma_{11}^{\times} & \eta_p \\ \rho_z & 0 \end{pmatrix}\begin{pmatrix} P^{\times}B \\ -YP^{\times}B + G - \widetilde{G} \end{pmatrix}\begin{pmatrix} C, -CPX - F + \widetilde{F} \end{pmatrix}.$$

Let $\hat{\Gamma}^{-1} : M^{\times} \oplus (M \cap M^{\times}) \to M \oplus K$ be given by

$$\hat{\Gamma}^{-1} = \begin{pmatrix} \Gamma_{11}^{\times} & \eta_p \\ \rho_z & 0 \end{pmatrix},$$

i.e. $\hat{\Gamma}$ is the coupling operator of $\tau_+ \oplus \tau_0$. Then

$$W(\lambda)^{-1} =$$
$$= I - \begin{pmatrix} C, -CPX - F + \widetilde{F} \end{pmatrix} \hat{\Gamma}^{-1} (\lambda - \hat{\Gamma}\widetilde{A}^{\times}\hat{\Gamma}^{-1})^{-1} \hat{\Gamma}\, \hat{\Gamma}^{-1} \begin{pmatrix} P^{\times}B \\ -YP^{\times}B + G \end{pmatrix},$$

which is equal to (4.15) once we have established

$$\widetilde{A}^{\times}\hat{\Gamma}^{-1} = \hat{\Gamma}^{-1} \begin{pmatrix} A^{\times}|_{\operatorname{Im} P^{\times}} & 0 \\ V_-^{\times} & \widetilde{S} \end{pmatrix}.$$

Noting that $\widetilde{G}\widetilde{F} = 0$ we have

$$\widetilde{A}^{\times}\hat{\Gamma}^{-1} = \begin{pmatrix} A|_M & 0 \\ 0 & T \end{pmatrix} \hat{\Gamma}^{-1} + \begin{pmatrix} 0 & V_- \\ 0 & \widetilde{T} - T \end{pmatrix} \hat{\Gamma}^{-1} - \hat{\Gamma}^{-1} \begin{pmatrix} P^{\times}B \\ -YP^{\times}B + G \end{pmatrix} \begin{pmatrix} C, -CPX - F \end{pmatrix} \hat{\Gamma}^{-1} +$$
$$+ \hat{\Gamma}^{-1} \begin{pmatrix} 0 \\ \widetilde{G} \end{pmatrix} \begin{pmatrix} C, -CPX - F \end{pmatrix} \hat{\Gamma}^{-1} - \hat{\Gamma}^{-1} \begin{pmatrix} P^{\times}B \\ -YP^{\times}B + G \end{pmatrix} \begin{pmatrix} 0, \widetilde{F} \end{pmatrix} \hat{\Gamma}^{-1}.$$

By (4.24) this equals

$$\hat{\Gamma}^{-1} \begin{pmatrix} A^{\times}|_{\operatorname{Im} P^{\times}} & 0 \\ 0 & S \end{pmatrix} + \begin{pmatrix} V_-\rho_Z & 0 \\ (\widetilde{T} - T)\rho_Z & 0 \end{pmatrix} + \hat{\Gamma}^{-1} \begin{pmatrix} 0 \\ \widetilde{G} \end{pmatrix} \begin{pmatrix} C(\Gamma_{11}^{\times} - PX\rho_Z) - F\rho_Z, C\eta_p \end{pmatrix} +$$
$$- \begin{pmatrix} (\Gamma_{11}^{\times} - \eta_p Y)P^{\times}B + \eta_p G \\ \rho_Z P^{\times}B \end{pmatrix} \begin{pmatrix} \widetilde{F}\rho_Z, 0 \end{pmatrix}.$$

From (4.6) one sees that $\widetilde{G}C\eta_p = (\widetilde{S} - S)\eta_p$, and from (4.5) and (4.19) that $\rho_Z P^{\times}B\widetilde{F} = \widetilde{T} - T$. Using also the definitions of V_- and $V_-^{\times}$ we obtain that

$$\widetilde{A}^{\times}\hat{\Gamma}^{-1} = \hat{\Gamma}^{-1} \begin{pmatrix} A^{\times}|_{\operatorname{Im} P} & 0 \\ 0 & S \end{pmatrix} + \begin{pmatrix} V_-\rho_Z & 0 \\ (\widetilde{T} - T)\rho_Z & 0 \end{pmatrix} +$$
$$+ \hat{\Gamma}^{-1} \begin{pmatrix} 0 & 0 \\ V_-^{\times} & \widetilde{S} - S \end{pmatrix} - \begin{pmatrix} V_-\rho_Z & 0 \\ (\widetilde{T} - T)\rho_Z & 0 \end{pmatrix} =$$
$$= \hat{\Gamma}^{-1} \begin{pmatrix} A^{\times}|_{\operatorname{Im} P^{\times}} & 0 \\ V_-^{\times} & \widetilde{S} \end{pmatrix}. \quad \square$$

Note by the way that it follows that the Wiener-Hopf factorization of the function $\widetilde{W}(\lambda)$ corresponding to $\tau_+ \oplus \tau_0$ is given by $\widetilde{W}(\lambda) = W_-(\lambda)D(\lambda)$.

5. FACTORIZATION INDICES FOR MATRIX POLYNOMIALS

In the case of matrix polynomials we are interested in Wiener-Hopf factorization with middle term $\operatorname{diag}(\lambda^{\kappa_j})_{j=1}^m$. We shall give formulas for the indices κ_j here.

Let $L(\lambda) = \lambda^\ell A_\ell + \dots + \lambda A_1 + I$ be an $m\times m$-matrix polynomial with invertible leading coefficient A_ℓ. Also, let γ be a contour with 0 inside γ and ∞ outside γ, and assume $L(\lambda)$ is invertible for $\lambda \in \gamma$. Choose a standard pair (X,T) for L, and let P be the spectral projection of T corresponding to γ_+, and set $M = \operatorname{Im} P$. Finally introduce the numbers

$$q_j = \operatorname{rank} \operatorname{col}(XT|_M^i)_{i=0}^j,$$

and let k be the smallest integer for which $q_{k-1} = \dim M$.

THEOREM 5.1. *With respect to* γ $L(\lambda)$ *has a Wiener-Hopf factorization*

$$L(\lambda) = H_-(\lambda)\operatorname{diag}(\lambda^{\kappa_j})_{j=1}^m H_+(\lambda),$$

where $\kappa_j = 0$ *for* $j = 1,\dots,\dim M - q_{k-2}$ *and for* $j > \dim M - q_{k-2}$ *we have*

$$\kappa_j = \#\{s \mid q_{k-s} - q_{k-s-1} \le j-1, s = 0,\dots,k-1\}$$

PROOF. Introduce $\tilde\gamma = \{\frac{1}{\lambda} \mid \lambda \in \gamma\}$. A $\tilde\gamma_+$-spectral triple for $L(\frac{1}{\lambda})^{-1}$ is

$$\tau_+ = \left\{(XT^{-1}|_M, T^{-1}|_M), \left(\begin{pmatrix} 0 & I_m & & \\ & \ddots & \ddots & \\ & & \ddots & I_m \\ & & & 0\end{pmatrix}, \begin{pmatrix} 0\\ \vdots \\ 0 \\ I_m\end{pmatrix}\right), \begin{pmatrix} XT|_M^{\ell-1} \\ \vdots \\ X|_M\end{pmatrix}\right\}$$

(see Section 3). Choose ε_1 inside $\tilde\gamma$, ε_2 outside $\tilde\gamma$. A Wiener-Hopf factorization of $L(\frac{1}{\lambda})^{-1}$ is given by

$$(5.1)\qquad L(\tfrac{1}{\lambda})^{-1} = W_-(\lambda)\operatorname{diag}\left(\left(\frac{\lambda-\varepsilon_1}{\lambda-\varepsilon_2}\right)^{\kappa_j}\right)_{j=1}^m W_+(\lambda).$$

Since 0 is inside $\tilde\gamma$ we have that $M^\times = (0)$, so $\kappa_j \ge 0$ for all j, and $\kappa_j = \omega_{m-j+1}$ for the positive κ_j's (see Theorem 4.1). However, we already computed the ω_j's corresponding to a triple like τ_+, see (3.15). This translates to the above formulas for the κ_j's.

Starting from (5.1) we have

$$L(\lambda) = W_+(\tfrac{1}{\lambda})^{-1}\operatorname{diag}\left(\left(\frac{1-\lambda\varepsilon_2}{\lambda}\right)^{\kappa_j}\right)_{j=1}^m \operatorname{diag}(\lambda^{\kappa_j})_{j=1}^m \operatorname{diag}\left(\left(\frac{1}{1-\lambda\varepsilon_1}\right)^{\kappa_j}\right)_{j=1}^m W_-(\tfrac{1}{\lambda})^{-1}.$$

Putting

$$H_-(\lambda) = W_+(\tfrac{1}{\lambda})^{-1}\,\mathrm{diag}\left(\left(\frac{1-\lambda\varepsilon_2}{\lambda}\right)^{\kappa_j}\right)_{j=1}^{m},$$

$$H_+(\lambda) = \mathrm{diag}\left(\left(\frac{1}{1-\lambda\varepsilon_1}\right)^{\kappa_j}\right)_{j=1}^{m} W_-(\tfrac{1}{\lambda})^{-1}$$

one verifies that H_- and its inverse are analytic outside γ and H_+ and its inverse are analytic inside γ. This proves the theorem. □

Formulas similar to this one for the right Wiener-Hopf factorization indices have been obtained earlier, see [12], also [11], Section 4.7.

REFERENCES

[1] Ball, J.A., Gohberg, I., Rodman, L.: Minimal factorization of meromorphic matrix functions in terms of local data, Integral Equations and Operator Theory 10 (1987), 309-348.

[2] Ball, J.A., Ran, A.C.M.: Global inverse spectral problems for rational matrix functions, Linear Algebra Appl. 86 (1987), 237-382.

[3] Ball, J.A., Ran, A.C.M.: Local inverse spectral problems for rational matrix functions, Integral Equations and Operator Theory 10 (1987), 349-415.

[4] Bart, H., Gohberg, I., Kaashoek, M.A.: Minimal factorization of matrix and operator functions, OT1, Birkhäuser Verlag (Basel) 1979.

[5] Bart, H., Gohberg, I., Kaashoek, M.A.: Explicit Wiener-Hopf factorization and realization, in: Constructive methods of Wiener-Hopf factorization (Eds. I. Gohberg, M.A. Kaashoek), OT21, Birkhäuser Verlag (Basel) 1986, 235-316.

[6] Bart, H., Gohberg, I., Kaashoek, M.A.: The coupling method for solving integral equations, in: Topics in Operator Theory, Systems and Networks (Eds. H. Dym, I. Gohberg), OT12, Birkhäuser Verlag (Basel) 1984, 39-73.

[7] Brunovský, P.: A classification of linear controllable systems, Kybernetika (Praha) 3 (1970), 173-187.

[8] Gohberg, I., Kaashoek, M.A.: An inverse spectral problem for rational matrix functions and minimal divisibility, Integral Equations and Operator Theory 10 (1987), 437-465.

[9] Gohberg, I., Kaashoek, M.A., Lerer, L., Rodman, L.: Minimal divisors of rational matrix functions with prescribed zero and pole structure, in: Topics in Operator Theory, Systems and Networks (Eds. H. Dym, I. Gohberg), OT12, Birkhäuser Verlag (Basel) 1984, 241-275.

[10] Gohberg, I., Lancaster, P., Rodman, L.: Invariant subspaces of matrices with applications, J. Wiley & Sons (New York etc.) 1986.

[11] Gohberg, I., Lancaster, P., Rodman, L.: Matrix polynomials, Academic Press (New York) 1982.

[12] Gohberg, I., Lerer, L., Rodman, L.: Factorization indices for matrix polynomials, Bull. Amer. Math. Soc. 84 (1978), 275-277.

[13] Gohberg, I., Rodman, L.: On the spectral structure of monic matrix polynomials and the extension problem, Linear Algebra Appl. 24 (1979), 157-172.

[14] Kaashoek, M.A., Ran, A.C.M.: Symmetric Wiener-Hopf factorization of self-adjoint rational matrix functions and realization, in: Constructive methods of Wiener-Hopf factorization (Eds. I. Gohberg, M.A. Kaashoek), OT21, Birkhäuser Verlag (Basel) 1986, 373-409.

[15] Rosenbrock, H.H.: State space and Multivariate Theory, Nelson, London 1970.

I. Gohberg
School of Mathematical Sciences
Raymond and Beverly Sackler
Faculty of Exact Sciences
Tel Aviv University
Ramat Aviv, Israel

M.A. Kaashoek, A.C.M. Ran
Fakulteit Wiskunde en Informatica
Vrije Universiteit
De Boelelaan 1081
1081 HV Amsterdam
The Netherlands

Operator Theory:
Advances and Applications, Vol. 33

REGULAR RATIONAL MATRIX FUNCTIONS WITH PRESCRIBED POLE AND ZERO STRUCTURE

I. Gohberg and M.A. Kaashoek

The problem to construct all regular rational matrix functions with a prescribed pole and zero structure is solved explicitly. Also the necessary and sufficient condition for the existence of a solution is derived. The proofs use an appropriate Möbius transformation to reduce the problem to the case when the functions are regular at infinity.

0. INTRODUCTION

This paper concerns the problem to reconstruct (if possible) a regular rational $m \times m$ matrix function given its pole and zero structure. To state more precisely the problem we need to define what is meant by the given pole and zero structure.

First let us consider the case when the rational matrix function W has no pole and zero at infinity, i.e., $D := W(\infty)$ is well-defined and invertible. In that case $W(\lambda)$ and its inverse $W(\lambda)^{-1}$ can be represented in the following form:

$$W(\lambda) = D + C(\lambda - A)^{-1}B, \tag{0.1}$$

$$W(\lambda)^{-1} = D^{-1} - D^{-1}C(\lambda - A^{\times})^{-1}BD^{-1}, \tag{0.2}$$

where A is a square matrix of which the order n may be much larger than the order m of $W(\lambda)$, the matrices B and C have sizes $n \times m$ and $m \times n$, respectively, and $A^{\times} = A - BD^{-1}C$. Now assume that the realizations (0.1) and (0.2) are minimal, i.e., the order n of A is as small as possible. Then the pair (C, A) is called a right pole pair for W and $(A^{\times}, BD^{-1})$ is a left zero pair for W. In [6], Section I.2, the following inverse problem has been solved. Let (C_p, A_p) and (A_z, B_z) be two given pairs of matrices with sizes $m \times n_p$, $n_p \times n_p$, $n_z \times n_z$ and $n_z \times m$, respectively, and such that

$$\bigcap_{\nu=1}^{n_p} \operatorname{Ker} C_p A_p^{\nu-1} = \{0\}, \qquad \operatorname{Im}[B_z \;\; A_z B_z \cdots A_z^{n_z-1} B_z] = \mathbb{C}^{n_z}. \tag{0.3}$$

Then (C_p, A_p) is a right pole pair and (A_z, B_z) is a left zero pair for a regular $m \times m$ rational matrix function W with $W(\infty) = I$ if and only if $n_p = n_z$ and there exists an invertible matrix S such that

$$SA_p - A_z S = B_z C_p. \tag{0.4}$$

In this case W is given by the formula $W(\lambda) = I + C_p(\lambda - A_p)^{-1}SB_z$, and there is a one-one correspondence between the invertible matrices S for which (0.4) holds and the rational matrix functions W with $W(\infty) = I$ for which (C_p, A_p) is a right pole pair and (A_z, B_z) is a left zero pair.

In the present paper we generalize this result for the case when W has singularities at infinity. First of all in this case the realizations (0.1) and (0.2) cannot be used, and they will be replaced by

$$W(\lambda) = D + (\lambda - \alpha)C(\lambda G - A)^{-1}B, \tag{0.5}$$

$$W(\lambda)^{-1} = D^{-1} - (\lambda - \alpha)D^{-1}C(\lambda G^{\times} - A^{\times})^{-1}BD^{-1}. \tag{0.6}$$

Here A, B and C are as before, G is a square matrix of the same size as A, and

$$A^{\times} = A + \alpha BD^{-1}C, \qquad G^{\times} = G + BD^{-1}C \tag{0.7}$$

with $D = W(\alpha)$. Such a representation is always possible. By choosing appropriate bases we may decompose $\lambda G - A$ and C in the following way:

$$\lambda G - A = \begin{bmatrix} \lambda I - T_F & 0 \\ 0 & \lambda T_\infty - I \end{bmatrix} \qquad C = [\, X_F \quad X_\infty \,],$$

where T_∞ has only zero as an eigenvalue, and (using different bases if necessary) we may write

$$\lambda G^{\times} - A^{\times} = \begin{bmatrix} \lambda I - T_F^{\times} & 0 \\ 0 & \lambda T_\infty^{\times} - I \end{bmatrix}, \qquad BD^{-1} = \begin{bmatrix} Y_F \\ Y_\infty \end{bmatrix},$$

where $T_\infty^{\times}$ has also $\lambda = 0$ as the only eigenvalue. If the realizations (0.5) and (0.6) are minimal, i.e., the size of $\lambda G - A$ (and hence of $\lambda G^{\times} - A^{\times}$) is as small as possible, then the pair (X_F, T_F) is called a *right pole pair for* W *on* $\mathbb{C}$, the pair (X_∞, T_∞) a *right pole pair for* W *at infinity*, the pair $(T_F^{\times}, Y_F)$ a *left zero pair for* W *on* $\mathbb{C}$, and $(T_\infty^{\times}, Y_\infty)$ a *left zero pair for* W *at infinity*.

In the present paper we solve the following inverse problem. Given four pairs of matrices:

$$(C_{p,F}, A_{p,F}), \qquad (C_{p,\infty}, A_{p,\infty}), \tag{0.8}$$

$$(A_{z,F}, B_{z,F}), \qquad (A_{z,\infty}, B_{z,\infty}), \tag{0.9}$$

construct, if possible, a regular $m \times m$ rational matrix function W such that the pairs (0.8) are right pole pairs for W on $\mathbb{C}$ and at infinity, respectively, and the pairs (0.9)) are left zero pairs for W on $\mathbb{C}$ and at infinity, respectively. Of course we must require that $A_{p,\infty}$ and $A_{z,\infty}$ have only $\lambda = 0$ as an eigenvalue. It is also natural to require that the pairs (0.8) are *right admissible* (which means that $C_{p,F}$ and $C_{p,\infty}$ have m rows, and products of the type appearing in the first identity of (0.3) make sense) and have a *null kernel* (which means that for both pairs (0.8) the analog of the first identity in (0.3) holds true), and similarly the pairs (0.9) have to be *left admissible* (the dual of right admissible) and full range (i.e., for both pairs (0.9) the analog of the second identity in (0.3) holds true). We shall prove the following theorem.

THEOREM 0.1. *Let the pairs (0.8) be right admissible null kernel pairs, let the pair (0.9) be left admissible full range pairs, and assume that $A_{p,\infty}$ and $A_{z,\infty}$ have only $\lambda = 0$ as an eigenvalue. Furthermore, let α be a given complex number and D a given invertible $m \times m$ matrix. In order that there exists a regular rational $m \times m$ matrix function such that*

(i) *the pairs (0.8) are right pole pairs for W on $\mathbb{C}$ and at infinity, respectively.*

(ii) *the pairs (0.9) are left zero pairs for W on $\mathbb{C}$ and at infinity, respectively.*

(iii) $W(\alpha) = D$,

it is necessary and sufficient that α is neither an eigenvalue of $A_{p,F}$ nor of $A_{z,F}$ and the equations

(0.10) $$\Gamma_F A_{p,F} - A_{z,F}\Gamma_F = B_{z,F}C_{p,F},$$

(0.11) $$\Gamma_\infty A_{p,\infty} - A_{z,\infty}\Gamma_\infty = B_{z,\infty}C_{p,\infty},$$

have solutions Γ_F and Γ_∞ such that

(0.12) $$\det \begin{bmatrix} \Gamma_F & \Gamma_{12} \\ \Gamma_{21} & \Gamma_\infty \end{bmatrix} \neq 0,$$

where Γ_{12} and Γ_{21} are the unique solutions of the equations

(0.13) $$A_{z,F}\Gamma_{12}A_{p,\infty} - \Gamma_{12} = B_{z,F}C_{p,\infty},$$

(0.14) $$A_{z,\infty}\Gamma_{21}A_{p,F} - \Gamma_{21} = B_{z,\infty}C_{p,F}.$$

In this case one such W is given by

(0.15) $$\begin{aligned} W(\lambda) = D + (\lambda - \alpha)[C_{p,F}C_{p,\infty}]\times \\ \times \begin{bmatrix} (\lambda - A_{p,F})^{-1} & 0 \\ 0 & (I - \lambda A_{p,\infty})^{-1} \end{bmatrix} \begin{bmatrix} \Gamma_F & \Gamma_{12} \\ \Gamma_{21} & \Gamma_\infty \end{bmatrix}^{-1} \times \\ \times \begin{bmatrix} (A_{z,F} - \alpha)^{-1}B_{z,F}D \\ (I - \alpha A_{z,\infty})^{-1}B_{z,\infty}D \end{bmatrix}, \end{aligned}$$

and this function $W(\lambda)$ *has the following inverse:*

$$W(\lambda)^{-1} = D^{-1} - (\lambda - \alpha)[D^{-1}C_{p,F}(\alpha - A_{p,F})^{-1}C_{p,\infty}(I - \alpha A_{p,\infty})^{-1}]\times \tag{0.16}$$
$$\times \begin{bmatrix} \Gamma_F & \Gamma_{12} \\ \Gamma_{21} & \Gamma_\infty \end{bmatrix}^{-1} \begin{bmatrix} (A_{z,F} - \lambda)^{-1} & 0 \\ 0 & (I - \lambda A_{z,\infty})^{-1} \end{bmatrix} \begin{bmatrix} B_{z,F} \\ B_{z,\infty} \end{bmatrix}.$$

Furthermore, the map $\{\Gamma_F, \Gamma_\infty\} \to W(\lambda)$ *gives a one-one correspondence between the solutions of* (0.10) *and* (0.11) *satisfying* (0.12) *and the* $m \times m$ *rational matrix function* W *for which* (i), (ii) *and* (iii) *hold true.*

The theorem will be proved by using an appropriate Möbius transformation which reduces the problem to the case when W is regular at infinity. Note that the equations (0.13) and (0.14) are uniquely solvable indeed; in fact, their unique solutions are given by

$$\Gamma_{12} = -\sum_{\nu=0}^{n_p-1} A_{z,F}^{\nu} B_{z,F} C_{p,\infty} A_{p,\infty}^{\nu}, \tag{0.17}$$

$$\Gamma_{21} = -\sum_{\nu=0}^{n_z-1} A_{z,\infty}^{\nu} B_{z,\infty} C_{p,F} A_{p,F}^{\nu}, \tag{0.18}$$

where n_p and n_z are the orders of the matrices $A_{p,\infty}$ and $A_{z,\infty}$, respectively.

Another solution of the above inverse problem, based on realizations of $W(\lambda)$ and $W(\lambda)^{-1}$ in the form

$$W(\lambda) = C(\lambda - A)^{-1}B + D + \lambda E(I - \lambda G)^{-1}F,$$

$$W(\lambda)^{-1} = c(\lambda - a)^{-1}b + d + \lambda e(I - \lambda g)^{-1}f,$$

is due to J.A. Ball, N. Cohen and A.C.M. Ran and appears (together with interesting applications) also in the present volume.

This paper consists of four sections. The first contains preliminaries about minimal realizations. In the second section we define pole pairs and zero pairs in terms of realizations and make the connection with the definitions employed in the present introduction. In the third section we describe the behaviour of pole and zero pairs under Möbius transformations. Section 4 consists of the proof of Theorem 0.1.

1. PRELIMINARIES ABOUT REALIZATION

Let F be an $m \times m$ rational matrix function. Assume that F is *strictly proper*, i.e., F is analytic at infinity and $F(\infty) = 0$, the $m \times m$ zero matrix. A system $\theta = [A, B, C]$ of linear operators $A: X \to X$, B: $\mathbb{C}^m \to X$ and $C: X \to \mathbb{C}^m$ is called a *realization* of F if

$$F(\lambda) = C(\lambda - A)^{-1}B. \tag{1.1}$$

The space X is a finite dimensional vector space over $\mathbb{C}$, which is called the *state space* of the realization θ. Here and in the sequel an $m \times m$ matrix is identified with the operator on $\mathbb{C}^m$ given by the canonical action of the matrix. Realization always exist (see, e.g., [4], Section 2.2).

A realization θ of F is said to be a *minimal realization* of F if among all realizations of F the state space dimension of θ is as small as possible. It is well-known (see, e.g., [4], Section 4.2) that a realization $\theta = [A, B, C]$ of F is a minimal realization if and only if

$$\bigcap_{\nu=0}^{n-1} \operatorname{Ker} CA^{\nu} = \{0\}, \qquad \bigvee_{\nu=0}^{n-1} \operatorname{Im} A^{\nu}B = X. \tag{1.2}$$

Here X is the state space of θ, the number n is equal to $\dim X$, and the symbol $\bigvee_{\nu=0}^{n-1} Z_\nu$ denotes the linear hull of the set $\bigcup_{\nu=0}^{n-1} Z_\nu$. If for $\theta = [A, B, C]$ the two identities in (1.2) hold, then θ will be called a *minimal system*.

We shall use the fact (see, e.g., [4], Section 3.2) that two minimal realizations $\theta_1 = [A_1, B_1, C_1]$ and $\theta_2 = [A_2, B_2, C_2]$ of F are *similar*, i.e., there exists an invertible operator $S: X_1 \to X_2$, where X_1 and X_2 are the state spaces of θ_1 and θ_2, respectively, such that

$$A_1 = S^{-1}A_2S, \qquad B_1 = S^{-1}B_2, \qquad C_1 = C_2S. \tag{1.3}$$

We shall refer to this result as the *state space isomorphism theorem*.

2. LOCAL POLE AND ZERO STRUCTURE

Throughout this section W is a regular $m \times m$ rational matrix function. To introduce the local pole and zero structure of W we need the following definitions (cf., [5], Section 2). A pair (C_p, A_p) is called a *right admissible pair* with *main space* X_p if X_p is a finite dimensional vector space over $\mathbb{C}$ and $A_p: X_p \to X_p$, $C_p: X_p \to \mathbb{C}^m$ are linear operators. Such a pair (C_p, A_p) is said to be a *null kernel pair* if

$$\bigcap_{\nu=0}^{n-1} \operatorname{Ker} C_pA_p^{\nu} = \{0\}, \tag{2.1}$$

where $n = \dim X_p$. Two right admissible pairs $(C_p^{(1)}, A_p^{(1)})$ and $(C_p^{(2)}, A_p^{(2)})$ are called *similar* if there exists an invertible operator S such that

$$C_p^{(1)} = C_p^{(2)} S, \qquad A_p^{(1)} = S^{-1} A_p^{(2)} S. \tag{2.2}$$

Let σ be a non-empty set of complex numbers. A right admissible pair (C_p, A_p) with main space X_p is called a *right pole pair for W on σ* if all the eigenvalues of A_p belong to σ and there exists a linear operator $B_p\colon \mathbb{C}^m \to X_p$ such that $\theta_p = [A_p, B_p, C_p]$ is a minimal system and the function

$$W(\lambda) - C_p(\lambda - A_p)^{-1} B_p \tag{2.3}$$

is analytic on σ. From the first identity, in (1.2) and the condition of minimality in the definition of a pole pair it follows that such a pair is always a null kernel pair.

To construct a right pole pair for W on σ, let W_- be the sum of the principal parts of the poles of W in σ. Note that W_- is a strictly proper rational matrix function. Thus W_- has a minimal realization $\theta = [A, B, C]$, say (see Section 1). The minimality of θ implies (cf., [4], Theorem 3.3) that the eigenvalues of A coincide with the poles of W_- and hence the set $\sigma(A)$ of all eigenvalues of A lies in σ. Furthermore

$$W(\lambda) - C(\lambda - A)^{-1} B = W(\lambda) - W_-(\lambda)$$

is analytic on σ. Thus (C, A) is a right pole pair for W on σ.

LEMMA 2.1. *Two right poles pairs of W on σ are similar and the similarity transformation is uniquely determined by the pairs.*

Lemma 2.1 may be found in Section 2 of [5], where it is proved for W's that are analytic at infinity and have $W(\infty) = I$. However the proof given in [5] carries over to the case considered here.

A right admissible pair (C_∞, A_∞) is called a *right pole pair for W at infinity* if the pair (C_∞, A_∞) is a right pole pair for $W^{\#}(\lambda) := W(\lambda^{-1})$ on $\{0\}$. In that case (C_∞, A_∞) is a null kernel pair and A_∞ has precisely one eigenvalue, namely $\lambda = 0$.

A *zero* of W is defined to be a pole of $W(\cdot)^{-1}$. To deal with the zeros we need the following terminology (cf., [5], Section 2). A pair (A_z, B_z) is called a *left admissible pair* with *main space* X_z if X_z is a finite dimensional vector space over $\mathbb{C}$ and $A_z\colon X_z \to X_z$, $B_z\colon \mathbb{C}^m \to X$ are linear operators. Such a pair (A_z, B_z) is said to be a *full range pair* if

$$\bigvee_{\nu=0}^{n-1} \operatorname{Im} A_z^\nu B_z = X_z, \tag{2.4}$$

where $n = \dim X_z$. Two left admissible pairs $(A_z^{(1)}, B_z^{(1)})$ and $(A_z^{(2)}, B_z^{(2)})$ are called *similar* if there exists an invertible operator S such that

$$B_z^{(1)} = S^{-1} B_z^{(2)}, \qquad A_z^{(1)} = S^{-1} A_z^{(2)} S. \tag{2.5}$$

Let σ be a non-empty subset of $\mathbb{C}$. A left admissible pair (A_z, B_z) with main space X_z is called a *left zero pair for* W *on* σ if all the eigenvalues of A_z are in σ and there exists a linear operator $C_z: X_z \to \mathbb{C}^m$ such that the system $\theta_z = [A_z, B_z, C_z]$ is minimal and the function

$$W(\lambda)^{-1} - C_z(\lambda - A_z)^{-1}B_z \tag{2.6}$$

is analytic on σ. From the second identity in (1.2) and the condition of minimality on θ_z it follows that a left zero pair is always a full range pair.

The construction of a left zero pair is analogous to the construction of a right pole pair. Also the following analogue of Lemma 2.1 holds true.

LEMMA 2.2. *Two left zero pairs for* W *on* σ *are similar and the similarity transformation is uniquely determined by the pairs.*

A left admissible pair (A_∞, B_∞) is called a left zero pair for W at infinity if the pair (A_∞, B_∞) is a left zero pair for $W^{\#}(\lambda) := W(\lambda^{-1})$ on $\{0\}$. In that case (A_∞, B_∞) is a full range pair and A_∞ has precisely one eigenvalue, namely $\lambda = 0$.

In Section 4 we shall need the following two propositions. Their proofs may be found in [5].

PROPOSITION 2.3. *For* $\nu = 1, 2$ *let* (C_ν, A_ν) *be a right admissible pair and let the eigenvalues of* A_ν *belong to the set* σ_ν. *Assume* $\sigma_1 \cap \sigma_2 = \emptyset$. *Then the pair*

$$\left([\, C_1 \quad C_2 \,], \quad \begin{bmatrix} A_1 & 0 \\ 0 & A_2 \end{bmatrix} \right)$$

is a right pole pair for W *on* $\sigma_1 \cup \sigma_2$ *if and only if* (C_1, A_1) *and* (C_2, A_2) *are right pole pairs for* W *on* σ_1 *and* σ_2, *respectively.*

PROPOSITION 2.4. *For* $\nu = 1, 2$ *let* (A_ν, B_ν) *be a left admissible pair and let the eigenvalues of* A_ν *belong to the set* σ_ν. *Assume* $\sigma_1 \cap \sigma_2 = \emptyset$. *Then the pair*

$$\left(\begin{bmatrix} A_1 & 0 \\ 0 & A_2 \end{bmatrix}, \begin{bmatrix} B_1 \\ B_2 \end{bmatrix} \right)$$

is a left zero pair for W *on* $\sigma_1 \cup \sigma_2$ *if and only if* (A_1, B_1) *and* (A_2, B_2) *are left pole pairs for* W *on* σ_1 *and* σ_2, *respectively.*

We conclude this section with the remark that the definitions of pole and zero pair employed in the introduction and the ones introduced in the present section are the same. To see this, let $\alpha \in \mathbb{C}$ be neither a pole nor a zero of $W(\lambda)$ and assume that

$$W(\lambda) = D + (\lambda - \alpha)[\, X_1 \quad X_2 \,] \begin{bmatrix} \lambda I_1 - T_1 & 0 \\ 0 & \lambda T_2 - I_2 \end{bmatrix}^{-1} \begin{bmatrix} Y_1 \\ Y_2 \end{bmatrix}, \tag{2.7}$$

$$W(\lambda)^{-1} = D^{-1} - (\lambda - \alpha)[\, X_1^\times \quad X_2^\times \,] \begin{bmatrix} \lambda I_1^\times - T_1^\times & 0 \\ 0 & \lambda T_2^\times - I_2^\times \end{bmatrix}^{-1} \begin{bmatrix} Y_1^\times \\ Y_2^\times \end{bmatrix}, \tag{2.8}$$

where T_2 and $T_2^\times$ have only $\lambda = 0$ as an eigenvalue. Now assume additionally that among all representations of $W(\lambda)$ of the form (2.7) the size of

$$\begin{bmatrix} \lambda I_1 - T_1 & 0 \\ 0 & \lambda T_2 - I_2 \end{bmatrix} \tag{2.9}$$

is as small as possible. We claim that in that case (X_1, T_1) and (X_2, T_2) are right pole pairs for W on $\mathbb{C}$ and at infinity, respectively. To prove this note that (2.7) implies that

$$\begin{aligned} W(\lambda) = {} & D + X_1Y_1 + \alpha X_2Y_2 + X_1(\lambda - T_1)^{-1}(T_1 - \alpha)Y_1 + \\ & + \lambda X_2(I_2 - \lambda T_2)^{-1}(\alpha T_2 - I_2)Y_2, \end{aligned} \tag{2.10}$$

$$\begin{aligned} W(\lambda^{-1}) = {} & D + X_1Y_1 + \alpha X_2Y_2 + \lambda X_1(I_1 - \lambda T_1)^{-1}(T_1 - \alpha)Y_1 + \\ & + X_2(\lambda - T_2)^{-1}(\alpha T_2 - I_2)Y_2. \end{aligned} \tag{2.11}$$

Since $\lambda = 0$ is the only eigenvalue of T_2, the function

$$W(\lambda) - X_1(\lambda - T_1)^{-1}(T_1 - \alpha)Y_1$$

is analytic on $\mathbb{C}$. The condition of minimality on the size of the matrix (2.9) implies that the system $[T_1, (T_1 - \alpha)Y_1, X_1]$ is minimal. It follows that (X_1, T_1) is a right pole pair for W on $\mathbb{C}$. Also the system $[T_2, (\alpha T_2 - I_2)Y_2, X_2]$ is minimal and

$$W(\lambda^{-1}) - X_2(\lambda - T_2)^{-1}(\alpha T_2 - I_2)Y_2$$

is analytic in $\lambda = 0$. We conclude that (X_2, T_2) is a right pole pair for $W^{\#}$ on $\{0\}$ and hence also for W at infinity. In a similar way one shows that $(T_1^\times, Y_1^\times)$ and $(T_2^\times, Y_2^\times)$ are left zero pairs for W on $\mathbb{C}$ and at infinity, provided that among all representations of $W(\lambda)^{-1}$ of the form (2.8) the size of

$$\begin{bmatrix} \lambda I_1^\times - T_1^\times & 0 \\ 0 & \lambda T_2^\times - I_2^\times \end{bmatrix}$$

is as small as possible.

3. MÖBIUS TRANSFORMATIONS

Consider the Möbius transformation

$$\varphi(\lambda) = \frac{k\lambda + \ell}{r\lambda + s}. \tag{3.1}$$

Throughout this section we shall assume that $\delta_\varphi := ks - \ell r \neq 0$. We consider φ as a map of the Riemann sphere $\mathbb{C} \cup \{\infty\}$ onto itself. The inverse map φ^{-1} is given by

$$\varphi^{-1}(\lambda) = \frac{-s\lambda + \ell}{r\lambda - k}. \tag{3.2}$$

In what follows W is a regular $m \times m$ rational matrix function and σ is a nonempty subset of $\mathbb{C}$.

LEMMA 3.1. *Let (C_p, A_p) be a right pole pair for W on σ, and let (A_z, B_z) be a left zero pair for W on σ. Put*

$$\tilde{W}(\lambda) = W(\varphi(\lambda)), \qquad \tilde{\sigma} = \varphi^{-1}[\sigma], \tag{3.3}$$

where φ is the Möbius transformation given by (3.1). *Assume that $r\lambda - k \neq 0$ for all $\lambda \in \sigma$. Then $\tilde{\sigma} \subset \mathbb{C}$, the pair $(\delta_\varphi C_p(rA_p - k)^{-1}, \varphi^{-1}(A_p))$ is a right pole pair for $\tilde{W}$ on $\tilde{\sigma}$, and the pair $(\varphi^{-1}(A_z), (rA_z - k)^{-1}B_z)$ is a left zero pair for $\tilde{W}$ on $\tilde{\sigma}$.*

PROOF. We shall prove the statement about pole pairs. The proof of the statement about zero pairs follows an analogous reasoning. Since A_p has all its eigenvalues in σ, the condition $r\lambda - k \neq 0$ for all $\lambda \in \sigma$ implies that $rA_p - k$ is invertible. Hence $(rA_p - k)^{-1}$ and

$$\varphi^{-1}(A_p) = (-sA + k)(rA_p - k)^{-1}$$

are well-defined. Obviously, $\tilde{\sigma} \subset \mathbb{C}$.

Choose B_p such that $\theta = [A_p, B_p, C_p]$ is a minimal system and $W(\lambda) - C_p(\lambda - A_p)^{-1}B_p$ is analytic on σ. Then

$$W(\varphi(\lambda)) - C_p(\varphi(\lambda) - A_p)^{-1}B_p \tag{3.4}$$

is analytic on $\tilde{\sigma}$. A simple computation (cf. Theorem 1.9 in [4]) shows that

$$\begin{aligned} C_p(\varphi(\lambda) - A_p)^{-1}B_p &= rC_p(k - rA_p)^{-1}B_p + \\ &\quad + \delta_\varphi C_p(rA_p - k)^{-1}(\lambda - \varphi^{-1}(A_p))^{-1}(rA_p - k)^{-1}B_p, \end{aligned}$$

where $\delta_\varphi = ks - \ell r$. By the spectral mapping theorem the eigenvalues of $\varphi^{-1}(A_p)$ are in the set $\varphi^{-1}[\sigma\} = \tilde{\sigma}$. Furthermore, from [4], Theorem 3.5 we known that the system

$$[\varphi^{-1}(A_p), (rA_p - k)^{-1}B_p, \delta_\varphi C_p(rA_p - k)^{-1}]$$

is a minimal system. We conclude that $(\delta_\varphi C_p(rA_p - k)^{-1}, \varphi^{-1}(A_p))$ is a right pole pair for $\tilde{W}$ on $\tilde{\sigma}$.□

4. PROOF OF THE MAIN THEOREM

In this section we prove Theorem 0.1. First we establish the necessity of the conditions in Theorem 0.1 and derive the formulas for $W(\lambda)$ and $W(\lambda)^{-1}$. Next we prove the sufficiency and we conclude the one-one correspondence given by the map $\{\Gamma_F, \Gamma_\infty\} \to W$. Throughout this section we work with the following pairs:

$$(C_{p,F}, A_{p,F}), \qquad (C_{p,\infty}, A_{p,\infty}), \tag{4.1}$$

$$(A_{z,F}, B_{z,F}), \qquad (A_{z,\infty}, B_{z,\infty}). \tag{4.2}$$

The pairs in (4.1) are right admissible null kernel pairs and those in (4.2) are left admissible full range pairs. By σ_F we denote the set consisting of all eigenvalues of $A_{p,F}$ and $A_{z,F}$. Further, we put $\sigma_\infty := \{0\}$.

Assume that α is neither an eigenvalue of $A_{p,F}$ nor of $A_{z,F}$. Thus $\alpha \notin \sigma_F$. Consider the Möbius transformation

$$\varphi(\lambda) = \frac{\alpha\lambda + 1}{\lambda}, \quad \varphi^{-1}(\lambda) = \frac{1}{\lambda - \alpha}, \qquad \delta_\varphi = -1, \tag{4.3}$$

$$\psi(\lambda) = \frac{\lambda}{\alpha\lambda + 1}, \quad \psi^{-1}(\lambda) = \frac{\lambda}{1 - \alpha\lambda}, \qquad \delta_\psi = 1. \tag{4.4}$$

Note that $\lambda - \alpha \neq 0$ for all $\lambda \in \sigma_F$. Also $1 - \alpha\lambda \neq 0$ for $\lambda = 0$. It follows that

$$\tilde{\sigma}_F := \varphi^{-1}[\sigma_F] \subset \mathbb{C}, \qquad \tilde{\sigma}_\infty := \psi^{-1}[\{0\}] = \{0\}. \tag{4.5}$$

Since $\tilde{\sigma}_\infty = \varphi^{-1}[\{\infty\}]$, the set $\tilde{\sigma}_F$ and $\tilde{\sigma}_\infty$ are disjoint.

4.1. Necessity and formulas. Let $W(\lambda)$ be a regular $m \times m$ rational matrix function such that the pairs (4.1) are right pole pairs for W on $\mathbb{C}$ and at infinity, respectively, and the pairs (4.2) are left zero pairs for W on $\mathbb{C}$ and at infinity, respectively. Furthermore, let $\alpha \in \mathbb{C}$ be a regular point of W (i.e., the point α is neither a pole nor a zero of W), and put $D = W(\alpha)$. Then E is invertible and α is neither an eigenvalue of $A_{p,F}$ nor of $A_{z,F}$. Now use the Möbius transformations (4.3) and (4.4), and put $\tilde{W}(\lambda) = W(\varphi(\lambda))D^{-1}$. Then $\tilde{W}$ is analytic at infinity, and $\tilde{W}(\infty) = I$. Furthermore,

$$\tilde{W}(\lambda) = W(\varphi(\lambda))D^{-1} = W^{\#}(\psi(\lambda))D^{-1}, \tag{4.6}$$

where $W^{\#}(\lambda) = W(\lambda^{-1})$. All poles and zeros of W are in $\sigma_F \cup \{\infty\}$. Thus all poles and zeros of $\tilde{W}$ are in

$$\varphi^{-1}[\sigma_F \cup \{0\}] = \tilde{\sigma}_F \cup \{0\} = \tilde{\sigma}_F \cup \tilde{\sigma}_\infty.$$

Now apply four times Lemma 3.1, first for W and φ and next for $W^{\#}$ and ψ. We conclude that $(-C_{p,F}(A_{p,F}-\alpha)^{-1},(A_{p,F}-\alpha)^{-1})$ is a right pole pair for $\tilde{W}$ on $\tilde{\sigma}_F$, the pair $((A_{z,F}-\alpha)^{-1},(A_{z,F}-\alpha)^{-1}B_{z,F})$ is a left zero pair for $\tilde{W}$ on $\tilde{\sigma}_F$, the pair $(C_{p,\infty}(I-\alpha A_{p,\infty})^{-1},A_{p,\infty}(I-\alpha A_{p,\infty})^{-1})$ is a right pole pair for $\tilde{W}$ on $\{0\}$, and the pair $(A_{z,\infty}(I-\alpha A_{z,\infty})^{-1},(I-\alpha A_{z,\infty})^{-1}B_{z,\infty})$ is a left zero pair for $\tilde{W}$ on $\{0\}$. Next, put

$$\theta_p=\left(\begin{bmatrix} -C_{p,F}(A_{p,F}-\alpha)^{-1} & C_{p,\infty}(I-\alpha A_{p,\infty})^{-1}\end{bmatrix},\right.$$
$$\left.\begin{bmatrix}(A_{p,F}-\alpha)^{-1} & 0\\ 0 & A_{p,\infty}(I-\alpha A_{p,\infty})^{-1}\end{bmatrix}\right); \tag{4.7}$$

$$\theta_z=\left(\begin{bmatrix}(A_{z,F}-\alpha)^{-1} & 0\\ 0 & A_{z,\infty}(I-\alpha A_{z,\infty})^{-1}\end{bmatrix},\begin{bmatrix}(A_{z,F}-\alpha)^{-1}B_{z,F}\\ (I-\alpha A_{z,\infty})^{-1}B_{z,\infty}\end{bmatrix}\right). \tag{4.8}$$

By proposition 2.3 and 2.4 the pair θ_p is a right pole pair for $\tilde{W}$ on $\mathbb{C}$ and θ_z is a left zero pair for $\tilde{W}$ on $\mathbb{C}$. Now use that $\tilde{W}(\infty)=I$. This allows us to apply Theorem I.2.1 in [6] (cf., [5], Theorem 4.1). Thus there exist a unique invertible $S=[S_{ij}]_{i,j=1}^2$ such that

$$\begin{aligned}&\begin{bmatrix}S_{11} & S_{12}\\ S_{21} & S_{22}\end{bmatrix}\begin{bmatrix}(A_{p,F}-\alpha)^{-1} & 0\\ 0 & A_{p,\infty}(I-\alpha A_{p,\infty})^{-1}\end{bmatrix}-\\ &-\begin{bmatrix}(A_{z,F}-\alpha)^{-1} & 0\\ 0 & A_{z,\infty}(I-\alpha A_{z,\infty})^{-1}\end{bmatrix}\begin{bmatrix}S_{11} & S_{12}\\ S_{21} & S_{22}\end{bmatrix}=\\ &=\begin{bmatrix}(A_{z,F}-\alpha)^{-1}B_{z,F}\\ (I-\alpha A_{z,\infty})^{-1}B_{z,\infty}\end{bmatrix}\begin{bmatrix} -C_{p,F}(A_{p,F}-\alpha)^{-1} & C_{p,\infty}(I-\alpha A_{p,\infty})^{-1}\end{bmatrix},\end{aligned} \tag{4.9}$$

and

$$\begin{aligned}\tilde{W}(\lambda)&=I+\begin{bmatrix} -C_{p,F}(A_{p,F}-\alpha)^{-1} & C_{p,\infty}(I-\alpha A_{p,\infty})^{-1}\end{bmatrix}\times\\ &\times\begin{bmatrix}\lambda-(A_{p,F}-\alpha)^{-1} & 0\\ 0 & \lambda-A_{p,\infty}(I-\alpha A_{p,\infty})^{-1}\end{bmatrix}^{-1}\times\\ &\times\begin{bmatrix}S_{11} & S_{12}\\ S_{21} & S_{22}\end{bmatrix}^{-1}\begin{bmatrix}(A_{z,F}-\alpha)^{-1}B_{z,F}\\ (I-\alpha A_{z,\infty})^{-1}B_{z,\infty}\end{bmatrix}.\end{aligned} \tag{4.10}$$

Furthermore,

(4.11)

$$\tilde{W}(\lambda)^{-1} = I - [\ -C_{p,F}(A_{p,F} - \alpha)^{-1} \quad C_{p,\infty}(I - \alpha A_{p,\infty})^{-1}\ ]\times$$
$$\times \begin{bmatrix} S_{11} & S_{12} \\ S_{21} & S_{22} \end{bmatrix}^{-1} \begin{bmatrix} \lambda - A_{z,F} - \alpha)^{-1} & 0 \\ 0 & \lambda - A_{z,\infty}(I - \alpha A_{z,\infty})^{-1} \end{bmatrix}^{-1} \times$$
$$\times \begin{bmatrix} (A_{z,F} - \alpha)^{-1} B_{z,F} \\ (I - \alpha A_{z,\infty})^{-1} B_{z,\infty} \end{bmatrix}.$$

The identity (4.9) is equivalent to the following four identities:

$$S_{11}A_{p,F} - A_{z,F}S_{11} = B_{z,F}C_{p,F}, \tag{4.12a}$$

$$S_{22}A_{p,\infty} - A_{z,\infty}S_{22} = B_{z,\infty}C_{p,\infty}, \tag{4.12b}$$

$$A_{z,F}S_{12}A_{p,\infty} - S_{12} = B_{z,F}C_{p,\infty}, \tag{4.12c}$$

$$A_{z,\infty}S_{21}A_{p,F} - S_{21} = B_{z,\infty}C_{p,F}. \tag{4.12d}$$

Since $A_{p,\infty}$ and $A_{z,\infty}$ have only $\lambda = 0$ as an eigenvalue, the operators S_{12} and S_{21} are uniquely determined by (4.12c) and (4.12d). In fact, $S_{12} = \Gamma_{12}$ and $S_{21} = \Gamma_{21}$, where Γ_{12} and Γ_{21} are given by (0.12) and (0.18). Put $\Gamma_F = S_{11}$ and $\Gamma_\infty = S_{22}$. Then Γ_F and Γ_∞ are solutions of the equations (0.10) and (0.11). Since S is invertible, also (0.12) is satisfied. This proves the necessity of the conditions in Theorem 0.1.

From (4.10) and the identities

$$W(\lambda) = \tilde{W}((\lambda - \alpha)^{-1})D, \tag{4.13}$$

$$\left[\frac{1}{\lambda - \alpha}(A_{p,F} - \alpha) - I\right]^{-1} = -(\lambda - \alpha)(\lambda - A_{p,F})^{-1}, \tag{4.14}$$

$$\left[\frac{1}{\lambda - \alpha}(I - \alpha A_{p,\infty}\right]^{-1} = (\lambda - \alpha)(I - \lambda A_{p,\infty})^{-1}, \tag{4.15}$$

one obtains formula (0.15) for $W(\lambda)$. In a similar way (replace in (4.14) and (4.15) the operator $A_{p,F}$ by $A_{z,F}$ and $A_{p,\infty}$ by $A_{z,\infty}$, and use (4.11)) one obtains formula (0.16) for $W(\lambda)^{-1}$.

4.2. Proof of sufficiency. In this subsection α is a given complex number such that α is neither an eigenvalue of $A_{p,F}$ nor of $A_{z,F}$. Furthermore we assume that the equations (0.10) and (0.11) have solutions Γ_F and Γ_∞ such that (0.12) holds. As in (0.12) the operators Γ_{12} and Γ_{21} are the unique solutions of (0.13) and (0.14). Let θ_p and θ_z be the pairs defined by (4.7) and (4.8). Since the pairs (4.1) are right admissible null kernel pairs, it is straightforward to check that θ_p is a right admissible null kernel pair. Similarly, since the pairs (4.2) are left admissible full range pairs, the same is true for θ_z. From the identities (0.10), (0.11), (0.13) and (0.14) it follows that (4.9) holds true for

$$\begin{bmatrix} S_{11} & S_{12} \\ S_{21} & S_{22} \end{bmatrix} = \begin{bmatrix} \Gamma_F & \Gamma_{12} \\ \Gamma_{21} & \Gamma_\infty \end{bmatrix}. \tag{4.16}$$

Next define $\tilde{W}$ by (4.10) with $[S_{ij}]_{i,j=1}^2$ given by (4.16). The Theorem I.2.1 in [6] implies that θ_p is a right pole pair and θ_z is a left zero pair both for $\tilde{W}$ on $\mathbb{C}$. Since $(A_{p,F}-\alpha)^{-1}$ and $A_{p,\infty}(I-\alpha A_{p,\infty})^{-1}$ have no common eigenvalue, it follows that the pair $(-C_{p,F}(A_{p,F}-\alpha)^{-1},(A_{p,F}-\alpha)^{-1})$ is a right pole pair for $\tilde{W}$ on $\tilde{\sigma}_F$ and $(C_{p,\infty}(I-\alpha A_{p,\infty})^{-1}, A_{p,\infty}(I-\alpha A_{p,\infty})^{-1})$ is a right pole pair for $\tilde{W}$ on $\tilde{\sigma}_\infty$ (see Proposition 2.3). Similarly, since $(A_{z,F}-\alpha)^{-1}$ and $A_{z,\infty}(I-\alpha A_{z,\infty})^{-1}$ have no common eigenvalue, the pair $((A_{z,F}-\alpha)^{-1},(A_{z,F}-\alpha)^{-1}B_{z,F})$ is a left pole pair for $\tilde{W}$ on $\tilde{\sigma}_F$ and $(A_{z,\infty}(I-\alpha A_{z,\infty})^{-1},(I-\alpha A_{z,\infty})^{-1}B_{z,\infty})$ is a left pole pair for $\tilde{W}$ on $\tilde{\sigma}_\infty$. Now define

$$W(\lambda) := \tilde{W}(\varphi^{-1}(\lambda))D, \tag{4.17}$$

where φ is as in (4.3). Note that $W(\infty)=\tilde{W}(\infty)D=D$. Furthermore,

$$W^{\#}(\lambda)=W(\lambda^{-1})=\tilde{W}(\varphi^{-1}(\lambda^{-1}))D=\tilde{W}(\psi^{-1}(\lambda))D, \tag{4.18}$$

with E as in (4.4). Formulas (4.17) and (4.18) allow us to apply Lemma 3.1, first for $\tilde{W}$ and φ^{-1}, and next for $\tilde{W}$ and ψ^{-1}. It follows that for W defined by (4.17) the statements (i), (ii) and (iii) in Theorem 0.1 are proved.

4.3. The one-one correspondence. Let the pairs (4.1) be right pole pairs for W on $\mathbb{C}$ and at infinity, respectively, and let the pairs (4.2) be left zero pairs for W on $\mathbb{C}$ and at infinity. Furthermore, assume that $W(\alpha)=D$ is invertible. In Subsection 4.1 we have shown that $W(\lambda)$ can be written in the form (0.15), where Γ_{12} and Γ_{21} are the unique solutions of (0.13) and (0.14) and the operators Γ_F and Γ_∞ are solutions of (0.10) and (0.11). It remains to prove that Γ_F and Γ_∞ are uniquely determined by W. To do this, consider again

$$\tilde{W}(\lambda) := W(\varphi(\lambda))D^{-1}.$$

From the representation (0.15) and the identities (4.14) and (4.15) (and their analogs for $A_{z,F}$ and $A_{z,\infty}$) it follows that $\tilde{W}(\lambda)$ is given by (4.10) provided one takes

$$\begin{bmatrix} S_{11} & S_{12} \\ S_{21} & S_{22} \end{bmatrix} = \begin{bmatrix} \Gamma_F & \Gamma_{12} \\ \Gamma_{21} & \Gamma_\infty \end{bmatrix}. \tag{4.18}$$

For $[S_{ij}]_{i,j=1}^{2}$ given by (4.18) the identities (0.10), (0.11), (0.13) and (0.14) imply that (4.9) holds true. But then Theorem I.2.1 in [6] guarantees that $[S_{ij}]_{i,j=1}^{2}$ is uniquely determined by $\tilde{W}$, which implies that Γ_F and Γ_∞ are uniquely determined by W. This completes the proof of Theorem 0.1.

The operators Γ_F and Γ_∞ appearing in Theorem 0.1 are coupling operators in the sense of [2,3]. These operators appear also in [1], Theorem 3.1.

REFERENCES

[1] Ball, J.A., Cohen, N., Ran, A.C.M.: Inverse spectral problems for regular improper rational matrix functions, this volume.

[2] Ball, J.A., Ran, A.C.M.: Global inverse spectral problems for rational matrix functions, Linear Algebra and Applications 86 (1987), 237–282.

[3] Ball, J.A., Ran, A.C.M.: Local inverse spectral problems for rational matrix functions, Integral Equations and Operator Theory 10 (1987), 349–415.

[4] Bart, H., Gohberg, I., Kaashoek, M.A.: Minimal factorization of matrix and operator functions, OT 1, Birkhäuser Verlag, Basel, 1979.

[5] Gohberg, I., Kaashoek, M.A.: An inverse spectral problem for rational matrix functions and minimal divisibility, Integral Equations and Operator Theory 10 (1987), 437–465.

[6] Gohberg, I., Kaashoek, M.A., Lerer, L., Rodman, L.: Minimal divisors of rational matrix functions with prescribed zero and pole structure, in: Topics in Operator Theory, Systems and Networks (Eds. H. Dym and I. Gohberg), OT12, Birkhäuser Verlag, Basel, 1984, pp. 241–275.

I. Gohberg
Raymond and Beverly Sackler
Faculty of Exact Sciences
School of Mathematical Sciences
Tel-Aviv University
Ramat-Aviv, Israel

M.A. Kaashoek
Department of Mathematics and Computer Science
Vrije Universiteit
Amsterdam
The Netherlands

Operator Theory:
Advances and Applications, Vol. 33

INVERSE SPECTRAL PROBLEMS FOR REGULAR IMPROPER RATIONAL MATRIX FUNCTIONS

Joseph A. Ball, Nir Cohen and André C.M. Ran

We consider the problem of constructing a regular rational $n \times n$ matrix function $W(z) = C(zI - A)^{-1}B + D + zE(I - zG)^{-1}F$ such that $W\mathcal{R}_n^+ = W_1\mathcal{R}_n^+$ and $W\mathcal{R}_n^- = W_2\mathcal{R}_n^-$. Here $\mathcal{R}_n^+$ (respectively $\mathcal{R}_n^-$) is the space of rational $\mathbb{C}^n$-valued functions analytic inside (respectively outside) a smooth closed contour in the complex plane, and realizations $W_j(z) = C_j(zI - A_j)^{-1}B_j + D_j + zE_j(I - zG_j)^{-1}F_j$ are given for $j = 1,2$. The special case where the contour Γ is the unit circle and $W_1(\infty) = W_2(\infty) = I$ was studied recently in [BR3]. As applications we consider the model reduction problem from linear systems theory for both the discrete time and continuous time settings and the special case of matrix polynomials.

0. INTRODUCTION

Let Γ be a Cauchy contour in the complex plane bounding an inside region Ω^+ and an outside region Ω^- in the extended complex plane. For convenience we assume $0 \in \Omega^+$ and $\infty \in \Omega^-$. We let $\mathcal{R}_n$ be the linear space of all rational n-dimensional column vector functions with no poles on Γ. Let $\mathcal{R}_n^+$ denote such functions which have no poles in Ω^+ and $\mathcal{R}_n^-$ be such functions with no poles in Ω^- and with value 0 at ∞. By the Mittag-Leffler expansion one easily sees the validity of the direct sum decomposition

$$\mathcal{R}_n = \mathcal{R}_n^- \dotplus \mathcal{R}_n^+. \tag{0.1}$$

If W is a regular rational $n \times n$ matrix function with no poles on Γ then $W\mathcal{R}_n = \mathcal{R}_n$ and hence also from (0.1)

$$\mathcal{R}_n = W\mathcal{R}_n^- \dotplus W\mathcal{R}_n^+.$$

One of our main results is a converse of this statement.

THEOREM 0.1 (see Theorem 3.1). *Suppose* W_1 *and* W_2 *are given regular rational* $n \times n$ *matrix functions with no poles on* Γ. *Then there exists a regular rational matrix function* W *such that*

$$WR_n^- = W_1R_n^-, \ WR_n^+ = W_2R_n^+$$

if and only if

$$R_n = W_1R_n^- \dotplus W_2R_n^+.$$

Moreover, given minimal realizations

$$W_j(z) = C_j(zI - A_j)^{-1}B_j + D_j + zE_j(I - zG_j)^{-1}F_j$$

for $W_j(z)$ *and minimal realizations*

$$W_j^{-1}(z) = c_j(zI - a_j)^{-1}b_j + d_j + ze_j(I - zg_j)^{-1}f_j$$

for $W_j^{-1}(z)$ $(j = 1,2)$, *we get in terms of these data explicit realization formulas for* $W(z)$

$$W(z) = C(zI - A)^{-1}B + D + zE(I - zG)^{-1}F$$

and for $W^{-1}(z)$

$$W^{-1}(z) = c(zI - a)^{-1}b + d + ze(I - zg)^{-1}f.$$

An intermediate step in the proof of Theorem 0.1 (which is also of independent interest) is to derive an explicit representation of the subspaces $M^\times := W_1R_n^-$ and $M := W_2R_n^+$ in terms of the matrices in the realizations for W_j and W_j^{-1}. The function W is then found by computing the subspace $L := zM^\times \cap M$ and then demanding that $W\cdot\mathbb{C}^n = L$. The space L is the analogue of the "wandering subspace" appearing in the proof of the main theorem in [BH2], and appearing originally in the proof of the Beurling-Lax Theorem of Halmos [H].

Our main source of inspiration is the article [BR3] where Theorem 0.1 was proved for the case when Γ is the unit circle and $W_1(\infty) = W_2(\infty) = I$. This paper generalizes [BR3] in two respects: the contour Γ is much more general (Ω_+ and Ω_- are not even required to be connected) and we allow singularities at infinity. The generalization to arbitrary contours is rather straightforward; one works with Cauchy's Theorem and contour integral formulas rather than with Fourier coefficients. The second generalization however has its cost; we must work with more complicated realizations as in Theorem 0.1, and the final formulas are somewhat less explicit.

The result in Theorem 0.1 can also be seen as an inverse spectral problem: one reconstructs the function W from spectral data inside Ω_+ and spectral data inside Ω_-. It is in fact this point of view which will be taken in the paper. Inverse spectral problems of various kinds have been studied extensively in the recent past. We mention inverse spectral problems for matrix polynomials ([GLR], [C2]), for rational matrix valued functions ([GKLR], [GKvS], [BR3,4]), for analytic matrix valued functions ([GR1]) and meromorphic matrix functions ([GR2]). Some of these results can also be obtained from our main theorems. In particular, in Section four we shall give an application of our result to matrix polynomials, thereby reproving some results from [GLR].

In Sections one and two we identify the precise data, in terms of the matrices appearing in the realizations for W and W^{-1}, required to describe the *singular subspases* WR_n^- and WR_n^+. In Section three we state and prove our main result, as described in Theorem 0.1. Possibly the final formulas for W and W^{-1} as presented in Theorem 3.1 can be checked directly without going through the machinery concerning singular subspaces developed in the first two sections. However, we think the result is better motivated by showing how it can be derived from the connection with singular subspaces. In Section four we restrict our attention to matrix polynomials in order to show how the theorems can be used to derive results obtained earlier by [GLR] in a completely different way. Finally, in Section five we give an application to model reduction (see [BR1,2], [Gl]); both discrete time systems and continuous time systems are considered here. We prove our results here by seeing model reduction as an inverse spectral problem. This approach was already taken in [BR3] for discrete time systems, however there were some restrictions which are now removed, and also the formulas here are different and do not require Cholesky factorization of a certain matrix as in [BR3]. Also the case of continuous time systems now can be seen as an iverse spectral problem and fall under the analysis of Section three. From the point of view of [BH1] the model reduction problem can be viewed as the problem of computing the J-inner factor in the J-inner-outer factorization of a given matrix function. In our earlier approach [BR1,2] we first computed the outer factor by solving a Wiener-Hopf factorization problem and then found the J-inner factor by muliplying by the inverse of the outer; in the present inverse spectral approach we find the J-inner factor directly from its spectral data inside the contour.

1. SINGULAR SUBSPACES

Let Γ be a closed smooth contour in the finite complex plane which is the positivily (negatively) oriented boundary of the domain Ω^+ (Ω^-) in the Riemann sphere which we consider fixed throughout the following discussion. We assume $0 \in \Omega^+$ and $\infty \in \Omega^-$; these restrictions will be relaxed later. Denote by R_n the linear space of all n-component column vector functions with rational components having no poles on Γ. We let R_n^+ denote such functions having no poles in Ω^+, and R_n^- the space of such functions having no poles in Ω^- and having value 0 at ∞. As is well known (see [CG]), R_n has the direct sum decomposition

$$R_n = R_n^- \dot{+} R_n^+$$

and the projection operator P_+ mapping R_n onto R_n^+ along R_n^- is given by the Cauchy integral formula

$$(P_+ h)(z) = \frac{1}{2\pi i} \int_\Gamma \frac{h(\tau)}{\tau - z} d\tau, \quad z \in \Omega^+.$$

Similarly, if $P_- = I - P_+$ is the projection onto R_n^- along R_n^+ and $h \in R_n$ the values of $(P_- h)(z)$ for $z \in \Omega^-$ are given by

$$(P_- h)(z) = \frac{-1}{2\pi i} \int_\Gamma \frac{h(\tau)}{\tau - z} d\tau, \quad z \in \Omega^-.$$

On the curve Γ itself, P_+ and P_- have the forms

$$P_+ = \tfrac{1}{2}[I + S_\Gamma], \quad P_- = \tfrac{1}{2}[I - S_\Gamma]$$

where S_Γ is the singular integral operator on Γ

$$(S_\Gamma h)(t) = \frac{1}{\pi i} \int_\Gamma \frac{h(\tau)}{\tau - t} d\tau$$

$$:= \lim_{\varepsilon \to 0} \frac{1}{\pi i} \int_{\Gamma_\varepsilon} \frac{h(\tau)}{\tau - t} d\tau, \quad t \in \Gamma$$

where for $\varepsilon > 0$, $\Gamma_\varepsilon = \{\tau \in \Gamma: |\tau - t| \geq \varepsilon\}$.

We now introduce a rational $n \times n$ matrix function W which is analytic and invertible at each point of Γ; thus multiplication by W maps the space R_n of rational column vector functions bijectively onto itself. We shall be interested in studying the subspaces

$$S^+(W) = WR_n^+$$

and

$$S^-(W) = WR_n^-.$$

We refer to these subspaces as the *singular subspaces* of W (with respect to Ω^+ and Ω^- respectively). Note that, since multiplication by W commutes with scalar multiplication, $S^+(W)$ is invariant under pointwise multiplication by functions in R^+ (scalar rational functions with no poles in $\Gamma \cup \Omega^+$), and hence in particular under multiplication by z. Similarly $S^-(W)$ is invariant under R^-, and hence in particular under multiplication by z^{-1}. In the sequel we will often abuse notation and write simply z (or z^{-1}) for the operator of multiplication by z (or z^{-1}). The first order of business is to obtain some properties for subspaces of R_n of the form $S^+(W)$ or $S^-(W)$.

LEMMA 1.1. *Suppose* W *is a rational* n × n *matrix function which is regular on* Γ, *and* $M = S^+(W)$. *Define subspaces* M_p *and* M_z *by*

$$M_p = P_-M, \quad M_z = M \cap R_n^+. \tag{1.1}$$

Then

(1) M_z *has a finite dimensional complement* N *in* R_n^+ ($R_n^+ = N \dot{+} M_z$),

(2) M_p *is a finite dimensional subspace of* R_n^-,

(3) $zM_z \subset M_z$, $P_-(zM_p) \subset M_p$,

(4) *if* $T : M_p \to N$ *is the uniquely determined linear transformation such that*

$$M = (I + T)M_p \dot{+} M_z \tag{1.2}$$

then

$$(TP_-z - zT - P_+z)M_p \subset M_z.$$

PROOF. See [BR1].

The analogous result for $S^-(W)$ is as follows.

LEMMA 1.2. *Let* W *be a rational* $n \times n$ *matrix function which is regular on* Γ *and let* $M^\times = S^-(W)$. *Define subspaces* $M_p^\times$ *and* $M_z^\times$ *by*

(1.3) $$M_p^\times = P_+ M^\times, \quad M_z^\times = M^\times \cap R_n^-.$$

Then

(1) $M_z^\times$ *has a finite dimensional complement* $N^\times$ *in* R_n^- $(R_n^- = N^\times \dot{+} M_z^\times)$,

(2) $M_p^\times$ *is a finite dimensional subspace of* R_n^+ ,

(3) $z^{-1}M_z^\times \subset M_z^\times$, $P_+(z^{-1}M_p^\times) \subset M_p^\times$,

(4) *if* $T^\times : M_p^\times \to N^\times$ *is the uniquely determined linear transformation such that*

$$M^\times = M_z^\times \dot{+} (I + T^\times)M_p^\times$$

then

$$(T^\times P_+ z^{-1} - z^{-1}T^\times - P_- z^{-1})M_p^\times \subset M_z^\times.$$

We are also interested in when can a pair of subspaces $(M^\times, M)$ of R_n be of the form $M^\times = S^-(W)$ and $M = S^+(W)$ for the same rational matrix function W. The following provides a necessary condition.

LEMMA 1.3. *Suppose* W *is a rational* $n \times n$ *matrix function regular on* Γ *and* $M = S^+(W)$, $M^\times = S^-(W)$. *Then*

(1) M *satisfies* (1) - (4) *in Lemma* 1.1.

(2) $M^\times$ *satisfies* (1) - (4) *in Lemma* 1.2.

(3) $R_n = M^\times \dot{+} M$.

Moreover,

$$\dim zM^\times \cap M = n.$$

If $V : \mathbb{C}^n \to zM^\times \cap M$ *is a linear isomorphism and we define the* $n \times n$ *matrix function* V(z) *by*

$$V(z)\,x = (Vx)(z), \quad x \in \mathbb{C}^n,$$

then $W(z) = V(z)X$ *where* X *is a nonsingular* $n \times n$ *(constant) matrix.*

Thus W *is determined by* $(S^-(W), S^+(W))$ *up to a constant invertible right factor.*

2. SINGULAR SUBSPACES IN TERMS OF LOCAL SPECTRAL DATA

Let the contour Γ, the regions $\Omega_{\pm}$ and the rational $n \times n$ matrix function $W(z)$ be as in §1. As W has no poles on Γ, by standard realization results (see [C]) one can represent W by a realization of the form

$$W(z) = C(zI_p - A)^{-1}B + D + zE(I_r - zG)^{-1}F \tag{2.1}$$

where A through G are matrices and

$$\sigma(A) \subset \Omega_+,\ \sigma(G) \subset \Omega_-^{-1} \equiv \{z \in \mathbb{C} : z^{-1} \in \Omega_-\}. \tag{2.2}$$

It follows that $P_+ W(z) = D + zE(I_r - zG)^{-1}F$ and $P_- W(z) = C(zI_p - A)^{-1}B$. It is known ([C]) that in general $p + r \geq \delta(W)$, where δ denotes the McMillan degree ([M]), and that realizations (2.1 - 2.2) for W exist with $p + r = \delta(W)$. Such realizations are called *minimal*; and the equation $p + r = \delta(W)$ is known to be equivalent to the four conditions

(2.3)
a) $C(zI - A)^{-1}: \mathbb{C}^p \to R_n^-$ is 1 : 1
b) $E(I - zG)^{-1}: \mathbb{C}^r \to R_n^+$ is 1 : 1
c) $R(zI - A)^{-1}B : R_n^+ \to \mathbb{C}^p$ is onto
d) $R(I - zG)^{-1}F : R_n^- \to \mathbb{C}^r$ is onto.

Here R is the linear mapping

$$R[h] = \frac{1}{2\pi i} \oint_\Gamma h(t)dt$$

which assigns to every (scalar, vector or matrix) rational function with poles off Γ a constant (scalar, vector or matrix) of the same dimension. This constant may be interpreted as the entrywise-sum of residues of h inside Ω_+; or equivalently as minus the entrywise-sum of residues of h inside Ω_-, if $\infty \in \Omega_-$ is included. We refer to a) or b) as an *observability* condition on the pair (C,A) or (E,G); and to c) or d) as a *controllability* condition on the pair (A,B) or (G,F). Many equivalent formulations of such conditions are known (see e.g. [K]). A realization (2.1-3) for W is unique up to a similarity transformation $(A,B,C,D,E,F,G)\to(S^{-1}AS,S^{-1}B,CS,D,ET,T^{-1}F,T^{-1}GT)$ with $S\in GL(p)$ and $T\in GL(r)$.

By our assumptions, $W^{-1}(z)$ also exists and has no poles on Γ, hence one can find for W^{-1} a realization of the form

$$W^{-1}(z) = c(I_q z - a)^{-1}b + d + ze(I_s - zg)^{-1}f \tag{2.4}$$

where a through g are matrices and

$$\sigma(a) \subset \Omega_-, \ \sigma(g) \subset \Omega_-^{-1} \equiv \{z\in\mathbb{C}: z^{-1}\in\Omega_-\}, \tag{2.5}$$

and with the minimality condition $\delta(W^{-1}) = \delta(W) = q + s$, i.e. such that (c,a) and (e,g) are observable and (a,b), (g,f) are controllable;

$$\begin{aligned} &\text{a)}\quad c(zI - a)^{-1} : \mathbb{C}^q \to R_n^- \text{ is } 1:1\\ &\text{b)}\quad e(I - zg)^{-1} : \mathbb{C}^s \to R_n^+ \text{ is } 1:1\\ &\text{c)}\quad \ddot{R}(zI - a)^{-1}b : R_n^- \to \mathbb{C}^q \text{ is onto}\\ &\text{d)}\quad R(I - zg)^{-1}f : R_n^+ \to \mathbb{C}^s \text{ is onto.} \end{aligned} \tag{2.6}$$

There is a close connection between realizations and singular subspaces of the form $M = WR_n^+$ and $M^\times = WR_n^-$. In fact, given a pair of realizations (2.1 - 6) for W and W^{-1}, one can determine the pieces M_p and M_z of M (see §1) from (C,A,a,b); similarly, the pieces $M_p^\times$ and $M_z^\times$ of $M^\times$ are determined from (E,G,g,f).

LEMMA 2.1. *Suppose (2.1 - 6) is a given pair of realizations for* W *and* W^{-1}. *Let* $M = S^+(W) = WR_n^+$ *and let* $M_p = P_- M$, $M_z = M \cap R_n^+$. *Then*

$$M_p = \{C(zI_p - A)^{-1}x : x\in\mathbb{C}^p\}$$

$$M_z = \{h\in R_n^+ : R[(zI_q - a)^{-1}bh(z)] = 0\}.$$

Similarly, let $M^\times = S^-(W) = WR_n^-$, $M_p^\times = P_+ M^\times$, $M_z^\times = M^\times \cap R_n^-$. *Then*

$$M_p^\times = \{E(I_r - zG)^{-1}x : x\in\mathbb{C}^q\}$$

$$M_z^\times = \{h\in R_n^- : R[(I_s - zg)^{-1}fh(z)] = 0\}.$$

PROOF. We shall only derive the forms of M_p and M_z; the derivation of $M_p^\times$ and $M_z^\times$ then follows by a similar argument. If $h = P_-Wk$ belongs to M_p (where $k\in R_n^+$) then by (2.1) $P_-\{[D + zE(I - zG)^{-1}F]k(z)\} = 0$, so for $z\in\Omega_+$

$$(2.7)\qquad \begin{aligned} h(z) &= (I - P_+)Wk(z) = (I - P_+)[C(zI - A)^{-1}BP_+ k(z)] = \\ &= \frac{1}{2\pi i}\oint_\Gamma C\{(zI - A)^{-1} - (tI - A)^{-1}\}(t - z)^{-1}Bk(t)dt = \\ &= \frac{1}{2\pi i}\oint_\Gamma C(zI - A)^{-1}(tI - A)^{-1}Bk(t)dt = \\ &= C(zI - A)^{-1}x \end{aligned}$$

where $x = \mathcal{R}[(tI - A)^{-1}Bk(t)] \in \mathbb{C}^p$.

Conversely, assume that $h(z) = C(zI - A)^{-1}x$ for some $x \in \mathbb{C}^p$.

By the controllability condition (2.3 - b) the equation

$$x = \mathcal{R}[(tI - A)^{-1}Bk(t)],\quad k \in \mathcal{R}_n^+$$

has a solution k(z). Reversing the order of computation in (2.7) we see that $h \equiv P_- Wk$, so $h \in M_p$. We have shown that $M_p = C(zI - A)^{-1}\mathbb{C}^p$.

To prove the assertion about M_z, let $h = Wk$ be in M_z, i.e. let both h,k be in $\mathcal{R}_n^+$. Then in (2.4) we shall have $P_-\{[d + ze(I - zg)^{-1}]h(z)\} = 0$, hence

$$0 = P_- k(z) = P_- W^{-1}h(z) = P_-[c(zI - a)^{-1}bh(z)]$$

and by a computation analogous to (2.7) we get

$$0 = e(zI - a)^{-1}\mathcal{R}[(tI - a)^{-1}bh(t)]$$

hence from the observability condition (2.6 - a) we deduce that $\mathcal{R}[(tI - a)^{-1}bh(t)] = 0$. The converse implication, i.e. that $\mathcal{R}[(tI - a)^{-1}bh(t)] = 0$ for $h \in \mathcal{R}_n^+$ implies $h \in M_z$, follows by reversing the steps in the above argument. □

Knowledge of (C,A,a,b), hence of M_p and M_z (via Lemma 2.1) does not suffice to determine the singular space $M = W\mathcal{R}_n^+$. As explained in §1, M is in general determined from a triple (M_p, M_z, T), where T is some operator from M_p into a direct complement N of M_z in $\mathcal{R}_n^+$. Being an input-output object, T is not fit for state space considerations, and will be effectively replaced by the $q \times p$ matrix

$$(2.8)\qquad S = \mathcal{R}[(zIq - a)^{-1}bTC(zIp - A)^{-1}].$$

From the discussion above it is clear (knowing the minimal realizations (2.1) and (2.3)) how to associate with the space $M = W\mathcal{R}_n^+$ a quintuple (C,A,a,b,S). Actually, such a quintuple determines M uniquely, as follows:

1) (C,A) and (a,b) determine M_p and M_z via Lemma 2.1.

2) S determines T.

3) (M_p, M_z, T) determine M via Lemma 1.1.

To check that indeed S determines T, fix some direct complement N of M_z in R_n^+. Our minimality assumptions (2.3) and (2.6) imply that $C(zI - A)^{-1} : \mathbb{C}^p \to M_p$ and $R(zI - A)^{-1}b : N \to \mathbb{C}^q$ are both bijective. Therefore (2.8) establishes an isomorphism between all the $q \times p$ matrices S and all the operators $T : M_p \to N$, as required.

Notice that replacing N by another complement N' of M_z changes the operator T to another one $T' : M_p \to N'$; however the net result $M_z + (I + T')M_p$ will still be the same space $M = M_z + (I + T)M_p$.

An equivalent, slightly more direct description of M in terms of (C,A,a,b,S), which can be derived from the description $M = M_z + (I + T)M_p$, is

$$(2.9) \qquad M = \{C(zI_p - A)^{-1}x + h(z) : x \in \mathbb{C}^p,\ h \in R_n^+,\ Sx = R[zI_q - a)^{-1}bh(z)]\}.$$

This leads us to define a collection of matrices (C,A,a,b,S) to be a *canonical set of spectral dat (c.s.s.d) for W over the region* Ω_+ if

1) (C,A) is observable and $\sigma(A) \subset \Omega_+$

2) (a,b) is controllable and $\sigma(a) \subset \Omega_+$

3) the singular subspace $M = S^+(W)$ for W over Ω_+ is given by (2.9).

Let us say that two data sets (C_1,A_1,a_1,b_1,S_1) and (C_2,A_2,a_2,b_2,S_2) are *similar* if and only if there are nonsingular matrices X and Y such that

$$(2.10) \qquad \begin{aligned} C_2 &= C_1X, \quad XA_2 = A_1X \\ Ya_2 &= a_1Y, \quad b_2 = Yb_1 \\ S_1 &= YS_2X. \end{aligned}$$

It is easy to check that if (C_1,A_1,a_1,b_1,S_1) is a c.s.s.d. for W on Ω_+, then (C_2,A_2,a_2,b_2,S_2) is also a c.s.s.d. for W on Ω_+ if and only if $(C_1, A_1, a_1, b_1, S_1)$ and $(C_2, A_2, a_2, b_2, S_2)$ are similar.

We next note a necessary condition (in addition to conditions (1) and (2) in the definition) for a quintuple of matrices (C,A,a,b,S) to be a c.s.s.d. for some rational matrix function W over Ω_+.

PROPOSITION 2.2. *Suppose* (C,A,a,b,S) *is a c.s.s.d. over* Ω_+ *for some rational* $n \times n$ *matrix function* W *which is regular on* Γ. *Then* S *Satisfies the Sylvester equation*

(2.11) $$SA - aS = bC.$$

PROOF. From (C,A,a,b,S) compute T from (2.8). From Lemma 1.1 we know that T satisfies

$$(TP_-z - zT - P_+z)M_p \subset M_z,$$

so, by the characterization of M_z and M_p in Lemma 2.1, for all $x \in M_p$

(2.12) $$\begin{aligned} 0 &= R[(zI_q - a)^{-1}b(TP_-z - zT - P_+z)C(zI_p - A)^{-1}x] \\ &= X - Y - Z \end{aligned}$$

where

$$X = R[(zI_q - a)^{-1}bTP_-zC(zI_p - A)^{-1}x]$$

$$Y = R[(zI_q - a)^{-1}bzTC(zI_p - A)^{-1}x]$$

and

$$Z = R[(zI_q - a)^{-1}bP_+zC(zI_p - A)^{-1}x].$$

To compute X, note that

$$\begin{aligned} &P_-zC(zI - A)^{-1}x \\ &= P_-(C + C(zI - A)^{-1}A)x \\ &= C(zI - A)^{-1}Ax \end{aligned}$$

where we used $CAx \in R_n^+$ and $C(zI - A)^{-1}Ax \in R_n^-$ (since $\sigma(A) \subset \Omega_+$) for the last step. From (2.8) we thus see that

$$X = SA.$$

To analyze Y, note that

$$(zI_q - a)^{-1}bz = b + a(zI_q - a)^{-1}b$$

so

$$Y = R[(b + a(zI_q - a)^{-1}b)TC(zI_p - A)^{-1}x]$$
$$= R[a(zI_q - a)^{-1}bTC(zI_p - A)^{-1}x]$$

by Cauchy's Theorem, since $bT[C(zI_p - A)^{-1}x]$ is analytic on Ω_+ by definition of T. By (2.8) again we see from this that

$$Y = aS.$$

Finally, to analyze Z, note that

$$P_+zC(zI_p - A)^{-1}x = P_+(Cx + C(zI_p - A)^{-1}Ax) = Cx$$

and hence, by the Riesz functional calculus for a we get

$$Z = bCx.$$

Thus equation (2.8) holding for all x is equivalent to the desired identity

$$SA - aS = bC. \qquad \square$$

Similar results hold with respect to Ω_-. Care must be taken in order to allow W to have a pole or zero at ∞. We say that the quintuple of matrices $(E,G,g,f,S^\times)$ is a *canonical set of spectral data (c.s.s.d.) for* W *over* Ω_- if

1) (E,G) is observable and $\sigma(G) \subset \Omega_-^{-1}$
2) (g,f) is controllable and $\sigma(g) \subset \Omega_-^{-1}$
3) the singular subspace $M^\times = S^-(W)$ for W over Ω_- is given by

$$M^\times = \{h + E(I_r - zG)^{-1}x: \quad h \in R_n^-,\ x \in \mathbb{C}^r,\ S^\times x = R[(I_s - zg)^{-1}fh(z)]\}. \tag{2.13}$$

One way to get a c.s.s.d. for W over Ω_-, by Lemma 2.2, is to suppose W and W^{-1} have representations as in (2.2) from which we get matrices E,G,g,f satisfying (1) and (2) above, let $T^\times: M_p^\times \to N^\times$ be as in (4) of Lemma 1.2, and then define $S^\times$ by

$$S^\times x = R[(I_s - zg)^{-1}fT^\times E(I_r - zG)^{-1}x]$$

for $x \in \mathbb{C}^r$. Conversely, (2.13) may be used to solve for $T^\times$ from $S^\times$. The definition of similarity between two such quintuples is as in (2.10). It is easy to see that, given a c.s.s.d. $(E_1,G_1,g_1,f_1,S_1^\times)$ for W over Ω_-, the quintuple $(E_2,G_2,g_2,f_2,S_2^\times)$ is another c.s.s.d. for W over Ω_- if and only if these two quintuples are similar. Condition (4) in Lemma 1.2 together with the connection (2.14) between $S^\times$ and $T^\times$ implies the following analogue of Proposition 2.2; as the proof is completely analogous, it will be omitted.

PROPOSITION 2.3. *Suppose* $(E,G,g,f,S^\times)$ *is a c.s.s.d. over* Ω_- *for some rational* $n \times n$ *matrix function* W *which is regular on* Γ. *Then* $S^\times$ *satisfies the Sylvester equation*

$$(2.15) \qquad S^\times G - gS^\times = fE.$$

Let us say that a collection of matrices $\{(C,A),(a,b),S\}$ is an admissible triple over Ω_+ if (C,A) is observable, (a,b) is controllable, and both $\sigma(A)$ and $\sigma(a)$ are contained in Ω_+. Similarly we say that the collection of matrices $\{(E,G),(g,f),S^\times\}$ is an admissible *triple over* Ω_- if (E,G) is observable, (g,f) is controllable and both $\sigma(G)$ and $\sigma(g)$ are contained in Ω_-^{-1}. We have seen that the validity of the Sylvester equation (2.11) is necessary for the admissible triple over Ω_+ $\{(C,A),(a,b),S\}$ to be a c.s.s.d. over Ω_+ for some rational matrix function W_1. This turns out to be also sufficient. Similarly, the validity of the Sylvester equetion (2.15) turns out to be necessary and sufficient for the admissible triple over Ω_- $\{(E,G),(g,f),S^\times\}$ to be a c.s.s.d. over Ω_- for some rational matrix function W_2. It turns out however that the validity of (2.11) and (2.15) put together does not guarantee that $W_1 = W_2$, that is, that (C,A,a,b,S) be a c.s.s.d. over Ω_+ and $(E,G,g,f,S^\times)$ be a c.s.s.d. over Ω_- for the *same* function W. We now derive an additional necessary condition. Sufficiency of all these necessary conditions put together will be shown in the next section.

THEOREM 2.4. *Suppose* (C,A,a,b,S) *is a c.s.s.d. over* Ω_+ *and* $(E,G,g,f,S^\times)$ *is a c.s.s.d. over* Ω_- *for the same rational* $n \times n$ *matrix function* W *which is regular on* Γ. *Define matrices* S_1, S_2 *and then S by*

$$S_1 = \frac{1}{2\pi i} \int_\Gamma (I - tg)^{-1} fC(tI - A)^{-1}\, dt$$

$$S_2 = \frac{1}{2\pi i} \int_\Gamma (tI - a)^{-1} bE(I - tG)^{-1}\, dt$$

and

$$\mathcal{S} = \begin{bmatrix} S^\times & S_1 \\ S_2 & S \end{bmatrix}.$$

Then $\mathcal{S}$ is nonsingular.

PROOF. We show that the invertibility of $\mathcal{S}$ is equivalent to the matching condition

(2.16) $\quad \mathcal{R}_n = M^\times \dot{+} M$

for the singular subspaces $M^\times = \mathcal{S}^-(W)$ and $M = \mathcal{S}^+(W)$. The invertibility of $\mathcal{S}$ will then follow from Lemma 1.3. We write M in the form

$$M = (I + T)M_p + M_z$$

where

$$M_p = \{h(z) = C(zI_p - A)^{-1}x : x \in \mathbb{C}^p\}$$

$$M_z = \{h \in \mathcal{R}_n^+ : \mathcal{R}[(zI_q - a)^{-1}bh(z)] = 0\}$$

and where $T : M_p \to N$ is defined by (2.8):

$$Sx = \mathcal{R}[(zI_q - a)^{-1}bTC(zI_p - A)^{-1}x],\ x \in \mathbb{C}^p.$$

Similarly, we write $M^\times$ in the form

$$M^\times = M_z^\times + (I + T^\times)M_p^\times$$

where

$$M_p^\times = \{h(z) = E(I_r - zG)^{-1}x : x \in \mathbb{C}^r\}$$

$$M_z^\times = \{h \in \mathcal{R}_n^- : \mathcal{R}[(I_s - zg)^{-1}f\, h(z)] = 0\}$$

and where $T^\times : M_p^\times \to N^\times$ is defined by (2.10):

$$S^\times x = \mathcal{R}[(I_s - zg)^{-1}fT^\times E(I_r - zG)^{-1}x].$$

Introduce the operator $F : M_p^\times \oplus M_p \to N^\times \oplus N$

$$F = \begin{bmatrix} T^{\times} & P_{N^{\times}}|M_p \\ P_N|M_p^{\times} & T \end{bmatrix}$$

Here $P_{N^{\times}}$ is the projection of R_n^- onto $N^{\times}$ along $M_z^{\times}$ and P_N is the projection of R_n^+ onto N along M_z. By a direct computation as in [BR3] (proof of Theorem 3.1) one can verify that $M^{\times}$ and M satisfy the matching condition (2.16) if and only if the operator F: $M_p^{\times} \oplus M_p \to N^{\times} \oplus N$ is invertible. Next by inspection one sees that

$$S = R\left[\begin{bmatrix} (I_s-zg)^{-1}f & 0 \\ 0 & (zI_q-a)^{-1}b \end{bmatrix} F \begin{bmatrix} E(I_r-Gz)^{-1} & 0 \\ 0 & C(zI_p-A)^{-1} \end{bmatrix}\right] \tag{2.17}$$

By controllability of (g,f) and (a,b) and since $N^{\times}$ and N are complementary to the kernels of $R\cdot(I_s-zg)^{-1}f : R_n^- \to \mathbb{C}^s$ and $R\cdot(sI_q-a)^{-1}b : R_n^+ \to \mathbb{C}^q$, we see that

$$R\cdot\begin{bmatrix} (I_s-zg)^{-1}f & 0 \\ 0 & (sI_q-a)^{-1}b \end{bmatrix}$$

maps $N^{\times} \oplus N$ bijectively to $\mathbb{C}^s \oplus \mathbb{C}^q$. Similarly, by the observability of (E,G) and (C,A) and since $M_p^{\times}$ and M_p are by definition the images of $E(I_r - zG)^{-1}|\mathbb{C}^r$ and of $C(zI_p - A)^{-1}|\mathbb{C}^p$ respectively, we see that $E(I_r - zG)^{-1} \oplus C(zI_p - A)^{-1}$ maps $\mathbb{C}^r \oplus \mathbb{C}^p$ bijectively to $M_p^{\times} \oplus M_p$. With these observations we see from (2.17) that invertibility of F is in turn equivalent to invertibility of S. Thus Theorem 2.4 follows. □

It remains to see how the coupling matrix S can be computed for a given matrix function W(z). This is explained in the following Theorem.

THEOREM 2.5 a) *Let* W(z) *be a matrix function of the form*

$$W(z) = C(zI - A)^{-1}B + D + zE(I - zG)^{-1}F$$

with inverse given by

$$W(z)^{-1} = e(zI - a)^{-1}b + d + ze(I - zg)^{-1}f$$

where $\sigma(A)$ *and* $\sigma(a)$ *are contained in* Ω_+ *and* $\sigma(G)$ *and* $\sigma(g)$ *are contained in* Ω_-^{-1}, *and where the controllability and observability assumptions* (2.6) *hold. Then* (C,A,a,b,S) *is a c.s.s.d. for* W *over* Ω_+ *where* S *is uniquely determined by the formula*

(i) $SR[(zI - A)^{-1}Bh(z)]$

$= R[(zI - a)^{-1}\{bD + aS_2F + bC(zI - A)^{-1}B\}h(z)]$

for all $h \in R_n^+$

(b) *Suppose* $W(z)$ *is given by*

$$W(z) = I + \tilde{C}(zI - \tilde{A})^{-1}\tilde{B}$$

with inverse

$$W(z)^{-1} = I - \tilde{C}(zI - \tilde{A}^{\times})^{-1}\tilde{B}$$

where $\tilde{A}^{\times} := \tilde{A} - \tilde{B}\tilde{C}$. *Assume neither* $\tilde{A}$ *nor* $\tilde{A}^{\times}$ *have spectrum on* Γ *and that* $(\tilde{C},\tilde{A})$ *is observable and that* $(\tilde{A},\tilde{B})$ *is controllable. Let* P *(respectively* $P^{\times}$*) be the spectral projection for* $\tilde{A}$ *(respectively* $\tilde{A}^{\times}$*) associated with* Ω_+. *Then a c.s.s.d. for* W *over* Ω_+ *is* $(\tilde{C}P|\mathrm{Im}P,\ \tilde{A}|\mathrm{Im}P,\ \tilde{A}^{\times}|\mathrm{Im}P,\ P^{\times}\tilde{B},\ P^{\times}|\mathrm{Im}P)$.

PROOF of (a). By assumption (2.6) it is clear that (i) determines an operator (or matrix) S uniquely (if any exists). As for existence, equation (i) follows easily from (2.11) combined with

$$SB = bD + aS_2F. \tag{2.18}$$

To derive (2.18) we use that constant $\mathbb{C}^n$-functions are in R_+^n, and hence

$$W(z)\mathbb{C}^n \subset S^+(W).$$

For $x \in \mathbb{C}^n$, $P_-(W(z)x) = C(zI - A)^{-1}Bx$ while $P_+(W(z)x) = [D + zE(I - zG)^{-1}F]x$. By the characterization (2.9) of $M = S^+(W)$ we conclude

$$SBx = R[(zI - a)^{-1}b\{D + zE(I - zG)^{-1}F\}x]. \tag{2.19}$$

From the identity

$$z(zI - a)^{-1} = I + a(zI - a)^{-1}$$

we get

$$\mathcal{R}[(zI - a)^{-1}bzE(I - zG)^{-1}Fx]$$
$$= \mathcal{R}[bE(I - zG)^{-1}Fx] + \mathcal{R}[a(zI - a)^{-1}bE(I - zG)^{-1}Fx].$$

The first term vanishes since $(I - zG)^{-1}$ is analytic in Ω_+, and the second one equals aS_2Fx by definition. Thus (2.18) follows.

PROOF of (b). When $W(z) = \tilde{C}(zI - \tilde{A})^{-1}\tilde{B} + I$, we may rewrite $W(z)$ in the form

$$W(z) = C(zI - A)^{-1}B + D + zE(I - zG)^{-1}F$$

where

$$C = \tilde{C}P|_{\mathrm{Im}P},\ A = \tilde{A}|_{\mathrm{Im}P},\ B = P\tilde{B},$$
$$D = I - \tilde{C}\tilde{A}^{-1}(I - P)\tilde{B} \tag{2.20}$$
$$E = -\tilde{C}\tilde{A}^{-1}(I - P)|_{\mathrm{Im}(I - P)}$$
$$F = \tilde{A}^{-1}(I - P)\tilde{B}$$
$$G = \tilde{A}^{-1}(I - P)|_{\mathrm{Im}(I - P)}\ .$$

Here while $\tilde{A}$ may not be invertible, $\tilde{A}|_{\mathrm{Im}(I - P)}$ is invertible since by assumption $0 \in \Omega_+$, and $\tilde{A}^{-1}(I - P)|_{\mathrm{Im}(I - P)}$ refers to the inverse of $\tilde{A}|_{\mathrm{Im}(I - P)}$. Similarly

$$W(z)^{-1} = c(I - za)^{-1}b + d + ze(I - zg)^{-1}f$$

where

$$c = -\tilde{C}P^{\times}|_{\mathrm{Im}P^{\times}}\ ,\ a = \tilde{A}^{\times}|_{\mathrm{Im}P^{\times}}\ ,\ b = P^{\times}\tilde{B},$$
$$d = I + \tilde{C}\tilde{A}^{\times^{-1}}(I - P^{\times})\tilde{B}$$
$$e = \tilde{C}\tilde{A}^{\times^{-1}}(I - P^{\times})|_{\mathrm{Im}(I - P^{\times})}$$
$$f = \tilde{A}^{\times^{-1}}(I - P^{\times})\tilde{B}$$
$$g = \tilde{A}^{\times^{-1}}(I - P^{\times})|_{\mathrm{Im}(I - P^{\times})}.$$

Thus $C(zI - A)^{-1} = \tilde{C}(zI - \tilde{A})^{-1}P|_{\mathrm{Im}P}$ and $(zI - a)^{-1}b = (zI - \tilde{A}^{\times})^{-1}P^{\times}\tilde{B}|_{\mathrm{Im}P^{\times}}$. If $k(z) \in \mathcal{R}_n^+$, from (2.7) we know that

$$P_-[Wk](z) = \tilde{C}(zI - \tilde{A})^{-1}Px$$

where

$$x = \mathcal{R}[(tI - \tilde{A})^{-1}\tilde{B}k(t)] = Px.$$

Thus, by definition

$$\begin{aligned} Sx &= R[P^{\times}(zI - \tilde{A}^{\times})^{-1}\tilde{B}\{W(z)k(z) - P_{-}[Wk](z)\}] \\ &= R[P^{\times}(zI - \tilde{A}^{\times})^{-1}\tilde{B}\tilde{C}(zI - \tilde{A})^{-1}\{\tilde{B}k(z) - x\}] \\ &\quad + R[P^{\times}(zI - \tilde{A}^{\times})^{-1}\tilde{B}k(z)] \\ &= R[P^{\times}(zI - \tilde{A}^{\times})^{-1}[(zI - \tilde{A}^{\times}) - (zI - \tilde{A})] \\ &\quad (zI - \tilde{A})^{-1}\{\tilde{B}k(z) - x\}] \\ &\quad + R[P^{\times}(zI - \tilde{A}^{\times})^{-1}\tilde{B}k(z)] \\ &\doteq (1) + (2) + (3) + (4) + (5) \end{aligned}$$

where

$$\begin{aligned} (1) &= R[P^{\times}(zI - \tilde{A})^{-1}\tilde{B}k(z)] \\ (2) &= -R[P^{\times}(zI - \tilde{A}^{\times})^{-1}\tilde{B}k(z)] \\ (3) &= -R[P^{\times}(zI - \tilde{A})^{-1}x] \\ (4) &= R[P^{\times}(zI - \tilde{A}^{\times})^{-1}x] \\ (5) &= R[P^{\times}(zI - \tilde{A}^{\times})^{-1}\tilde{B}k(z)]. \end{aligned}$$

Note that (2)+(5)= 0, (1)= $P^{\times}x$ by definition (3)= $-P^{\times}x$ since $x = Px$ by the functional calculus for $\tilde{A}$ and (4)= $P^{\times}x$ by the functional calculus for $\tilde{A}^{\times}$. Thus the sum collapses to $P^{\times}$ and we have verified that $Sx = P^{\times}x$ for all $x \in \operatorname{Im}P$ as desired. □

The corresponding result for Ω_{-} is as follows.

THEOREM 2.6 a) *Let* $W(z)$ *and* $W(z)^{-1}$ *be as in* (a) *in Theorem* 2.5. *Then a c.s.s.d. for* W *over* Ω_{-} *is given by* $(E,G,g,f,S^{\times})$ *where* $S^{\times}$ *is uniquely determined by the formula*

$$\begin{aligned} &S^{\times}\cdot R[(I - zG)^{-1}Fh(z)] \\ &\quad = R[(I - zg)^{-1}\{gS_{1}B + fD + fE(I - zG)^{-1}F\}h(z)] \end{aligned}$$

for all $h \in R_{n}^{-}$.

(b) *Let* $W(z)$ *and* $W(z)^{-1}$ *be as in* (b) *in Theorem* 2.5. *Then a c.s.s.d. for* W *over* Ω_{-} *is* $(-\tilde{C}\tilde{A}^{-1}(I-P)|_{\operatorname{Im}(I-P)}$, $\tilde{A}^{-1}(I-P)|_{\operatorname{Im}(I-P)}$, $\tilde{A}^{\times^{-1}}(I-P^{\times})|_{\operatorname{Im}(I-P^{\times})}$, $\tilde{B}\tilde{A}^{\times^{-1}}(I-P^{\times})|_{\operatorname{Im}(I-P^{\times})}$,

$(I-P^\times)|_{\mathrm{Im}(I-P)})$.

PROOF of (a). The proof is similar to the proof of (a) in Theorem 2.5. This time use the Sylvester equation (2.15) for $S^\times$ together with the identity

(2.22) $$S^\times F = gS_1B + fD.$$

One derives (2.22) by observing that $z^{-1}W(z)\mathbb{C}^n \subset S^-(W) =: M^\times$ and then using the characterization (2.13) of $M^\times$. Note that

$$P_+(W(z)z^{-1}y) = E(I - zG)^{-1}Fy$$

and

$$P_-(W(z)z^{-1}y) = [z^{-1}C(zI - A)^{-1}B + z^{-1}D]y$$

and use the identity

$$z^{-1}(I - zg)^{-1} = z^{-1}I + g(I - zg)^{-1}$$

together with the residue Theorem.

PROOF of (b). Proceed as in the proof of (b) in Theorem 2.5. In this case, for $k \in R_n^-$ and $z \in \Omega_-$,

$$\begin{aligned} P_-(Wk)(z) &= -R_t[(\tilde{C}(zI - \tilde{A})^{-1}\tilde{B} + I)k(z)(t - z)^{-1}] \\ &= -R_t[\tilde{C}(zI - \tilde{A})^{-1}(I - P)\tilde{B}(t - z)^{-1}k(z)] \end{aligned}$$

so

$$\begin{aligned} &(I - P_-)(Wk)(z) \\ &\quad = -R_t[\tilde{C}\{(zI - \tilde{A})^{-1} - (tI - \tilde{A})^{-1}\}(t - z)^{-1}(I - P)\tilde{B}k(t)] \\ &\quad = -R_t[\tilde{C}(zI - \tilde{A})^{-1}(tI - \tilde{A})^{-1}(I - P)\tilde{B}k(t)] \\ &\quad = \tilde{C}(zI - \tilde{A})^{-1}(I - P)y \end{aligned}$$

where

$$y = -R[(tI - \tilde{A})^{-1}\tilde{B}k(t)] = (I - P)y.$$

From (2.21),

$$\begin{aligned} (I - zg)^{-1}f &= (I - z\tilde{A}^{\times^{-1}})^{-1}\tilde{A}^{\times^{-1}}(I - P^\times)\tilde{B} \\ &= -(zI - \tilde{A}^\times)^{-1}(I - P^\times)\tilde{B}. \end{aligned}$$

Thus, by definition of $S^\times$,

$$\begin{aligned} S^{\times}y &= R[(I - zg)^{-1}fP_{-}[Wk](z)] \\ &= R[-(zI - \tilde{A}^{\times})^{-1}(I - P^{\times})\tilde{B}\cdot\{k(z) + \\ &\qquad C(zI - \tilde{A})^{-1}\tilde{B}k(z) - \tilde{C}(zI - \tilde{A})^{-1}y\}] \\ &= R[-(zI - \tilde{A}^{\times})^{-1}(I - P^{\times})\tilde{B}k(z)] \\ &\qquad -R[(I - P^{\times})(zI - \tilde{A}^{\times})^{-1}[(zI - \tilde{A}^{\times}) - (zI - \tilde{A})] \\ &\qquad (zI - \tilde{A})^{-1}\{\tilde{B}k(z) - y\}] \\ &= 1 + 2 + 3 + 4 + 5 \end{aligned}$$

where

$$\begin{aligned} 1 &= -R[(zI - \tilde{A}^{\times})^{-1}(I - P^{\times})\tilde{B}k(z)] \\ 2 &= -R[(I - P^{\times})(zI - \tilde{A})^{-1}\tilde{B}K(z)] \\ 3 &= R[(I - P^{\times})(zI - \tilde{A}^{\times})^{-1}\tilde{B}k(z)] \\ 4 &= R[(I - P^{\times})(zI - \tilde{A})^{-1}y] \\ 5 &= -R[(I - P^{\times})(zI - \tilde{A}^{\times})^{-1}y]. \end{aligned}$$

Note that $1 + 3 = 0$, $2 = (I - P^{\times})y$ by definition and each of 4 and 5 is zero since the integrands are analytic on Ω_+. We conclude that $S^{\times}y = (I - P^{\times})y$ for all $y \in \mathrm{Im}(I - P)$ as asserted. □

3. THE INVERSE SPECTRAL PROBLEM

In this section we consider the following inverse spectral problem. We are given an admissible triple $\{(C,A),(a,b),S\}$ over Ω_+ and an admissible triple $\{(E,G),(g,f),S^{\times}\}$ over Ω_-. The problem is to decide if there is a rational $n \times n$ matrix function W regular on Γ which has (C,A,a,b,S) as a c.s.s.d. over Ω_+ and $(E,G,g,f,S^{\times})$ as a c.s.s.d. over Ω_-, and if so, exhibit the function W as explicitly as possible. In the previous Section we obtained a number of necessary conditions for the existence of such a W. In this Section we show that these necessary conditions together are also sufficient. The result is as follows.

THEOREM 3.1 *Let there be given an admissable triple over* Ω_+ $\{(C,A),(a,b),S\}$ *and an admissable triple over* Ω_- $\{(E,G),(g,f),S^{\times}\}$. *Then there exists a rational matrix function* W *regular on* Γ *such that* (C,A,a,b,S) *is a c.s.s.d. over* Ω_+ *for* W *and* $(E,G,g,f,S^{\times})$ *is a c.s.s.d. over* Ω_- *for* W *if and only if*

(3.1) $SA - aS = bC$

(3.2) $S^{\times}G - gS^{\times} = fE$

and

(3.3) $\mathcal{S} = \begin{bmatrix} S^{\times} & S_1 \\ S_2 & S \end{bmatrix}$ *is invertible,*

where

(3.3a) $S_1 = \frac{1}{2\pi i} \int_{\Gamma} (I_s - tg)^{-1} fC(tI - A)^{-1} dt$

and

(3.3b) $S_2 = \frac{1}{2\pi i} \int_{\Gamma} (tI_q - a)^{-1} bE(I - tG)^{-1} dt$.

Suppose conditions (3.1) - (3.3) *hold. Choose an invertible* n × n *matrix* Δ *and a point* $z_0 \in \mathbb{C}$ *not in* $\sigma(A) \cup \sigma(a) \cup \sigma(G)^{-1} \cup \sigma(g)^{-1}$. *Then the unique solution* W(z) *of the above inverse spectral problem which also satisfies* $W(z_0) = \Delta$ *is given by*

(3.4) $$W(z) = [zE(I - zG)^{-1}(I - z_0G), C(zI - A)^{-1}(z_0I - A)] \cdot$$

$$\cdot \mathcal{S}^{-1} \cdot \begin{bmatrix} (I - z_0g)^{-1}f \\ (z_0I - a)^{-1}b \end{bmatrix} \Delta$$

$$+\{-[z_0E, C]\mathcal{S}^{-1} \begin{bmatrix} (I - z_0g)^{-1}f \\ (z_0I - a)^{-1}b \end{bmatrix} + I\} \Delta$$

with inverse given by

(3.5) $$W(z)^{-1} = \Delta^{-1}[E(I - z_0G)^{-1}, -C(z_0I - A)^{-1}] \cdot \mathcal{S}^{-1} \cdot$$

$$\cdot \begin{bmatrix} -z(I - zg)^{-1}(I - z_0g)g \\ (zI - a)^{-1}(z_0I - a)b \end{bmatrix}$$

$$+ \Delta^{-1}\{[E(I - z_0G)^{-1}, -C(z_0I - A)^{-1}]\mathcal{S}^{-1} \begin{bmatrix} z_0f \\ -b \end{bmatrix} + I\}.$$

If both G *and* g *are invertible, then the unique solution* W(z) *satisfying* $W(\infty) = \Delta$ *is given by*

$$(3.4a)\qquad W(z) = \Delta + [EG^{-1}(zI - G^{-1})^{-1}, C(zI - A)^{-1}]S^{-1}\begin{bmatrix} -g^{-1}f \\ b \end{bmatrix}\Delta$$

with inverse given by

$$(3.5a)\qquad W(z)^{-1} = \Delta^{-1} + \Delta^{-1}[EG^{-1}, C]S^{-1}\begin{bmatrix} (zI - g^{-1})^{-1}g^{-1}f \\ -(zI - a)^{-1}b \end{bmatrix}$$

Most of the remaining part of this section is devoted to the proof of Theorem 3.1, and is divided into several lemmas. Throughout the proof we shall fix an admissible data set $\{(C,A),(a,b),S\}$ and $\{(E,G),(g,f),S^{\times}\}$. If this data comes from a rational matrix function W(z), Propositions 2.2, 2.3 and Theorem 2.4 show that properties (3.1 - 3.3) are fulfilled. Conversely, given (3.1 - 3.3) then, motivated by (2.9) and (2.13) let us define two subspaces of R_n

$$M = \{C(zI-A)^{-1}x + h(z) \colon x \in \mathbb{C}^p,\ h \in R_n^+,\ Sx = [(zI-a)^{-1}bh(z)]\}$$

$$M^{\times} = \{E(I - zG)^{-1}x + h(z) \colon x \in \mathbb{C}^r,\ h \in R_n^-,\ S^{\times}x = R[(I - zg)^{-1}fh(z)]\}.$$

If we want M and $M^{\times}$ to be WR_n^+ and WR_n^- for some W, we certainly expect the invariance properties $R^+M \subset M$, $zR^-M^{\times} \subset M^{\times}$ to hold, as well as the direct sum decomposition $R_n = M^{\times} \dotplus M$. The first step in the proof is to verify these properties.

LEMMA 3.2. *(3.1) implies that* $R^+M \subset M$; *(3.2) implies that* $zR^-M^{\times} \subset M^{\times}$; *and (3.3) implies that* $R_n = M \dotplus M^{\times}$.

PROOF. Actually the implications in the Lemma are two-sided, but we shall need only one direction here (the other direction was proved in §2). The fact that (3.3) $\Longleftrightarrow$ $R_n = M \dotplus M^{\times}$ was given in Theorem 2.4. As to the two remaining assertions, we shall only prove the assertion about M, whereby the other assertion follows by a symmetry of the argument. Given $k \in R^+$, we shall show that $kM \subset M$.

First we show that $kM_z \subset M_z$. By the Riesz functional calculus $[k(z)I - k(a)](zI - a)^{-1}$ is analytic on Ω_+, hence for all $h \in R_n^+$ we have by Cauchy's theorem

(3.6) $$R\{[(k(z)I - k(a)](zI - a)^{-1}bh(z)\} = 0$$

or equivalently

$$R[(zI - a)^{-1}bk(z)h(z)] = k(a)R[(zI - a)^{-1}bh(z)].$$

In particular, if $h \in M_z$ then the right hand side vanishes, hence the left hand side vanishes, which means that $kh \in M_z$, as required.

Next we show that $P_-kM_p \subset M_p$. By the Riesz functional calculus $[k(z)I - k(A)](zI - A)^{-1}$ is analytic on Ω_+, hence for $w \in \Omega_-$ we have by Cauchy's theorem

$$\frac{1}{2\pi i} \int_\Gamma C[k(t)I - k(A)](tI - A)^{-1}(t - w)^{-1}dt = 0.$$

Now for a typical element $\tilde{m}(z) = C(zI - A)^{-1}x$ of M_p $(x \in \mathbb{C}^p)$

$$P_-(k\tilde{m})(w) = -\frac{1}{2\pi i} \int_\Gamma k(t)C(tI - A)^{-1}(t - w)^{-1}x\, dt =$$

$$= -\frac{1}{2\pi i} \int_\Gamma C(tI - A)^{-1}(t - w)^{-1}k(A)x\, dt =$$

(3.7) $$= C(wI - A)^{-1}y$$

where $y = k(A)x$. The last equality follows from the functional calculus. This shows that $P_-(k\tilde{m}) \in M_p$ as required.

Now let $m(z) = \tilde{m}(z) + h(z)$ be a typical element in M, i.e. $\tilde{m}(z) = C(zI - A)^{-1}x$ as before and $h \in R_n^+$ fulfill

(3.8) $$Sx = R[(zI - a)^{-1}bh(z)].$$

We will show that $km \in M$. From (3.7) we have $P_-(k\tilde{m})(z) = C(zI - A)^{-1}y$ where $y = k(A)x$, hence $(km)(z) = C(zI - A)^{-1}y + h_1(z)$, where

(3.9) $$h_1(z) = (kh)(z) + (I - P_-)(k\tilde{m})(z) = (kh)(z) + C(zI - A)^{-1}(k(z) - k(A))x \in R_n^+.$$

To show that $km \in M$ we have to show that

(3.10) $$Sy = R[(zI - a)^{-1}bh_1(z)].$$

Indeed, by (3.1) we get the identity

$$S(zI - A)^{-1} - (zI - a)^{-1}S = (zI - a)^{-1}bC(zI - A)^{-1}$$

hence by the functional calculus

$$Sy = Sk(A)x = \frac{1}{2\pi i}\int_\Gamma k(t)S(tI - A)^{-1}x\,dt =$$

$$= \frac{1}{2\pi i}\int_\Gamma k(t)(tI - a)^{-1}Sx\,dt +$$

$$+ R[k(z)(zI - a)^{-1}bC(zI - A)^{-1}x] =$$

$$= k(a)Sx + R[k(z)(zI - a)^{-1}bC(zI - A)^{-1}x].$$

On the other hand from (3.9)

$$R[(zI - a)^{-1}bh_1(z)] = R[(zI - a)^{-1}bk(z)h(z)] +$$

$$+ R\{(zI - a)^{-1}bC(zI - A)^{-1}[k(z)I - k(A)]x\}.$$

Due to (3.6) and (3.8), the first integral on the right hand side of the last expression equals

$$k(a)R[(zI - a)^{-1}bh(z)] = k(a)Sx$$

hence we are left with

$$Sy - R[(zI - a)^{-1}bh_1(z)] = R[K(z)]y$$

where $K(z) \equiv (zI-a)^{-1}bC(zI-A)^{-1}$. Now K is analytic throughout Ω_- (and vanishes at ∞), hence its residues in Ω_+ must sum up to zero, hence $R[K(z)]y = 0$ and (3.10) is established. Therefor km must belong to M and the Lemma is proved. □

LEMMA 3.3. *Let* ψ *be defined by* $\psi \begin{bmatrix} -S^\times & gS_1 & f \\ & & \\ aS_2 & -S & b \end{bmatrix}$. *A vector function* h(z) *belongs to* $L = zM^\times \cap M$ *if and only if*

$$h(z) = C(zI - A)^{-1}x + \alpha + zE(I - zG)^{-1}y$$

where $\psi\begin{pmatrix} y \\ x \\ \alpha \end{pmatrix} = 0.$

PROOF. We compute as in the verification of identities (2.18) and (2.19) in the direct analysis. Let h belong to L. Since $h \in M$ we have

$$(3.11)\qquad h(z) = C(zI - A)^{-1}x + h_+(z)$$

where $x \in \mathbb{C}^p$, $h_+ \in R_n^+$ and

(3.12) $$Sx = R[(zI - a)^{-1}bh_+(z)].$$

Since $z^{-1}h \in M^\times$ we have

(3.12a) $$z^{-1}h(z) = E(I - zG)^{-1}y + h_-(z)$$

where $y \in \mathbb{C}^r$, $h_- \in R_n^-$ and

$$S^\times y = R[(I - zg)^{-1}fh_-(z)].$$

For $z \in \Omega_-$ we compute by the residue theorem that

$$P_-[z^{-1}h_+(z)] = -\frac{1}{2\pi i}\int_\Gamma \frac{h_+(t)dt}{t(t-z)} = \frac{h_+(0)}{z} \equiv \frac{\alpha}{z} .$$

Multiply (3.11) by $z^{-1}$and apply P_-:

$$h_-(z) = P_-[z^{-1}h(z)] = z^{-1}C(zI - A)^{-1}x + z^{-1}\alpha$$

hence by (3.12a)

$$h(z) = C(zI - A)^{-1}x + \alpha + zE(I - zG)^{-1})y$$

as required in the Lemma. By (3.11), then,

(3.13) $$h_+(z) = \alpha + zE(I - zG)^{-1}y.$$

It remains to show that $\psi\begin{pmatrix} y \\ x \\ \alpha \end{pmatrix} = 0$, and first we shall show that

(3.14) $$[aS_2 \ , \ -S \ , \ b]\begin{pmatrix} y \\ x \\ \alpha \end{pmatrix} = 0.$$

Indeed, from (3.12) and (3.13) we have

(3.15) $$Sx = R[(zI - a)^{-1}b\{\alpha + zE(I - zG)^{-1}y\}].$$

From the identity

$$z(zI - a)^{-1} = I + a(zI - a)^{-1}$$

we see that

$$R[(zI - a)^{-1}bzE(I - zG)^{-1}y] =$$
$$= R[bE(I - zG)^{-1}y] + R[a(zI - a)^{-1}bE(I - zG)^{-1}y].$$

The first term vanishes since $(I - zG)^{-1}$ is analytic on Ω_+ , and the second one equeals aS_2y by definition. Also by the residu theorem $R[(zI - a)^{-1}b\alpha] = b\alpha$, hence going back to (3.15) we get $Sx = b\alpha + aS_2y$ as required.

In a similar way one can show that

$$[-S^{\times}\ ,\ gS_1\ ,\ f]\begin{pmatrix} y \\ x \\ \alpha \end{pmatrix} = 0$$

and together with (3.14) this implies that indeed $\psi\begin{pmatrix} y \\ x \\ \alpha \end{pmatrix} = 0.$ □

LEMMA 3.4. *Choose a point* $z_0 \in \mathbb{C}$ *which is not in* $\sigma(A) \cup \sigma(a) \cup \sigma(G)^{-1} \cup \sigma(g)^{-1}$, *and let the matrix* T_{z_0} *be given by*

$$T_{z_0} = \begin{bmatrix} & \psi & \\ z_0E(I-z_0G)^{-1} & C(z_0I-A)^{-1} & I \end{bmatrix}$$

$$= \begin{bmatrix} -S^{\times} & gS_1 & f \\ aS_2 & -S & b \\ z_0E(I-z_0G)^{-1} & C(z_0I-A)^{-1} & I \end{bmatrix}.$$

Then T_{z_0} *is invertible with inverse given by*

$$T_{z_0}^{-1} = \begin{bmatrix} I-z_0G & 0 & 0 \\ 0 & z_0I-A & 0 \\ -z_0E & -C & I \end{bmatrix} \begin{bmatrix} S^{-1} & & 0 \\ & & 0 \\ 0 & 0 & I \end{bmatrix} \cdot$$

$$\cdot \begin{bmatrix} -(I-z_0g)^{-1} & 0 & (I-z_0g)^{-1}f \\ 0 & (a-z_0I)^{-1} & -(a-z_0I)^{-1}b \\ 0 & 0 & I \end{bmatrix}.$$

In particular, ψ *has full rank and hence* $\dim L = n$.

PROOF. First note that

$$T_{z_0} = \begin{bmatrix} -S^{\times}(I-z_0G) & gS_1(z_0I-A) & f \\ aS_2(I-z_0G) & -S(z_0I-A) & b \\ z_0I & C & I \end{bmatrix} D_1$$

where

$$D_1 = \begin{bmatrix} (I-z_0G)^{-1} & 0 & 0 \\ 0 & (z_0I-A)^{-1} & 0 \\ 0 & 0 & I \end{bmatrix}.$$

By direct computation one can verify that S_1 and S_2 satisfy the Sylvester equations

(3.16) $$S_1 - gS_1A = fC$$

and

(3.17) $$S_2 - aS_2G = bE.$$

(Moreover, the separation of $\sigma(A)$ from $\sigma(g)^{-1}$ and of $\sigma(a)$ from $\sigma(G)^{-1}$ guaranties that the solutions S_1 and S_2 of these equations are unique.) Using (3.16) and (3.17) together with the Sylvester equations (3.1) and (3.2) for S and $S^\times$, we get

$$T_{z_0} = \begin{bmatrix} -(I-z_0g)S^\times & -(I-z_0g)S_1 & f \\ (a-z_0I)S_2 & (a-z_0I)S & b \\ 0 & 0 & I \end{bmatrix} \begin{bmatrix} I & 0 & 0 \\ 0 & I & 0 \\ z_0E & C & I \end{bmatrix} D_1$$

$$= \begin{bmatrix} I & 0 & f \\ 0 & I & b \\ 0 & 0 & I \end{bmatrix} D_2 \begin{bmatrix} \multicolumn{2}{c}{S} & 0 \\ & & 0 \\ 0 & 0 & I \end{bmatrix} \begin{bmatrix} I & 0 & 0 \\ 0 & I & 0 \\ z_0E & C & I \end{bmatrix} D_1$$

where

$$D_2 = \begin{bmatrix} -(I-z_0g) & 0 & 0 \\ 0 & (a-z_0I) & 0 \\ 0 & 0 & I \end{bmatrix}.$$

Thus

$$T_{z_0}^{-1} = D_1^{-1} \begin{bmatrix} I & 0 & 0 \\ 0 & I & 0 \\ -z_0E & -C & I \end{bmatrix} \begin{bmatrix} \multicolumn{2}{c}{S^{-1}} & 0 \\ & & 0 \\ 0 & 0 & I \end{bmatrix} D_2^{-1} \begin{bmatrix} I & 0 & -f \\ 0 & I & -b \\ 0 & 0 & I \end{bmatrix}$$

$$\begin{bmatrix} I-z_0G & 0 & 0 \\ 0 & z_0I-A & 0 \\ -z_0E & -C & I \end{bmatrix} \begin{bmatrix} \multicolumn{2}{c}{S^{-1}} & 0 \\ & & 0 \\ 0 & 0 & I \end{bmatrix} \cdot$$

$$\begin{bmatrix} -(I-z_0g)^{-1} & 0 & (I-z_0g)^{-1}f \\ 0 & (a-z_0I)^{-1} & -(a-z_0I)^{-1}b \\ 0 & 0 & I \end{bmatrix}$$

as asserted. From the form of T_{z_0} we see that the surjectivity of

T_{z_0} forces that of ψ, and hence dim Ker $\psi = n$. By Lemma 3.3, we get from this that dim $L = n$. □

LEMMA 3.5. *Let* $W(z)$ *be defined by* (3.4), *where* $z_0 \notin \sigma(A) \cup \sigma(G)^{-1} \cup \sigma(a) \cup \sigma(g)^{-1}$ *and where* Δ *is invertible. Then* $W(z_0) = \Delta$ *is invertible (so* W *is regular), and*

$$WR_n^+ \subset M$$

$$WR_n^- \subset M^\times.$$

PROOF. By a direct computation using Lemma 3.4, one sees that

$$W(z) = C(zI - A)^{-1}B + D + zE(I - zG)^{-1}F$$

where

$$\begin{bmatrix} F \\ B \\ D \end{bmatrix} = T_{z_0}^{-1} \begin{bmatrix} 0 \\ 0 \\ \Delta \end{bmatrix}.$$

Thus in particular

$$\psi \begin{bmatrix} F \\ B \\ D \end{bmatrix} = \begin{bmatrix} 0 \\ 0 \end{bmatrix}. \tag{3.17a}$$

By the correspondence between Ker ψ and L given by Lemma 3.3, we see that $W\mathbb{C}^n = L$. Since $L \subset M$ and $R^+M \subset M$ by Lemma 3.2, we deduce that

$$WR_n^+ = W(R^+\mathbb{C}^n) = R^+W\mathbb{C}^n = R^+L \subset M.$$

Similarly, from $z^{-1}L \subset M^\times$ and $zR^-M^\times \subset M^\times$ by Lemma 3.2 again, we get

$$WR_n^- = W(zR^- \cdot z^{-1}\mathbb{C}^n) = zR^- \cdot W(z^{-1}\mathbb{C}^n) = zR^-(z^{-1}L) \subset M^\times$$

as asserted. One sees that $W(z_0) = \Delta$ by inspection. □

In the next Lemma we use the formula (3.5) for the inverse of $W(z)$. To motivate the formula, one can check that $(Wz)^{-T}$ must have $\{(b^T,a^T), (A^T,C^T), -S^T\}$ as a c.s.s.d. over Ω_+, $\{(f^T,g^T), (G^T,E^T), -S^{\times T}\}$ as c.s.s.d. over Ω_- and value at z_0 equal

Δ^{-T}. Then formula (3.5) for $W(z)^{-1}$ can be derived from this collection of data for W^{-1} in the same way that formula (3.4) for $W(z)$ was derived from the original set of data for W.

LEMMA 3.6. *Define a matrix function* $V(z)$ *by the right hand side of formula* (3.5). *Then* $V(z) = W(z)^{-1}$.

PROOF. By inspection, $V(z_0) = \Delta^{-1}$ while $W(z_0) = \Delta$. Thus to show that $V(z) = W(z)^{-1}$, it suffices to show that $V(z)W(z)$ is a constant. To do this we note first that

$$V(z) = c(zI - a)^{-1}b + d + ze(I - zg)^{-1}f$$

where

$$[e\ c\ d]\begin{bmatrix} S^{\times} & S_1A & z_0(I-z_0g)^{-1}f \\ S_2G & S & (z_0I-a)^{-1}b \\ E & C & I \end{bmatrix} = [0\ 0\ \Delta^{-T}]\ .$$

[Indeed, verify as in the proof of Lemma 3.4 that

$$\begin{bmatrix} S^{\times} & S_1A & z_0(I-z_0g)^{-1}f \\ S_2G & S & (z_0I-a)^{-1}b \\ E & C & I \end{bmatrix}^{-1}$$

$$= \begin{bmatrix} -(I-z_0G)^{-1} & 0 & 0 \\ 0 & (z_0I-A)^{-1} & 0 \\ E(I-z_0G)^{-1} & -C(z_0I-A)^{-1} & I \end{bmatrix} \cdot \begin{bmatrix} S^{-1} & & 0 \\ & & 0 \\ 0 & 0 & I \end{bmatrix} \cdot$$

$$\cdot \begin{bmatrix} -(I-z_0g) & 0 & z_0f \\ 0 & z_0I-a & -b \\ 0 & 0 & I \end{bmatrix}$$

and then verify that the two forms of $V(z)$ are equivalent by direct computation.] In particular, if we define a matrix Φ by

$$\Phi = \begin{bmatrix} S^{\times} & S_1A \\ S_2G & S \\ E & C \end{bmatrix}$$

then

$$[e\ c\ d]\Phi = [0\ 0]. \tag{3.17b}$$

Introduce the matrix functions $h(z) = (z^{-1}I - g) \oplus (zI - a) \oplus I_n$, and $H(z) = (z^{-1}I - G) \oplus (zI - A) \oplus I_n$ and the matrix

$$Q = \begin{bmatrix} fE & fC & f \\ bE & bC & b \\ E & C & I \end{bmatrix} \equiv \begin{bmatrix} f \\ b \\ I \end{bmatrix} [E\ C\ I] .$$

The Sylvester equations for S_1, S_2, S and $S^\times$ imply that

$$Q = \begin{bmatrix} S^\times G - gS^\times & S_1 - gS_1A & f \\ S_2 - aS_2G & SA - aS & b \\ E & C & I \end{bmatrix}$$

and term-by-term computation shows that

$$Q = \begin{bmatrix} I & 0 \\ 0 & I \\ 0 & 0 \end{bmatrix} \psi H(z) + h(z)\Phi \begin{bmatrix} I & 0 & 0 \\ 0 & I & 0 \end{bmatrix}$$

$$+ h(z) \begin{bmatrix} 0 & S_1 & 0 \\ S_2 & 0 & 0 \\ 0 & 0 & I \end{bmatrix} H(z).$$

where Ψ is as in Lemma 3.3.

Now from the definition of $V(z)$ and $W(z)$ we see that

$$(VW)(z) = xh^{-1}(z)\, QH^{-1}(z)X =$$

$$= x \left\{ h^{-1}(z) \begin{bmatrix} I & 0 \\ 0 & I \\ 0 & 0 \end{bmatrix} \Psi + \Phi \begin{bmatrix} I & 0 & 0 \\ 0 & I & 0 \end{bmatrix} H^{-1}(z) \right.$$

$$\left. + \begin{bmatrix} 0 & S_1 & 0 \\ S_2 & 0 & 0 \\ 0 & 0 & I \end{bmatrix} \right\} X$$

where

$$x = [e\ ,\ c\ ,\ d]\ ,\ X = \begin{bmatrix} F \\ B \\ D \end{bmatrix} .$$

Since $x\Phi = 0$ and $\psi X = 0$ we have

$$V(z)W(z) = x \begin{bmatrix} 0 & S_1 & 0 \\ S_1 & 0 & 0 \\ 0 & 0 & I \end{bmatrix} X \equiv \text{a constant.}$$

□

CONCLUSION OF THE PROOF OF THEOREM 3.1.

To show that W(z) as defined by (3.4) has $\{(C,A), (a,b), S\}$ as a c.s.s.d. over Ω_+ and $\{(E,G), (g,f), S^\times\}$ as a c.s.s.d. over Ω_-, by the definition of M and $M^\times$ given before and the direct analysis from Section 1 we need only show that $WR_n^+ = M$ and $WR_n^- = M^\times$. In Lemma 3.5 we have already observed the containments $WR_n^+ \subset M$, $WR_n^- \subset M^\times$. Since $\sigma(A) \cup \sigma(G)^{-1}$ does not intersect Γ, it is clear that W has no poles on Γ from the formula (3.4); similarly from Lemma 3.6 and formula (3.5) it is clear that $W(z)^{-1}$ is analytic on Γ. Thus W is regular on Γ and we conclude that

$$WR_n = R_n .$$

From Lemma 3.2 we know that $R_n = M \dotplus M^\times$. Thus

$$\begin{aligned} R_n &= M \dotplus M^\times \supset WR_n^+ \dotplus WR_n^- \\ &= W(R_n^+ \dotplus R_n^-) \\ &= WR_n = R_n . \end{aligned}$$

From the equality

$$M \dotplus M^\times = WR_n^+ \dotplus WR_n^-$$

together with the containments

$$M \supset WR_n^+ , \quad M^\times \supset WR_n^-$$

we get the desired equalities

$$M = WR_n^+ , \quad M^\times = WR_n^- .$$

The existence assertion in Theorem 3.1 together with the formulas (3.4) and (3.5) now follow.

Formulas (3.4a) and (3.5a) for the solution of the inverse spectral problem W(z) with value Δ at ∞ prescribed follow in a analogous way.

If $W(z) = C(zI - A)^{-1}B + D + zE(I - zG)^{-1}F$ and G is invertible, then alternatively we may write

$$W(z) = C(zI - A)^{-1}B - EG^{-1}(zI - G^{-1})^{-1}G^{-1}F + [D - EG^{-1}F]$$

and thus $W(\infty) = D - EG^{-1}F$. To solve for W(z) in the inverse spectral problem with $W(\infty) = \Delta$, we must impose the condition

$$D - EG^{-1}F = \Delta$$

in addition to

$$\Psi \begin{bmatrix} F \\ B \\ D \end{bmatrix} = \begin{bmatrix} 0 \\ 0 \end{bmatrix}$$

that is

$$\begin{bmatrix} -S^{\times} & gS_1 & f \\ aS_2 & -S & b \\ -EG^{-1} & 0 & I \end{bmatrix} \begin{bmatrix} F \\ B \\ D \end{bmatrix} = \begin{bmatrix} 0 \\ 0 \\ \Delta \end{bmatrix}. \tag{3.18}$$

Recall that S, $S^{\times}$, S_1 and S_2 satisfy Sylvester equations (3.1), (3.2), (3.16) and (3.17). From these identities one can verify by direct comportation that

$$\begin{bmatrix} -S^{\times} & gS_1 & f \\ aS_2 & -S & b \\ -EG^{-1} & 0 & I \end{bmatrix} = \begin{bmatrix} -g & 0 & f \\ 0 & I & b \\ 0 & 0 & I \end{bmatrix} \begin{bmatrix} S^{\times} & S_1 & 0 \\ S_2 & S & 0 \\ 0 & 0 & I \end{bmatrix} \cdot \begin{bmatrix} G^{-1} & 0 & 0 \\ 0 & -I & 0 \\ -EG^{-1} & 0 & I \end{bmatrix}.$$

One easily checks

$$\begin{bmatrix} G^{-1} & 0 & 0 \\ 0 & -I & 0 \\ -EG^{-1} & 0 & I \end{bmatrix}^{-1} = \begin{bmatrix} G & 0 & 0 \\ 0 & -I & 0 \\ E & 0 & I \end{bmatrix}$$

and

$$\begin{bmatrix} -g & 0 & f \\ 0 & I & b \\ 0 & 0 & I \end{bmatrix}^{-1} = \begin{bmatrix} -g^{-1} & 0 & g^{-1}f \\ 0 & I & -b \\ 0 & 0 & I \end{bmatrix}$$

and hence the solution of (3.18) is given by

$$\begin{bmatrix} F \\ B \\ D \end{bmatrix} = \begin{bmatrix} G & 0 & 0 \\ 0 & -I & 0 \\ E & 0 & I \end{bmatrix} \begin{bmatrix} S^{-1} & & 0 \\ & & 0 \\ 0 & 0 & I \end{bmatrix} \begin{bmatrix} -g^{-1} & 0 & g^{-1}f \\ 0 & I & -b \\ 0 & 0 & I \end{bmatrix} \begin{bmatrix} 0 \\ 0 \\ \Delta \end{bmatrix}$$

$$= \begin{bmatrix} \begin{bmatrix} G & 0 \\ 0 & -I \end{bmatrix} S^{-1} \begin{bmatrix} g^{-1}f \\ -b \end{bmatrix} \Delta \\ \hline \begin{bmatrix} E & 0 \end{bmatrix} S^{-1} \begin{bmatrix} g^{-1}f \\ -b \end{bmatrix} \Delta + \Delta \end{bmatrix}$$

Then

$$W(z) = \begin{bmatrix} zE(I - zG)^{-1}, C(zI - A)^{-1} \end{bmatrix} \begin{bmatrix} G & 0 \\ 0 & -I \end{bmatrix} S^{-1} \begin{bmatrix} g^{-1}f \\ -b \end{bmatrix} \Delta$$

$$+ \left\{ I + \begin{bmatrix} E & 0 \end{bmatrix} S^{-1} \begin{bmatrix} g^{-1}f \\ -b \end{bmatrix} \right\} \Delta$$

$$= \begin{bmatrix} zE(G^{-1} - zI)^{-1}, \ -C(zI - A)^{-1} \end{bmatrix} S^{-1} \begin{bmatrix} g^{-1}f \\ -b \end{bmatrix} \Delta$$

$$+ \Delta + \begin{bmatrix} E & 0 \end{bmatrix} S^{-1} \begin{bmatrix} g^{-1}f \\ -b \end{bmatrix} \Delta$$

$$= \begin{bmatrix} EG^{-1}(zI - G^{-1})^{-1}, \ C(zI - A)^{-1} \end{bmatrix} S^{-1} \begin{bmatrix} -g^{-1}f \\ b \end{bmatrix} \Delta + \Delta$$

as desired. The inverse can be computed in a similar way. □

It is interesting to note the following more "coordinate-free" formula for the solution W(z) of the inverse spectral problem in Theorem 3.1, where the point z_0 and the value Δ are not specified.

COROLLARY 3.7. *Suppose* $\{(C,A)(a,b),S\}$ *and* $\{(E,G)(g,f),S^{x}\}$ *be admissible triples of* Ω_+ *and* Ω_- *respectively, and assume that* (3.1), (3.2) *and* (3.3) *hold as in* Theorem 3.1, *with* S_1 *and* S_2 *given by* (3.3a) *and* (3.3b). *Define matrices* Ψ *and* Φ *by*

$$\Psi = \begin{bmatrix} -S^{\times} & gS_1 & f \\ aS_2 & -S & b \end{bmatrix}$$

and

$$\Phi = \begin{bmatrix} S^{\times} & S_1A \\ S_2G & S \\ E & C \end{bmatrix}$$

construct $r \times n$, $p \times n$ *and* $n \times n$ *matrices* F,B,D *such that the columns of* $\begin{bmatrix} F \\ B \\ D \end{bmatrix}$ *form a basis for the right kernel of* Ψ. *Construct* $n \times s$, $n \times g$ *and* $n \times n$ *matrices* e,c,d *so that the rows of* $[e \quad c \quad d]$ *form a basis for the left kernel of* Φ. *Then the matrix function*

$$W(z) = E(z^{-1}I - G)^{-1}F + D + C(zI - A)^{-1}B \tag{3.19}$$

is a solution of the inverse spectral problem in Theorem 3.1, *with inverse given by*

$$W(z)^{-1} = L^{-1}[e(z^{-1}I - g)^{-1}f + d + c(zI - a)^{-1}b] \tag{3.20}$$

where L *is the invertible constant* $n \times n$ *matrix given by*

$$L = eS_1B + cS_2F + dD.$$

Moreover, any solution arises in this way.

PROOF. Suppose $W(z)$ is defined as in (3.19) and $W'(z)$ is defined by (3.4) for some choice of z_0 not in $\sigma(A) \cup \sigma(a) \cup \sigma(G)^{-1} \cup \sigma(g)^{-1}$ and of invertible Δ. From (3.17a) we see that

$$W(z) = W'(z)\Gamma_1$$

for an invertible $n \times n$ (constant) matrix Γ_1. Thus W is a solution of the inverse spectral problem. If we define $V(z) = e(z^{-1}I - g)^{-1} + d + c(zI - a)^{-1}b$ where e,c,d are as in the statement of Corollary 3.7, then from (3.17b) we see that

$$V(z) = \Gamma_2W'(z)^{-1}$$

for an invertible constant $n \times n$ matrix Γ_2. Then $V(z)W(z) = \Gamma_2\Gamma_1$ is an invertible constant. A computation as in the proof of

Lemma 3.6 reveals the constant to be

$$V(z)W(z) = \begin{bmatrix} e & c & d \end{bmatrix} \begin{bmatrix} 0 & S_1 & 0 \\ S_2 & 0 & 0 \\ 0 & 0 & I \end{bmatrix} \begin{bmatrix} F \\ B \\ D \end{bmatrix} =: L$$

Corollary 3.7 now follows. □

It is also interesting to note that the four equations (3.1), (3.2), (3.16) and (3.17) for S, S^x, S_1 and S_2 can be summarized in a single generalized Sylvester equation for $\mathcal{S} = \begin{bmatrix} S^x & S_1 \\ S_2 & S \end{bmatrix}$

$$(3.21) \qquad \begin{bmatrix} I & 0 \\ 0 & a \end{bmatrix} \mathcal{S} \begin{bmatrix} G & 0 \\ 0 & I \end{bmatrix} - \begin{bmatrix} g & 0 \\ 0 & I \end{bmatrix} \mathcal{S} \begin{bmatrix} I & 0 \\ 0 & A \end{bmatrix} = \begin{bmatrix} f \\ -b \end{bmatrix} \begin{bmatrix} E & C \end{bmatrix} .$$

This suggests the following solution to a less restrictive inverse spectral problem; the special case where both G and g are invertible was obtained in [GKLR].

COROLLARY 3.8. *Let observable pairs* $(C,A),(E,G)$ *and controllable pairs* $(a,b),(g,f)$ *of matrices be given such that* $\sigma(A)$ *and* $\sigma(a)$ *are in* Ω_+ *while* $\sigma(G)$ *and* $\sigma(g)$ *are in* Ω_-^{-1}. *Then there exist a regular rational matrix function of the form*

$$W(z) = C(zI - A)^{-1}B + D + zE(I - zG)^{-1}F$$

with inverse of the form

$$W(z)^{-1} = c(zI - a)^{-1}b + d + ze(I - zg)^{-1}f$$

for some matrices B,D,F,c,d,e *if and only if there exists an invertible solution* $\mathcal{S} = \begin{bmatrix} S^x & S_1 \\ S_2 & S \end{bmatrix}$ *of the generalized Sylvester equation (3.21). If* W_1 *and* W_2 *are two solutions arising from the same* $\mathcal{S}$, *then* $W_2(z)^{-1}W_1(z)$ *is an invertible constant matrix functio function.*

PROOF. Any solution $\mathcal{S}$ of (3.21) is of the form $\mathcal{S} = \begin{bmatrix} S^x & S_2 \\ S_1 & S \end{bmatrix}$ where S, S^x, S_1 and S_2 are as in (3.1)-(3.3) in Theorem 3.1. If $\mathcal{S}$ in addition is invertible we may apply Theorem 3.1 to produce a $W(z)$ of the desired form. The uniqueness assertion follows from the uniqueness assertion in Theorem 3.1. □

REMARK. We note that the four Sylvester equations (3.1), (3.2), (3.16) and (3.17) together with the conditions

$$\Psi \begin{bmatrix} F \\ B \\ D \end{bmatrix} = \begin{bmatrix} 0 \\ 0 \end{bmatrix}$$

and

$$\begin{bmatrix} e & c & d \end{bmatrix} \Phi = \begin{bmatrix} 0 & 0 \end{bmatrix}$$

in Corollary 3.7 can be condensed into a single matrix pencil equation

$$\begin{bmatrix} -S & b & aS_2 \\ eS_1 & d & cS_1 \\ gS_2 & f & -S^x \end{bmatrix} \begin{bmatrix} A-zI & B & 0 \\ C & D & zE \\ 0 & F & zG-I \end{bmatrix} =$$

(3.22)

$$= \begin{bmatrix} zI-a & 0 & 0 \\ -c & I & -ze \\ 0 & 0 & I-zg \end{bmatrix} \begin{bmatrix} S & 0 & S_2 \\ 0 & I & 0 \\ S_1 & 0 & S^x \end{bmatrix}.$$

The pencil

$$\begin{bmatrix} A-zI & B & 0 \\ C & D & zE \\ 0 & F & zG-I \end{bmatrix}$$

has appeared in the literature before; it resembles the "zero pencil" for $W(z)$ (see [C3],[C4]). The identity (3.22) suggests a way to solve for a,b,c,d,e,f,g from A,B,C,D,E,F,G, i.e. to find a realization for $W(z)^{-1}$ from a known realization for $W(z)$. We leave this here as an open problem.

4. THE SPECIAL CASE OF MATRIX POLYNOMIALS

In this section we illustrate the theory of Section 3 for the special case of matrix polynomials and thereby pick up a couple of known results from [GLR]; other results from [GLR] (e.g. on factorization) can also be obtained in this way.

The first problem we consider is that of determining a monic matrix polynomial $P(z) = z^{\ell} I_n + z^{\ell-1} A_{\ell-1} + \ldots + zA_1 + A_0$ from its finite spectral data. We take the "finite spectral data" to be a zero pair (a,b) for $P(z)$ as defined in Section 2. Since a matrix polynomial has no finite poles, we see that a c.s.s.d.

for $P(z)$ over $\Omega_+ := \mathbb{C}$ must be simply

(4.1) $$\{(C,A),(a,b),S\} = \{(0,0),(a,b),0\}.$$

If $P(z)$ is a monic matrix polynomial, then in a neighbourhood of ∞ $P(z)$ behaves like $z^{\ell}I_n = zE(I - zG)^{-1}\tilde{F}$ where $E = [0,\ldots,0,I_n]$ and where G is the $n\ell \times n\ell$ matrix

$$G = \begin{bmatrix} 0 & & & \\ I_n & 0 & & \\ & \ddots & \ddots & \\ & & I_n & 0 \end{bmatrix}$$

and $\tilde{F} = [I_n,0,\ldots,0]^T$, while $P(z)^{-1}$ is analytic at ∞. Thus a c.s.s.d. for $P(z)$ over $\Omega_- := \{\infty\}$ is

(4.2) $$\left\{(E,G),(g,f),S^x\right\} = \\ = \left\{\left(\left[0,\ldots,0,I_n\right], \begin{bmatrix} 0 & & & \\ I_n & \ddots & & \\ & \ddots & \ddots & \\ & & I_n & 0 \end{bmatrix}\right), (0,0)0\right\}.$$

The problem of finding a monic matrix polynomial having prescribed finite spectral data is then the same problem as that addressed in Theorem 3.1 with prescribed admissible triples (4.1) over $\Omega_+ = \mathbb{C}$ and (4.2) over $\Omega_- = \{\infty\}$, but with normalization condition

(4.3) $$\lim_{z\to\infty} z^{-\ell}P(z) = I_n$$

rather than prescribing an invertible value Δ at a regular point z_0. The reader should note that Theorem 1.23 of [GLR] is the transposed version of the following result.

THEOREM 4.1 (see Theorem 1.23 in [GLR]). *Let* (a,b) *be a controllable pair. Then there exists a monic* $n \times n$ *matrix polynomial* $P(z)$ *of degree* ℓ *having* (a,b) *as a zero pair over* $\mathbb{C}$ *if and only if the* $n\ell \times n\ell$ *matrix*

(4.4) $$S_2 = [a^{\ell-1}b,\ldots,ab,b]$$

is invertible (so in particular a *has size* $n\ell \times n\ell$*). In this case the unique solution* $P(z)$ *is given by*

(4.5) $$P(z) = I_n z^{\ell} + \Sigma_{j=0}^{\ell-1} A_j z^j$$

where

(4.6) $$a^{\ell}b + a^{\ell-1}bA_{\ell-1} + \dots + abA_1 + bA_0 = 0.$$

PROOF. The existence criterion follows from Theorem 3.1 using the data (4.1) and (4.2), once we check that the associated matrix S coincides with S_2 in (4.4). From the special form of the data (4.1) and (4.2) we see immediately that S, S_1 and S^x are vacuous, so $S = S_2$ where S_2 is the unique solution of the Sylvester equation

$$S_2 - aS_2G = bE.$$

Since G is nilpotent of index ℓ, the solution S_2 is given by

$$\begin{aligned} S_2 &= \Sigma_{j=0}^{\ell-1} a^j bEG^j \\ &= [0,\dots,0,b] + a[0,\dots,b,0] + \dots + a^{\ell-1}[b,0,\dots,0] \\ &= [a^{\ell-1}b,\dots,ab,b]. \end{aligned}$$

Thus $S = S_2$ is as in (4.4) as desired. The required solution P(z) will have the form $P(z) = D + zE(I - zG)^{-1}F$ where D and F are to be determined. To obtain the solution satisfying the normalization (4.3) requires only a slight modification of the procedure developed in Section 3. From Lemma 3.3 we know that F,D must satisfy

(4.7) $$\Psi \begin{bmatrix} F \\ D \end{bmatrix} = 0.$$

where Ψ for the case at hand collapses to

$$\Psi = \left[aS_2, b\right].$$

Decompose the (unknown) matrix $F : \mathbb{C}^n \to \mathbb{C}^{n\ell}$ as $F = \left[A_{\ell}^T, A_{\ell-1}^T, \dots, A_1^T\right]^T$ and relabel the unknown matrix D as A_0. Then (4.7) can be written in more explicit form as

(4.8) $$\left[a^{\ell}b, a^{\ell-1}b, \dots, ab, b\right] \begin{bmatrix} A_{\ell} \\ A_{\ell-1} \\ \vdots \\ A_0 \end{bmatrix} = 0.$$

On the other hand, from the special form of E and G we compute

$$P(z) = D + zE(I - zG)^{-1}F =$$
$$= A_0 + zA_1 + \dots + z^{\ell-1}A_{\ell-1} + z^{\ell}A_{\ell}.$$

The requirement that P(z) be monic (i.e. 4.3) amounts to $A_\ell = I_n$. When this is plugged into (4.8) we get (4.6) as desired with P(z) as in (4.5). □

As a second illustration we consider the problem of constructing a comonic polynomial P(z) (i.e. a polynomial with $P(0) = I_n$) having a prescribed zero pair (a,b) over $\mathbb{C}$, where a is invertible. In this case the solution is determined only up to a unimodular right factor having the value I_n at 0. Without loss of generality we assume that (a,b) is controllable. We then define the *index of stabilization* of the pair (a,b) to be the first integer ℓ such that $\mathrm{Im}[b,ab,\dots,a^{\ell-1}b] = \mathbb{C}^n$. The following result should be compared to Theorem 7.16 of [GLR]; here there is a difference in the form of the solution since [GLR] takes the approach of Kronecker for analyzing the data at ∞ while we maintain the McMillan approach.

THEOREM 4.2. *Let (a,b) be a controllable pair with the matrix a invertible. Then the smallest integer ℓ for which there exists a comonic matrix polynomial P(z) of degree ℓ which has (a,b) as a zero pair over $\mathbb{C}$ is the index of stabilization of the pair (a,b). For this value of ℓ, any such polynomial P(z) has the form*

$$P(z) = I + zB_1 + \dots + z^{\ell}B_{\ell}$$

where

$$(4.9)\qquad \begin{bmatrix} g^{\ell}f, \dots, gf, f \\ b, \dots, a^{\ell-1}b, a^{\ell}b \end{bmatrix} \begin{bmatrix} I \\ B_1 \\ \vdots \\ B_\ell \end{bmatrix} = \begin{bmatrix} 0 \\ 0 \end{bmatrix}$$

and where (g,f) is chosen to be a controllable pair with g nilpotent for which the matrix

$$\hat{S} = \begin{bmatrix} g^{\ell-1}f, \dots, gf, f \\ b, \dots, a^{\ell-2}b, a^{\ell-1}b \end{bmatrix}$$

is invertible.

REMARK. For details on how to construct (g,f) in a way which guarantees that $\hat{S}$ is invertible, see [GR3].

PROOF. In this case it is easier to solve for $\tilde{P}(z) := z^{-\ell}P(z)$ rather than P(z). To formulate the problem in the framework of Theorem 3.1 we again take $\Omega_+ = \mathbb{C}$ and $\Omega_- = \{\infty\}$. Since P(z) is to be a polynomial of degree ℓ, we see that $\tilde{P}(z)$ is analytic at ∞, and hence a c.s.s.d over Ω_- for $\tilde{P}$ has the form

$$\{(E,G),(g,f),S^{x}\} = \{(0,0),(g,f),0\} \tag{4.10}$$

where g is nilpotent. Note that (g,f) is not specified by the data of the problem. Since P is to be a monic polynomial we deduce that we may take a pole pair for P over Ω_+ to be of the form (C,A) with

$$C = [0,\ldots,0,I_n]$$

and A equal to the $n\ell \times n\ell$ block Jordan matrix

$$A = \begin{bmatrix} 0 & & & \\ I_n & \ddots & & \\ & \ddots & \ddots & \\ & & I_n & 0 \end{bmatrix}$$

Since the function $z^{-\ell}I_n$ is analytic and invertible at all non-zero points in $\mathbb{C}$, we deduce that (a,b) is a zero pair over $\mathbb{C}$ for $\tilde{P}$ as well as for P. The coupling matrix S in this case is uniquely determined as the solution of the Sylvester equation

$$SA - aS = bC$$

or equivalently

$$a^{-1}SA - S = a^{-1}bC.$$

As in the proof of Theorem 4.1, we see that for A and C of the special form here we get

$$S = -\Sigma_{j=0}^{\ell-1}a^{-j}(a^{-1}bC)A^{j} = \left[-a^{-\ell}b,\ldots,-a^{-2}b,-a^{-1}b\right].$$

Thus a c.s.s.d for $\tilde{P}$ over Ω_+ has the form

$$\{(C,A),(a,b),S\} =$$

$$\left\{\left(\left[0,\ldots,0,I\right],\begin{bmatrix} 0 & & \\ I & \ddots & \\ & \ddots & \ddots \\ & & I & 0\end{bmatrix}\right),(a,b),\left[-a^{-\ell}b,\ldots,-a^{-1}b\right]\right\}$$

and is completely specified by the data of the problem. The criterium for the existence of a solution $\tilde{P}(z)$ from Theorem 3.1 is the invertibility of the matrix S; in this case S collapses to

$$S = \begin{bmatrix} S_1 \\ S \end{bmatrix}$$

where S is as in (4.11) and S_1 is the unique solution of

$$S_1 - gS_1A = fC.$$

From the special form of A and C we deduce that

$$S_1 = \left[g^{\ell-1}f, \ldots, gf, f\right]$$

and hence

$$S = \begin{bmatrix} S_1 \\ S \end{bmatrix} = \begin{bmatrix} g^{\ell-1}f, \ldots, & gf & , f \\ -a^{-\ell}b, \ldots, & -a^{-2}b, & -a^{-1}b \end{bmatrix} = \begin{bmatrix} I & 0 \\ 0 & -a^{-\ell} \end{bmatrix} \hat{S}.$$

Thus S is invertible if and only if $\hat{S}$ is invertible. A necessary condition for the invertibility of $\hat{S}$ is that the second row be surjective. Conversely, if the second row is surjective, one can construct admissible matrices g and f which makes $\hat{S}$ invertible; for details we refer to [C3,GR3]. This establishes the existence criterion.

To obtain the formula for P(z) we proceed as in the proof of Theorem 3.1. Necessarily $\tilde{P}(z)$ has the form $\tilde{P}(z) = D + C(zI - A)^{-1}B$ where D and B are to be determined. One condition on D and B is that

$$\Psi \begin{bmatrix} B \\ D \end{bmatrix} = \begin{bmatrix} 0 \\ 0 \end{bmatrix} \tag{4.12}$$

where in this case Ψ collapses to

$$\Psi = \begin{bmatrix} gS_1 & f \\ -S & b \end{bmatrix} = \begin{bmatrix} g^{\ell}f, & \ldots, & gf & ,f \\ a^{-\ell}b, & \ldots, & a^{-1}b, & b \end{bmatrix} =$$

$$= \begin{bmatrix} I & 0 \\ 0 & a^{-\ell} \end{bmatrix} \begin{bmatrix} g^{\ell}f, \ldots, & gf & , f \\ b, & \ldots, a^{\ell-1}b, & a^{\ell}b \end{bmatrix}$$

Thus, if we write $B = \left[B_0^T, \ldots, B_{\ell-1}^T\right]^T$ and $D = B_\ell$, (4.12) collapses to

$$(4.13)\qquad \begin{bmatrix} g^{\ell}f,\dots, & gf\ , & f \\ b\ ,\dots, & a^{\ell-1}b, & a^{\ell}b \end{bmatrix} \begin{bmatrix} B_0 \\ \vdots \\ B_{\ell-1} \\ B_{\ell} \end{bmatrix} = \begin{bmatrix} 0 \\ 0 \end{bmatrix}.$$

On the other hand, from the special form of A,B,C,D we see that

$$\tilde{P}(z) = D + C(zI - A)^{-1}B =$$

$$= B_{\ell} + [0,\dots,0,I] \begin{bmatrix} z^{-1}I & & & \\ z^{-2}I & \ddots & & \\ \vdots & \ddots & \ddots & \\ z^{-\ell}I & \dots z^{-2}I & & z^{-1}I \end{bmatrix} \begin{bmatrix} B_0 \\ B_1 \\ \vdots \\ B_{\ell-1} \end{bmatrix}$$

$$= B_{\ell} + z^{-1}B_{\ell-1} + \dots + z^{-\ell+1}B_1 + z^{-\ell}B_0$$

and hence

$$P(z) = z^{\ell}\tilde{P}(z) = B_0 + zB_1 + \dots + z^{\ell-1}B_{\ell-1} + z^{\ell}B_{\ell}.$$

The condition that P(z) be comonic then amounts to $B_0 = I$; with this substitution (4.13) reduces to (4.9). Note that the invertibility of $\hat{S}$ guarantees that (4.9) has a unique solution $B_1,\dots,B_{\ell}$ once an admissible (g,f) has been selected. □

5. APPLICATIONS

In general all the applications of inverse spectral problems for rational matrix functions discussed in [BR3] (viz. Wiener-Hopf factorization, J-inner-outer factorization, coprime factorization) can now be handled with a more general contour Γ and without a regularity assumption at ∞. Related results were obtained in [BR4] for the situation where the data is presented in a more detailed local form. It is also possible to analyze minimal divisors as in [BGR], but now with data at ∞ allowed. We discuss a few examples in detail.

5a. MODEL REDUCTION: DISCRETE TIME

This problem is discussed in [BR2,3]. We are given a stable (i.e. all poles inside the unit disk) $p \times q$ matrix function $G(z) = C(zI_n - A)^{-1}B$ where C is $p \times n$, A is $n \times n$ and B is $n \times q$ and $\sigma(A) \subset \mathcal{D} = \{z \in \mathbb{C} : |z| < 1\}$. We are given a tolerance level

σ such that $\sigma_{\ell+1}(G) < \sigma < \sigma_{\ell}(G)$ where $\sigma_j(G)$ $(1 \le j \le n)$ are the Hankel singular values of G (see [BR2] for details). The problem is to characterize all functions of the form $\hat{G}(z) + F(z)$, where $\hat{G}$ is rational of McMillan degree at most ℓ and where F is rational with no poles in $\mathcal{D}^-$ (the closure of $\mathcal{D}$) such that $\|G - \hat{G} - F\|_\infty \le \sigma$; here in general $\|H\|_\infty = \sup\{\|H(z)\| : |z| = 1\}$. It is known that there is a matrix function

$$\theta(z) = \begin{bmatrix} \theta_{11}(z) & \theta_{12}(z) \\ \theta_{21}(z) & \theta_{22}(z) \end{bmatrix}$$

such that all possible error functions are characterized by

$$G - \hat{G} - F = (\theta_{11}H + \theta_{12})(\theta_{21}H + \theta_{22})^{-1}$$

where H is an arbitrary rational $p \times q$ matrix function in the Schur class $S_{p\times q}$ (i.e. H has no poles in $\mathcal{D}^-$ and $\|H\|_\infty \le 1$). One of the main results from [BH1] is that the matrix function $\theta(z)$ is characterized as any rational matrix function for which

$$\theta H^2_{p+q} = LH^2_{p+q} \tag{5.1}$$

and

$$\theta(\bar{z}^{-1})^* \begin{bmatrix} I_p & 0 \\ 0 & -\sigma^2 I_q \end{bmatrix} \theta(z) = \begin{bmatrix} I_p & 0 \\ 0 & -I_q \end{bmatrix}. \tag{5.2}$$

Here $L(z) = \begin{bmatrix} I_p & G(z) \\ 0 & I_q \end{bmatrix}$ and H^2_{p+q} is the usual Hardy space of $\mathbb{C}^{p+q}$-valued functionsover the unit disk. From (5.2) and (5.1) we deduce that

$$\theta H^{2\perp}_{p+q} = J_\sigma^{-1}(\theta H^2_{p+q})^\perp = J_\sigma^{-1}(LH^2_{p+1})^\perp$$

or

$$\theta H^{2\perp}_{p+q} = J_\sigma^{-1} L^{-*} J_\sigma H^{2\perp}_{p+q}. \tag{5.3}$$

(where $L^{-*}(z) = L(\bar{z}^{-1})^{-*}$ and $J_\sigma = \begin{bmatrix} I & 0 \\ 0 & -\sigma^2 I_2 \end{bmatrix}$). Since we are dealing with rational functions we write (5.1) and (5.3) in the form

$$\theta R^+_{p+q} = LR^+_{p+q} \tag{5.4}$$

and

$$\theta R^-_{p+q} = J_\sigma^{-1} L^{-*} J_\sigma R^-_{p+q}. \tag{5.5}$$

Here R^{+}_{p+q} and R^{-}_{p+q} are as in Section 1 with Γ chosen to be the unit circle. Conversely (5.4) and (5.5) together with the normalization

$$(5.6)\qquad \theta(z_0)^* J_\sigma \theta(z_0) = J := \begin{bmatrix} I_p & 0 \\ 0 & -I_q \end{bmatrix}$$

for some z_0 on the unit circle imply (5.1) and (5.2). Thus the construction of θ reduces to an inverse spectral problem (5.4) and (5.5) with an added normalization condition (5.6). Note that (5.4) means that a c.s.s.d. for θ over $\Omega_+ = \mathcal{D}$ should agree with a c.s.s.d. for L over Ω_+, while (5.5) means that a c.s.s.d. for θ over Ω_- should agree with a c.s.s.d. for $J_\sigma^{-1} L^{-*} J_\sigma$ over Ω_-.

Explicit realization formulas for θ were obtained in [BR2] with the assumption that G(z) have no pole at 0, or equivalently, that A be invertible. We now use the machinery of this paper to obtain an explicit formula for θ for the general case where A is not necessarily invertible.

From $G(z) = C(zI - A)^{-1}B$ we get

$$L(z) = \begin{bmatrix} I & 0 \\ 0 & I \end{bmatrix} + \begin{bmatrix} C \\ 0 \end{bmatrix} (zI - A)^{-1}[0,B] =$$

$$= I_{p+q} + \underline{C}(zI - \underline{A})^{-1}\underline{B}$$

with

$$\underline{A}^{\times} := \underline{A} - \underline{B}\underline{C} = A.$$

Hence by Theorem 2.5b, since $\sigma(\underline{A}) = \sigma(\underline{A}^{\times}) \subset \Omega_+ = \mathcal{D}$, we conclude that a c.s.s.d. for L over Ω_+ (and hence also for θ over Ω_+) is

$$(5.7)\qquad \left\{\underline{C},\underline{A},\underline{a},\underline{b},S\right\} = \left\{\begin{bmatrix} C \\ 0 \end{bmatrix}, A, A, [0,B], I\right\}.$$

Next compute

$$J_\sigma^{-1} L^{-*}(z) J_\sigma =$$

$$= J_\sigma^{-1}\left\{\begin{bmatrix} I & 0 \\ 0 & I \end{bmatrix} + z\begin{bmatrix} 0 \\ B^* \end{bmatrix}(I - zA^*)^{-1}[C^*,0]\right\}^{-1} J_\sigma$$

$$= \begin{bmatrix} I & 0 \\ 0 & I \end{bmatrix} + \begin{bmatrix} 0 \\ \sigma^{-2}B^* \end{bmatrix}(I - zA^*)^{-1}[C^*,0].$$

By a computation as in the proof of Theorem 2.6b one sees that a c.s.s.d. for $J_\sigma^{-1}L^{-*}J_\sigma$ over Ω_- (and hence also for θ over Ω_-) is

(5.8) $$\left\{\underline{E},\underline{G},\underline{g},\underline{f},S^x\right\} = \left\{\begin{bmatrix} 0 \\ \sigma^{-2}B^* \end{bmatrix}, A^*, A^*, [C^*,0], I\right\}$$

From the Lyapunov equation (3.21) and (3.22) for S_1 and S_2 we get

$$S_1 - A^*S_1A = C^*C$$

and

$$S_2 - AS_2A^* = \sigma^{-2}BB^*$$

from which we get

$$S_1 = \sum_{j=0}^{\infty} A^{*j}C^*CA^j =: \hat{Q}$$

and

$$S_2 = \sigma^{-2} \sum_{j=0}^{\infty} A^jBB^*A^{*j} =: \sigma^{-2}\hat{P}$$

where $\hat{Q}$ and $\hat{P}$ are the observability and controllability gramians. Then the matrix S assumes the form

$$S = \begin{bmatrix} I & \hat{Q} \\ \sigma^{-2}\hat{P} & I \end{bmatrix}$$

$$= \begin{bmatrix} I & \hat{Q} \\ 0 & I \end{bmatrix}\begin{bmatrix} (I - \sigma^{-2}\hat{Q}\hat{P}) & 0 \\ 0 & I \end{bmatrix}\begin{bmatrix} I & 0 \\ \sigma^{-2}\hat{P} & I \end{bmatrix}$$

and hence invertibility of S is equivalent to that of $I - \sigma^{-2}\hat{Q}\hat{P}$; it is known (see [BR2]) that this is quaranteed by the condition $\sigma_{\ell+1}(G) < \sigma < \sigma_\ell(G)$. We can then calculate S^{-1} explicitly as

(5.9) $$S^{-1} = \begin{bmatrix} Z & -Z\hat{Q} \\ -\sigma^{-2}\hat{P}Z & Z^* \end{bmatrix}$$

where

$$Z = (I - \sigma^{-2}\hat{Q}\hat{P})^{-1}.$$

Note that the normalization (5.6) is satisfied if we demand that

(5.10) $$\theta(z_0) = \begin{bmatrix} I_p & 0 \\ 0 & \sigma^{-1}I_q \end{bmatrix}$$

for some z_0 on the unit circle. The formula for θ from the data (5.7), (5.8) and (5.10) is obtained simply by plugging into formula (3.4) and using (5.9) for S^{-1} and setting

$\Delta = \begin{bmatrix} I & 0 \\ 0 & \sigma^{-1}I \end{bmatrix}$ with $|z_0| = 1$. For the case where A is invertible, the result is even more explicit than that obtained in [BR2] since one is not required to perform a Cholesky factorization. Also the flexibility in the choice of the point z_0 on the unit circle may be useful for numerical implementation. The result is as follows.

THEOREM 5.1. *Suppose* $G(z) = C(zI_n - A)^{-1}B$ *is a minimal realization of the rational* $p \times q$ *matrix function where* $\sigma(A)$ *is contained in the unit disk. Choose* σ *so that* $\sigma_{\ell+1}(G) < \sigma < \sigma_\ell(G)$. *Then the collection of functions* $\hat{G} + F$ *such that*

(i) $\hat{G}$ *has McMillan degree* ℓ

(ii) F *is rational with no poles in the closed unit disk*

and

(iii) $\|G - \hat{G} - F\|_\infty \le \sigma$

is given by

$$\hat{G} + F = G - (\theta_{11}H + \theta_{12})(\theta_{21}H + \theta_{22})^{-1}$$

where H *is an arbitrary rational* $p \times q$ *matrix function with no poles in the unit disk and with* $\|H\|_\infty \le 1$, *and where* $\theta_{ij}(z)$ $(i,j = 1,2)$ *is given by*

$$\theta_{11}(z) = 1 + \sigma^{-2}C[I - (z_0I - A)(zI - A)^{-1}]\hat{P}Z(I - z_0A^*)^{-1}C^*$$

$$\theta_{12}(z) = \sigma^{-1}C[(z_0I - A)(zI - A)^{-1} - I]Z^*(z_0I - A)^{-1}B$$

$$\theta_{21}(z) = \sigma^{-2}B^*[z(I - z_0A^*)(I - zA^*)^{-1} - z_0I]Z(I - z_0A^*)^{-1}C^*$$

$$\theta_{22}(z) = \sigma^{-1}I - \sigma^{-3}B^*[z(I - z_0A^*)(I - zA^*)^{-1} - z_0I]Z\hat{Q}(z_0I - A)^{-1}B.$$

Here z_0 *can be any point in* $\mathbb{C}$ *with* $|z_0| = 1$.

5b. MODEL REDUCTION: CONTINUOUS TIME

This problem is exactly the same as in the discrete time version discussed above but with the contour Γ now taken to be the imaginary axis $\{iy : y \in \mathbb{R}\}$ and the "inside" region Ω_+ with respect to Γ taken to be the left half plane $\{z \in \mathbb{C} : \mathrm{Re}\, z < 0\}$. While Theorem 3.1 is stated only for finite contour which include zero in Ω_+, in fact the final result applies to an infinite contour Γ such as the imaginary axis. To see this, choose a finite contour Γ' which has zero in the associated region Ω'_+ and has the property that $\sigma(A)$ and $\sigma(a)$ are in Ω'_+ and $\sigma(G)$ and $\sigma(g)$ are in Ω'_-.

The continuous time model reduction problem is as follows (see [BR1], [BR2] and [G1] for details). We are given a rational matrix function $G(z) = C(zI - A)^{-1}B$ having all poles in the left half plane Ω_+; thus if we assume that (C,A,B) is a minimal realization of $G(z)$, we have that $\sigma(A) \subset \Omega_+$:= the left half plane. We are given a tolerance level σ such that $\sigma_{\ell+1}(G) < \sigma < \sigma_\ell(G)$ where $\sigma_j(G)$ $(1 \le j \le n)$ are the (continuous time) Hankel singular values of G. The problem is to characterize all functions of the form $\hat{G}(z) + F(z)$, where $\hat{G}$ is rational of McMillan degree at most ℓ and where F is rational with no poles in the closed right half plane. As in the discrete time case, it follows from results in [BH1] (after a linear fractional change of variable) that there is a $(p + q) \times (p + q)$ matrix function

$$\theta(z) = \begin{bmatrix} \theta_{11}(z) & \theta_{12}(z) \\ \theta_{21}(z) & \theta_{22}(z) \end{bmatrix}$$

such that the set of all possible error functions can be parametrized by the formula

$$G - \hat{G} - F = (\theta_{11}H + \theta_{12})(\theta_{21}H + \theta_{22})^{-1}$$

where H is rational and analytic on the left half plane with $\|H\|_\infty := \sup\{\|H(iy)\| : y \in \mathbb{R}\} \le 1$. Moreover the matrix function $\theta(z)$ can be shown to be the solution of the inverse spectral problem

$$\theta R^+_{p+q} = LR^+_{p+q}$$

$$\theta R^-_{p+q} = J_\sigma^{-1}L^{-*}J_\sigma R^-_{p+q}$$

with the normalization side condition

$$\theta(\infty) = \begin{bmatrix} I_p & 0 \\ 0 & \sigma I_q \end{bmatrix}$$

Here $L(z) = \begin{bmatrix} I_p & G(z) \\ 0 & I_q \end{bmatrix}$, $L^{-*}(z) := L(-\bar{z})^{-*} = \begin{bmatrix} I_p & 0 \\ -G(-\bar{z})^* & I_p \end{bmatrix}$,

$J_\sigma = \begin{bmatrix} I_p & 0 \\ 0 & -\sigma^2 I_q \end{bmatrix}$, R^+_{p+q} refers to rational $\mathbb{C}^{p+q}$-valued function analytic on the closed left half plane and R^-_{p+q} refers to such functions which are analytic on the closed right half plane and vanish at infinity. Using the realization $G(z) = C(zI - A)^{-1}B$, we get

$$L(z) = \begin{bmatrix} I & 0 \\ 0 & I \end{bmatrix} + \begin{bmatrix} C \\ 0 \end{bmatrix}(zI - A)^{-1}\begin{bmatrix} 0 & B \end{bmatrix}$$

and

$$J_\sigma^{-1} L^{-*}(z) J_\sigma = \begin{bmatrix} I & 0 \\ 0 & I \end{bmatrix} + \begin{bmatrix} 0 \\ -\sigma^{-2}B^* \end{bmatrix}(zI + A^*)^{-1}\begin{bmatrix} -C^*, 0 \end{bmatrix}.$$

By Theorem 2.5b), a c.s.s.d. for L over Ω_+ (and hence also for θ over Ω_+) is

(5.11) $$\{(\underline{C},\underline{A}),(\underline{a},\underline{b}),S\} = \{(\begin{bmatrix} C \\ 0 \end{bmatrix},A),(A,\begin{bmatrix} 0,B \end{bmatrix}),I\}$$

while a c.s.s.d. for $J_\sigma^{-1} L^{-*} J_\sigma$ over Ω_- (and hence also for θ over Ω_-) by Theorem 2.6b is

(5.12) $$\{(\underline{E},\underline{G}),(\underline{g},\underline{f}),S^x\} = \{(\begin{bmatrix} 0 \\ \sigma^{-2}B^*A^{-*} \end{bmatrix},-A^{-*}),(-A^{-*},\begin{bmatrix} A^{-*}C, 0 \end{bmatrix}),I\}.$$

The matrix S_1 is determined from the Lyapunov equation (3.21)

$$S_1 - \underline{g}S_1\underline{A} = \underline{f}\,\underline{C}$$

which for our case becomes

$$S_1 + A^{-*}S_1A = A^{-*}C^*C$$

which may be rewritten as

$$A^*S_1 + S_1A = C^*C.$$

This means that $S_1 = -\hat{Q}$ where $\hat{Q}$ is the (continuous time) observability gramian for the realization (A,B,C) of the transfer function G(z). Similarly the Lyapunov equation (3.22) for S_2 leads to

$$S_2 + AS_2A^{-*} = \sigma^{-2}BB^*A^{-*}$$

which implies that $S_2 = -\sigma^{-2}\hat{P}$ where $\hat{P}$ is the controllability gramian. Then the matrix $S = \begin{bmatrix} S^x & S_1 \\ S_2 & S \end{bmatrix}$ takes the form

$$S = \begin{bmatrix} I & -\hat{Q} \\ -\sigma^{-2}\hat{P} & I \end{bmatrix} = \begin{bmatrix} I & -\hat{Q} \\ 0 & I \end{bmatrix}\begin{bmatrix} I-\sigma^{-2}\hat{Q}\hat{P} & 0 \\ 0 & I \end{bmatrix}\begin{bmatrix} I & 0 \\ -\sigma^{-2}\hat{P} & I \end{bmatrix}.$$

The invertibility of S is equivalent to that of $I - \sigma^{-2}\hat{Q}\hat{P}$; this in turn is guaranteed by the hypothesis that $\sigma_{\ell+1}(G) < \sigma < \sigma_{\ell}(G)$. Then S^{-1} can be computed explicitly as

$$S^{-1} = \begin{bmatrix} Z & Z\hat{Q} \\ \sigma^{-2}\hat{P}Z & Z^{*} \end{bmatrix} \tag{5.13}$$

where $Z = (I - \sigma^{-2}\hat{P}\hat{Q})^{-1}$. We can now apply formula (3.4a) with $\Delta = \begin{bmatrix} I_p & 0 \\ 0 & \sigma^{-1}I_q \end{bmatrix}$ to produce the solution $\theta(z)$ of the inverse spectral problem having $\theta(\infty) = \begin{bmatrix} I_p & 0 \\ 0 & \sigma^{-1}I_q \end{bmatrix}$. The result is

$$\theta(z) = \begin{bmatrix} I & 0 \\ 0 & \sigma^{-1}I \end{bmatrix} +$$

$$+ \begin{bmatrix} 0 & C \\ -\sigma^{-2}B^{*} & 0 \end{bmatrix}\begin{bmatrix} (zI+A^{*})^{-1} & 0 \\ 0 & (zI-A)^{-1} \end{bmatrix}\begin{bmatrix} Z & Z\hat{Q} \\ \sigma^{-2}\hat{P}Z & Z^{*} \end{bmatrix}\begin{bmatrix} C^{*} & 0 \\ 0 & \sigma^{-1}B \end{bmatrix}$$

$$= \begin{bmatrix} I+\sigma^{-2}C(zI-A)^{-1}\hat{P}ZC^{*} & \sigma^{-1}C(zI-A)^{-1}Z^{*}B \\ -\sigma^{-2}B^{*}(zI+A^{*})^{-1}ZC^{*} & \sigma^{-1}I-\sigma^{-3}B(zI+A^{*})^{-1}Z\hat{Q}B \end{bmatrix}.$$

This result agrees with that obtained in [BR1] by a less direct factorization process, where it was also shown that this formula is also consistent with that obtained by Glover [Gl] by yet another approach. Also in [BR2] the formula was obtained by a linear fractional change of variable applied to the discrete time version.

REFERENCES

[BGR] Ball, J.A., Gohberg, I and Rodman, L.: Minimal factorization in terms of spectral data, Integral Equations and Operator Theory, 10 (1987), 309-348.

[BH1] Ball, J.A. and Helton, J.W.: A Beurling-Lax theorem for the Lie group U(m,n) which contains most classical interpolation theory, J. Operator Theory 9 (1983), 107-142.

[BH2] Ball, J.A. and Helton, J.W.: Beurling-Lax representations using classical Lie Groups with many applications II: GL(n,$\mathbb{C}$) and Wiener-Hopf factorization, Integral Equations and Operator Theory 7 (1984), 291-309.

[BR1] Ball, J.A. and Ran, A.C.M.: Hankel norm approximation of a rational matrix function in terms of its realization, in Modelling, Identification and Robust Control, C.I. Byrnes and A. Lindquist (eds.), North Holland, 1986, 285-296.

[BR2] Ball, J.A. and Ran, A.C.M.: Optimal Hankel norm model reductions and Wiener-Hopf factorization I: The canonical case, SIAM J. Control and Opt. 25 (1987), 362-382.

[BR3] Ball, J.A. and Ran, A.C.M.: Global inverse spectral problems for rational matrix functions, Linear Algebra and Applications 86 (1987), 237-282.

[BR4] Ball, J.A. and Ran, A.C.M.: Local inverse spectral problems for rational matrix functions, Integral Equations and Operator Theory, 10 (1987), 349-415.

[CG] Clancey, K, and Gohberg, I.: Factorization of Matrix Functions and Singular Integral Operators, OT3, Birkhäuser, Basel, 1981.

[C1] Cohen, N.: On minimal factorizations of rational matrix functions, Integral Equations and Operator Theory 6 (1983), 647-671.

[C2] Cohen, N.: Spectral analysis of regular matrix polynomials, Integral Equations and Operator Theory 6 (1983), 161-183.

[C3] Cohen, N.: On spectral analysis and factorization of rational matrix functions, Ph. D. Thesis, Weizmann Institute of Science, Rehovot, Isreal, 1984.

[C4] Cohen, N.: Realization of nonproper rational matrices, to appear: Int. J. Control.

[G1] Glover, K.: All optimal Hankel-norm approximations of linear multivariable systems and their L^∞-error bounds, Int. J. Control 39 (1984), 1115-1193.

[GLR] Gohberg, I., Lancaster, P. and Rodman, L.: Matrix polynomials, Academic Press, New York etc., 1982.

[GKLR] Gohberg, I., Kaashoek, M.A., Lerer, L. and Rodman, L.: Minimal divisors of rational matrix functions with prescribed zero and pole structure, in: Topics in Operator Theory Systems and Networks, H. Dym and I. Gohberg (eds.) OT12, Birkhäuser, Basel, 1984.

[GKvS] Gohberg, I., Kaashoek, M.A. and Van Schagen, F.: Rational matrix and operator functions with prescribed singularities, Integral Equations and Operator Theory 5 (1982), 673-717.

[GR1] Gohberg, I. and Rodman, L.: Analytic matrix functions with prescribed local data, J. de Analyse Mathematique 40 (1981), 90-128.

[GR2] Gohberg, I. and Rodman, L.: Interpolation and local data for meromorphic matrix and operator functions, Integral Equations and Operator Theory 9 (1986), 60-94.

[GR3] Gohberg, I. and Rodman, L.: On spectral analysis of nonmonic matrix and operator polynomials II. Dependence on the finite spectral data, Isreal J. Math. 30 (1978), 321-334.

[K] Kailath, T.: Linear systems, Prentice Hall Inc., Englewood Cliffs N.J., 1980.

[H] Halmos, P.: Shifts on Hilbert space, J. Reine Angew. Math. 208 (1961), 102-112.

[M] McMillan, B.: Introduction to formal realizability theory I.II. Bell Sys. Tech. J. 31 (1952), 541.

J.A. Ball
Department of Mathematics
Virginia Polytechnic Institute and State University
Blacksburg, Virginia 24061, USA

N. COHEN
Department of Mathematics
Michigan State University
Wells Hall, East Lansing
Michigan 48824-1027, USA

A.C.M. Ran
Faculteit Wiskunde en Informatica
Vrije Universiteit
de Boelelaan 1081
1081 HV Amsterdam, The Netherlands

Operator Theory:
Advances and Applications, Vol. 33

UNITARY RATIONAL MATRIX FUNCTIONS

Daniel Alpay and Israel Gohberg

This paper contains the theory of realization and minimal factorization of rational matrix valued functions unitary on the unit circle or on the imaginary line (in the framework of a, generally speaking, indefinite scalar product). The Blaschke-Potapov decomposition for unitary or J-inner rational matrix-valued functions is obtained as a corollary, and the inertia theorems are explained from the point of view of the theory of rational matrix-valued functions. The main results are also used to obtain decomposition theorems for selfadjoint matrix-valued functions.

Table of Contents

1. INTRODUCTION.

In this paper we analyze rational matrix valued functions which are unitary on the imaginary line or on the unit circle. The selfadjoint case is also considered. Such matrices appear in many problems of system theory and interpolation theory and have been investigated in many papers. We would like to mention the papers [BR], [DLS], [D], [GVKDM], [G], [GLR1] [KL] and [S], with which this paper has common points. We identify these connections in the text.

The main topic here is the description of realizations and factorizations of unitary matrix-valued functions in an indefinite inner product. Special attention is paid to minimal factorization theorems. The results of [BGK] on minimal factorizations of rational matrix-valued functions play an important role here. In case of unitary or J-inner rational functions the factorization theorem gives a simple way to obtain the Blaschke-Potapov decomposition. The results obtained are also applied to the analysis and additive minimal decompositions of selfadjoint matrix-valued functions. An interesting connection is found between the realization of unitary rational matrix functions and existing inertia theorems from linear algebra. Special realizations provided by the finite-dimensional de Branges spaces are also described.

Applications of some of the results to matrix orthogonal polynomials will be presented elsewhere.

This paper is an extended version of a part of a preprint [AG] which was in circulation for some time.

It is a pleasure to thank R.L. Ellis for his help in the preparation of the final version of this paper.

The paper consists of six sections. The second section concerns rational matrix functions $U(\lambda)$ which are unitary on the imaginary line with respect to an indefinite scalar product. These functions are defined by the equality $U(\lambda)JU(\lambda)^* = J, \lambda \in i\mathbb{R}$, for an appropriate choice of the signature matrix $J(J^* = J; J^2 = I)$. A special subsection is dedicated to the definite case $J = I$: in this case, the matrix function can be completely factorized. The third section contains results, parallel to those of the second section, for rational matrix functions J-unitary on the unit circle. Rational matrix functions which are selfadjoint on the unit circle or on the imaginary line are analyzed in the fourth and fifth sections. We use a method of reduction to the J-unitary case for a special J. Connections with the finite-dimensional de Branges spaces [dB] and a special realization, as they appear in [AD], are contained in the sixth section.

We now review some properties of matrix-valued rational function and set the notation; we refer to the monograph [BGK] for more information on these functions.

Let W be a rational matrix-valued function analytic at infinity. Then W admits realizations of the form

$$W(\lambda) = D + C(\lambda I_n - A)^{-1}B \tag{1.1}$$

where A, B, C, D are matrices of adequate size and where I_n denotes the identity in the space $C_{n\times n}$ of $n \times n$ matrices with complex entries. The smallest n in (1.1) is called the MacMillan degree of W, and if n is equal to the MacMillan degree of W, then the realization (1.1) is called minimal. The realization is minimal if and only if the following two conditions hold: the pair (C, A) is observable, that is,

$$\bigcap_{k=0}^{\infty} \ker CA^k = \{0\}$$

and the pair (A, B) is controllable, that is,

$$\text{Span} \bigcup_{k=0}^{\infty} \text{Im} BA^k = C_n$$

where C_n denotes the space of n dimensional column vectors with complex components.

The matrices A, B, C in (1.1) are in general not unique, but when the realization is minimal, they are unique up to similarity: if (1.1) is a minimal realization of W and $W(\lambda) = D + C_1 (\lambda I_n - A_1)^{-1} B_1$ is another minimal realization, then there exists a unique invertible matrix S such that

$$C = C_1 S, \quad A = S^{-1}A_1 S, \quad B = S^{-1}B_1$$

The matrix S is given by each of the formulas

$$S = \left[\text{col}\left(C_1 A_1^j\right)_0^{n-1}\right]^+ \left[\text{col}(CA^j)_0^{n-1}\right]$$

and

$$S = \left[\text{row} \left(A_1^j B_1 \right)_0^{n-1} \right] \left[\text{row} \left(A^j B \right)_0^{n-1} \right]^\dagger$$

where the symbol + indicates a left inverse and the symbol † indicates a right inverse.

Let $W(\lambda) = D + C(\lambda I_n - A)^{-1} B$ be a realization of the rational matrix function W and suppose D is invertible. Then

$$W^{-1}(\lambda) = D^{-1} - D^{-1} C \left(\lambda I_n - A^{\times}\right)^{-1} B D^{-1}$$

is a realization of the function W^{-1}, where

$$A^{\times} = A - BD^{-1}C$$

If one of these realizations is minimal, then the other is minimal also. The operator $A^{\times}$ plays an important role in the description of all minimal factorizations of W, i.e., of all representations $W = W_1 W_2$, where W_1 and W_2 are rational $C_{m\times m}$ valued functions invertible at infinity such that

$$\deg W = \deg W_1 + \deg W_2$$

where $\deg W$ denotes the MacMillan degree of W (see [BGK], p.83). We now recall this description. A projection π is called a supporting projection if

$$\operatorname{Ker} \pi = M \qquad \operatorname{Im} \pi = N$$

where M and N are such that

$$\begin{array}{ll} M + N = C_n, & M \cap N = \{0\} \\ AM \subset M, & A^{\times} N \subset N \end{array}$$

Let $D = D_1 D_2$ be a decomposition of D into two invertible matrices and π a supporting projection. Then $W = W_1 W_2$ is a minimal factorization of W, with

$$W_1(\lambda) = D_1 + C(\lambda I_n - A)^{-1} (I_n - \pi) B D_2^{-1}$$

$$W_2(\lambda) = D_2 + D_1^{-1} C \pi (\lambda I_n - A)^{-1} B$$

and any minimal factorization can be obtained in such a way.

Moreover, for a fixed decomposition $D = D_1 D_2$ there is a one to one correspondence between supporting projections and minimal factorizations of W. Uniqueness up to similarity of the minimal realization and the description of all the minimal factorizations are two results which we will often use in the paper.

Finally, $\mathbb{D}$ will denote the open unit disk, $i\mathbb{R}$ the imaginary line, and C_+ (resp. C_-) the open right (resp. left) half plane. The matrix A^* will denote the adjoint of the matrix A, and $F(\lambda)^*$ will denote $(F(\lambda))^*$.

2. RATIONAL FUNCTIONS J-UNITARY ON THE IMAGINARY LINE

In this section we study matrix-valued rational functions which are J-unitary on the imaginary line. The main topics are realizations and minimal factorizations of such functions.

From now on, the term line will stand for imaginary line.

2.1. Realization Theorem and Examples

Let us recall that a $C_{m\times m}$ valued function is J-unitary on the line for some signature matrix J $\left(J = J^* = J^{-1}\right)$ if

$$U(\lambda)JU(\lambda)^* = J \tag{2.1}$$

at all points λ of $i\mathbb{R}$ where it is defined.

From equation (2.1) it follows that

$$U(\lambda)JU^*(-\lambda^*) = J$$

for any point λ in C which is not a pole or a zero of U. In particular, if λ is a singular point of U (that is, a zero or a pole), then the point $-\lambda^*$ is a singular point of U. More precisely, using the Smith form at the point λ (see [GLR2] p. 193), we see that if λ is a singular point of U with partial multiplicities

$$k_1 \geq k_2 \geq \cdots \geq k_r > 0 \geq k_{r+1} \geq \cdots \geq k_s$$

then $-\lambda^*$ is also a singular point of U with partial multiplicities

$$-k_s \geq -k_{s-1} \geq \cdots \geq -k_{r+1} \geq 0 > -k_r \geq \cdots \geq -k_1$$

THEOREM 2.1. *Let U be a matrix-valued rational function analytic at infinity, and let $U(\lambda) = D + C\left(\lambda I_n - A\right)^{-1}B$ be a minimal realization of U. Then U is J- unitary on the imaginary line if and only if the following conditions hold:*

a) D is J- unitary

b) There exists an invertible hermitian solution H of the Lyapunov equation

$$A^*H + HA = -C^*JC \tag{2.2}$$

and

$$B = -H^{-1}C^*JD\ . \tag{2.3}$$

We note that property b) is equivalent to

b') There exists an invertible hermitian matrix H such that

$$H^{-1}A^* + AH^{-1} = -BJB^* \tag{2.4}$$

and

$$C = -DJB^*H \tag{2.5}$$

PROOF. It is clear that $D = U(\infty)$; hence D is J-unitary. Equation (2.1) may be rewritten as

$$U^{-1}(\lambda) = JU(\lambda)^*J \tag{2.6}$$

and, since a minimal realization of $U^{-1}(\lambda)$ is

$$U^{-1}(\lambda) = D^{-1} - D^{-1}C\left(\lambda I_n - A^\times\right)^{-1}BD^{-1}$$

with

$$A^\times = A - BD^{-1}C$$

(see [BGK], p. 63-64), equation (2.6) may be written as

$$D^{-1} - D^{-1}C\left(\lambda I_n - A^\times\right)^{-1}BD^{-1} = J\left(D^* + B^*(-\lambda I_n - A^*)^{-1}C^*\right)J \tag{2.7}$$

Equation (2.7) is an equality between two minimal realizations of a given rational function and thus ([BGK], p. 65) there exists a uniquely defined invertible matrix X such that

$$JB^*X = D^{-1}C \qquad X^{-1}C^*J = BD^{-1} \tag{2.8}$$

$$X^{-1}A^*X = -A^\times \tag{2.9}$$

These equations are also satisfied by X^*, and hence by the uniqueness of the similarity matrix, $X = -X^*$. Setting $X = -H$, we obtain equations (2.2) and (2.3) by a straightforward calculation from equations (2.8) and (2.9).

Conversely, let D be J-unitary and H an hermitian invertible matrix satisfying (2.2) for some matrices A, C, J, with J being a signature matrix. Then, with B as in (2.3) and $U(\lambda) = D + C\left(\lambda I_n - A\right)^{-1}B$, we have, for λ and ω in the resolvent set of A,

$$\begin{aligned}U(\lambda)JU(\omega)^* &= \left(D + C(\lambda I_n - A)^{-1}B\right)J\left(D^* + B^*(\omega^* I_n - A^*)^{-1}C^*\right)\\ &= DJD^* + C(\lambda I_n - A)^{-1}BJD^* + DJB^*(\omega^* I_n - A^*)^{-1}C^*\\ &\quad + C(\lambda I_n - A)^{-1}BJB^*(\omega^* I_n - A^*)^{-1}C^*\end{aligned}$$

Replacing DJD^* by J, BJB^* by $-\left(AH^{-1}-H^{-1}A^*\right)$, and BJD^* by $-H^{-1}C^*$, we obtain

$$U(\lambda)JU(\omega)^*=J-(\lambda+\omega^*)C\left(\lambda I_n-A\right)^{-1}H^{-1}\left(\omega^*I_n-A^*\right)^{-1}C^* \tag{2.10}$$

From (2.10) it follows that U is J-unitary on the line. □

Let U be a rational matrix valued function J-unitary on the line, with minimal realization $D+C\left(\lambda I_n-A\right)^{-1}B$. It follows from the proof of the theorem that the solution H of the Lyapunov equation (2.2) is uniquely defined by the given minimal realization of U. The matrix H will be called the *associated hermitian matrix* (associated with the given minimal realization).

There are explicit formulas for the similarity matrix (see [BGK], p. 66) and thus H is also given by each of the formulas

$$H=-\left[\operatorname{col}\left(JB^*(A^*)^j\right)_0^{n-1}\right]^+\left[\operatorname{col}\left(D^{-1}C(A^\times)^j\right)_0^{n-1}\right] \tag{2.11}$$

and

$$H=\left[\operatorname{row}\left((-A^*)^jC^*J\right)_0^{n-1}\right]\left[\operatorname{row}\left((A^\times)^jBD^{-1}\right)_0^{n-1}\right]^\dagger \tag{2.12}$$

where the symbol $+$ indicates a left inverse and the symbol $\dagger$ indicates a right inverse.

We remark that, for any J-unitary matrix D and any matrices A, B, C satisfying equations (2.2) and (2.5) for some not necessarily invertible hermitian matrix H, the equality $U(\lambda)=D+C\left(\lambda I_n-A\right)^{-1}B$ defines a rational function such that, for λ and ω in the resolvent set of A,

$$U(\omega)^*JU(\lambda)=J-(\lambda+\omega^*)B^*\left(\omega^*I_n-A^*\right)^{-1}H\left(\lambda I_n-A\right)^{-1}B \tag{2.13}$$

From this equality it follows that $U(\lambda)$ is J-unitary on the line. The realization of $U(\lambda)$ in general may not be minimal.

If D is J-unitary and A, B, C satisfy equations (2.3) and (2.4), where H^{-1} is replaced by some not necessarily invertible hermitian matrix Y, then the function $U(\lambda)=D+C\left(\lambda I_n-A\right)^{-1}B$ is also a realization, not necessarily minimal, of a rational function J-unitary on the line. This will follow from the fact that equation (2.10) is valid for $U(\lambda)$ with H^{-1} replaced by Y.

Finally, we note that when $H=I_n$, the function U is the Livsic-Brodskii characteristic operator function [Br] of the operator A (for $D=I_m$).

We now give some examples of rational functions J-unitary on the line.

EXAMPLE 2.1. *Let P be an element of $C_{m\times m}$ such that*

$$PJP^* = P$$

and let ω be a complex number with $Re\omega \neq 0$. Then the function

$$U(\lambda) = I_m - PJ + \left(\frac{\lambda - \omega}{\lambda + \omega^*}\right) PJ \tag{2.14}$$

is J-unitary on the line.

A typical example of P is given by

$$P = \frac{uu^*}{u^*Ju}$$

where u is not a J-neutral vector $(u^*Ju \neq 0)$.

EXAMPLE 2.2. *Let u_1 and u_2 be two vectors such that*

$$u_1^*Ju_1 = u_2^*Ju_2 = 0$$

$$u_1^*Ju_2 \neq 0$$

and define, for $i \neq j$,

$$W_{ij} = u_i \left(u_j^*Ju_i\right)^{-1} u_j^*J \ .$$

Let ω_1 and ω_2 be two points that are not purely imaginary. Then the function

$$U(\lambda) = I_m + \left(\frac{\lambda - \omega_2}{\lambda + \omega_1^*} - 1\right) W_{12} + \left(\frac{\lambda - \omega_1}{\lambda + \omega_2^*} - 1\right) W_{21} \tag{2.15}$$

is J-unitary on the line (see [AD]).

EXAMPLE 2.3. *Let α be a purely imaginary number, n a positive integer, and u a J-neutral vector. Then the function*

$$U(\lambda) = I_m + \frac{iuu^*J}{(\lambda - \alpha)^{2n}} \tag{2.16}$$

is J-unitary on the line.

Theorem 2.1 enables us to consider some inverse problems for rational functions which are J-unitary on the line. Let us first recall that if $U(\lambda) = D + C\left(\lambda I_n - A\right)^{-1} B$ is a minimal realization of a rational function, then the pair (C, A) (resp. (A, B)) characterizes the left (resp. right) pole structure of U (for details, see [GKLR]). The inverse problems we consider consist of reconstructing a rational function J-unitary on the line from its left or its right pole structure.

THEOREM 2.2. *Let (C, A) be an observable pair of matrices and let J be a signature matrix. Then there exists a rational function J-unitary on the line with minimal realization $D + C(\lambda I_n - A)^{-1} B$ if and only if the Lyapunov equation*

$$A^*H + HA = -C^*JC \tag{2.2}$$

has a solution H which is both invertible and hermitian. If such a solution H exists, possible choices of D and B are

$$D_0 = I_m \ , \qquad B_0 = -H^{-1}C^*J$$

Finally, for a given H, all other choices of B and D differ from B_0, D_0 by a right multiplicative J-unitary constant matrix.

PROOF. Let H be a solution of equation (2.2) that is both hermitian and invertible. We first check that the pair $\left(A, H^{-1}C^*\right)$ is controllable or equivalently that the pair $\left(CH^{-1}, A^*\right)$ is observable. Using the Lyapunov equation, we see that, for any k, there exist matrices $K_0, \ldots, K_{k-1}$ such that

$$CH^{-1}(A^*)^k = (-1)^k CA^k H^{-1} + K_0 CA^{k-1} H^{-1} + \cdots + K_{k-1} CH^{-1}$$

Thus, if f is such that $CH^{-1}(A^*)^k f = 0$ for $k = 0, 1, \ldots$, then $CA^k H^{-1} f = 0$ for $k = 0, 1, \ldots$, and the observability of the pair (C, A) forces f to be zero. Hence, the pair $\left(CH^{-1}, A^*\right)$ is observable and so the pair $\left(A, H^{-1}C^*\right)$ is a controllable pair of matrices.

From Theorem 2.1, it can be seen that $U_0(\lambda) = I_m - C(\lambda I_n - A)^{-1} H^{-1}C^*J$ is a minimal realization of a rational function J-unitary on the line, with associated hermitian matrix H.

Let B, D be another solution to the inverse problem for a given H, and let $U(\lambda) = D + C(\lambda I_n - A)^{-1} B$. Equation (2.12) leads to

$$U(\lambda)JU(\omega)^* = U_0(\lambda)JU_0(\omega)^*$$

Thus, U and U_0 differ by a right multiplicative J-unitary constant (which clearly has to be D), that is,

$$U(\lambda) = U_0(\lambda)D$$

and we easily deduce that $B = -H^{-1}C^*JD$ from the observability of the pair (C, A). □

The following theorem can be proved similarly.

THEOREM 2.3. *Let (A, B) be a controllable pair of matrices and let J be a signature matrix. Then there exists a rational function J-unitary on the line with minimal realization $D + C(\lambda I_n - A)^{-1} B$ if and only if the Lyapunov equation*

$$GA^* + AG = -BJB^* \tag{2.17}$$

has a solution which is both invertible and hermitian. When such a solution exists, possible choices of D and C are

$$D_0 = I_m , \qquad C_0 = -JB^*G^{-1}$$

For a given G, all other choices of D and C differ from D_0, C_0 by a left multiplicative J-unitary constant.

Results close to Theorem 2.1 and Theorem 2.3 may be found in different sources; see for example [BR], [GVKDM] and [G].

We conclude this subsection with the remark that inverse problems of a more general character are solved in the paper [GKLR]. In particular, the criteria of [GKLR] are in terms of Sylvester equations which are more general than (2.2). Links between Lyapunov equations and realizations are also discussed in [S].

2.2. The Associated Hermitian Matrix.

Here we study the invariants of the associated hermitian matrix.

LEMMA 2.1. *Let U be a rational function, analytic at infinity and J-unitary on the imaginary line. Let $U(\lambda) = D + C_i(\lambda I_n - A_i)^{-1} B_i, i = 1, 2$, be two minimal realizations of U, with associated hermitian matrices $H_i, i = 1, 2$. Then the two minimal realizations are similar, that is,*

$$C_1 = C_2 S , \qquad A_1 = S^{-1} A_2 S , \qquad B_1 = S^{-1} B_2 \tag{2.18}$$

for a unique invertible matrix S, and

$$H_1 = S^* H_2 S \tag{2.19}$$

In particular, the matrices H_1 and H_2 have the same number of positive and negative eigenvalues.

PROOF. The existence and uniqueness of the similarity matrix is well known (see [BGK], p. 65-66). Using equations (2.18), one checks that the equations

$$A_2^* H + H A_2 = -C_2^* J C_2 \tag{2.20}$$

$$C_2 = -DJB_2^*H \tag{2.21}$$

are satisfied by both H_2 and $(S^{-1})^* H_1 S^{-1}$. By the uniqueness of the associated hermitian matrix associated with the minimal realization $D + C_2 (\lambda I_n - A_2)^{-1} B_2$ we conclude that $H_1 = S^* H_2 S$. □

We remark that the similarity matrix S is a unitary mapping from C_n endowed with the inner product $[,]_{H_1}$ onto C_n endowed with the inner product $[,]_{H_2}$, where, for any hermitian matrix H,

$$[x, y]_H = \langle Hx, y \rangle$$

We recall (see [KL]) the following definition. Let $K(\lambda, \omega)$ be a $C_{m\times m}$ valued function defined for λ and ω in some set E and such that $K(\lambda, \omega)^* = K(\omega, \lambda)$. This function has k negative squares if for any positive integer r, any points $\omega_1, \ldots, \omega_r$ in E and any vectors $c_1, \ldots, c_r$ in C_m, the $C_{r\times r}$ hermitian matrix with ij entry

$$c_j^* K(\omega_j, \omega_i) c_i \tag{2.22}$$

has at most k negative eigenvalues and has exactly k negative eigenvalues for some choice of $r, \omega_1, \ldots, \omega_r, c_1, \ldots, c_r$. With this definition at hand, we can now state the following theorem, which gives a characterization of the number of negative eigenvalues of the associated matrix H.

THEOREM 2.4. *Let U be a rational matrix function J-unitary on the imaginary line and analytic at infinity, and let $U(\lambda) = D + C(\lambda I_n - A)^{-1} B$ be a minimal realization of U, with associated hermitian matrix H. Then the number of negative eigenvalues of the matrix H is equal to the number of negative squares of each of the functions*

$$K_U(\lambda, \omega) = \frac{J - U(\lambda) J U(\omega)^*}{(\lambda + \omega^*)} \quad \text{and} \quad \frac{J - U(\omega)^* J U(\lambda)}{(\lambda + \omega^*)} \tag{2.23}$$

Finally, let $\mathcal{K}(U)$ be the span of the functions $\lambda \to K_U(\lambda, \omega)c$, where ω is in the resolvent set of A and where c is in C_m. Then $\mathcal{K}(U)$ is a finite-dimensional space of rational functions analytic on the resolvent set of A, and the dimension of $\mathcal{K}(U)$ is equal to the MacMillan degree of U.

PROOF. From formula (2.10) we have that for λ and ω in the resolvent set of A,

$$K_U(\lambda, \omega) = C(\lambda I_n - A)^{-1} H^{-1} (\omega^* I_n - A^*)^{-1} C^* \tag{2.24}$$

Let r be a positive integer, let $\omega_1, \ldots, \omega_r$ be in the resolvent set of A, and let $c_1, \ldots, c_r$ be in C_m. Then the matrix equality

$$\left(c_j^* K_U(\omega_j, \omega_i) c_i\right)_{i,j=1,r} = X^* H^{-1} X \tag{2.25}$$

with

$$X = \left((\omega_1^* - A^*)^{-1}C^*c_1, \ldots, (\omega_r^* - A^*)^{-1}C^*c_r\right)$$

makes it clear that the function K_U has at most k_H negative squares, where k_H denotes the number of negative eigenvalues of the hermitian matrix H. The pair (C, A) is observable, and thus we can choose a basis of C_n of the form $x_i = \left(\omega_i^* - A_i^*\right)^{-1} C^*c_i$ $i = 1, \ldots, n$. In particular, $\det X \neq 0$ for $X = (x_1, \ldots, x_n)$ and the matrix $X^*H^{-1}X$ has exactly k_H negative squares, and thus K_U has k_H negative squares.

The case of the function $K_{U^*}(\omega, \lambda)$ is treated similarly, and relies on formula (2.13). Equation (2.24) implies that any finite linear combination of functions $K_U(\lambda, \omega)c$ is of the form

$$C\left(\lambda I_n - A\right)^{-1} f$$

where f is in C_n. Thus, $\mathcal{K}(U)$ is a finite-dimensional vector space of dimension at most n. From the observability of the pair (C, A), we see that $C\left(\lambda I_n - A\right)^{-1} f \equiv 0$ implies that $f = 0$ and then $\dim \mathcal{K}(U) = n$ follows. □

We will denote by $\nu(U)$ the number of negative squares of either of the functions defined in (2.23).

The last theorem of this subsection deals with the product of two J-unitary rational functions.

THEOREM 2.5. *Let $U_i, i = 1, 2$, be two rational J-unitary functions, analytic at infinity, with minimal realizations $U_i(\lambda) = D_i + C_i\left(\lambda I_{n_i} - A_i\right)^{-1} B_i$, $i = 1, 2$, and associated hermitian matrices $H_i, i = 1, 2$, and suppose the product U_1U_2 minimal. Then the hermitian matrix*

$$H = \begin{pmatrix} H_1 & 0 \\ 0 & H_2 \end{pmatrix} \tag{2.26}$$

is the associated hermitian matrix associated with the minimal realization $D + C\left(\lambda I_n - A\right)^{-1} B$ of the product U_1U_1, where $n = n_1 + n_2$,

$$D = D_1D_2, \quad C = (C_1 \quad D_1C), \quad B = \begin{pmatrix} B_1D_2 \\ B_2 \end{pmatrix} \tag{2.27}$$

and

$$A = \begin{pmatrix} A_1 & B_1C_2 \\ 0 & A_2 \end{pmatrix} \tag{2.28}$$

PROOF. It suffices to check that equations (2.2) and (2.3) are satisfied for A, B C, D and H as in the statement. This is an easy computation which is omitted. The second claim of the theorem is a consequence of Lemma 2.1. □

COROLLARY 2.2. *Let U_1 and U_2 be rational functions analytic at infinity and J-unitary on the line, and suppose the factorization $U = U_1U_2$ minimal. Then*

$$\nu(U_1U_2) = \nu(U_1) + \nu(U_2) \tag{2.29}$$

This additive property of the number of negative squares of the function defined in (2.23) was noticed by Sakhnovich in [S].

2.3. Factorizations of Rational Functions J-Unitary on the Line

In this subsection we study minimal factorizations of a rational function, J-unitary on the line and analytic at infinity, into factors both of which are J-unitary on the line. Such factorizations will be called J-unitary factorizations.

We first need to introduce certain hermitian forms on C_n. Let H be an invertible hermitian matrix in $C_{n\times n}$. We will denote by $[\ ,\ ]_H$ the hermitian form

$$[x,y]_H = \langle x, Hy\rangle \tag{2.30}$$

where $\langle\ ,\ \rangle$ denotes the usual euclidean inner product of C_n.

Two vectors x and y in C_n will be called H-orthogonal if $[x,y]_H = 0$. For any subspace M of C_n, $M^{[\perp]}$ will denote the subspace of all vectors H-orthogonal to M:

$$M^{[\perp]} = \{y \in C_n \ :\ [y,m]_H = 0 \quad \forall m \in M\}$$

The subspace M of C_n is called non-degenerate if $M \cap M^{[\perp]} = \{0\}$. In this case,

$$M\,[\dot{+}]\,M^{[\perp]} = C_n$$

where $[\dot{+}]$ denotes orthogonal direct sum.

When H is the hermitian matrix associated with some given minimal realization of a rational function J-unitary on the line, $[\ ,\ ]_H$ will be called the *associated inner product* (associated with the given minimal realization).

THEOREM 2.6. *Let U be a rational function analytic at infinity and J-unitary on the line. Let $U(\lambda) = D + C(\lambda I_n - A)^{-1}B$ be a minimal realization of U with associated hermitian matrix H. Finally, let M be an invariant subspace of A, non-degenerate in the associated inner product $[\ ,\ ]_H$, let π be the projection defined by*

$$\text{Ker } \pi = M \qquad \text{Im } \pi = M^{[\perp]}$$

and let $D = D_1D_2$ be a factorization of D into two J-unitary factors. Then the factorization $U = U_1U_2$, where

$$U_1(\lambda) = D_1 + C(\lambda I_n - A)^{-1}(I_n - \pi)BD_2^{-1} \tag{2.31}$$

$$U_2(\lambda) = D_2 + D_1^{-1}C\pi(\lambda I_n - A)^{-1}B \tag{2.32}$$

is a minimal J-unitary factorization of U.

Conversely, any minimal J-unitary factorization of U can be obtained in such a way, and with a fixed decomposition $D = D_1 D_2$ the correspondence between J-unitary factorizations of U and non-degenerate invariant subspaces of A is one to one.

PROOF. Let $A^{[*]}$ denote the adjoint of the operator A with respect to the associated inner product $[\ ,\ ]_H$. Then A^*, the adjoint of A in the usual metric of $\mathbf{C}_n$, and $A^{[*]}$ are linked by

$$A^{[*]} = H^{-1} A^* H$$

and hence equation (2.9) reads

$$A^{[*]} = -A^{\times}$$

Let M be an A-invariant subspace. Then $M^{[\perp]}$ is $A^{\times}$-invariant. When M is non-degenerate, the projection π is thus a supporting projection and, by Theorem 1.1 of [BGK], (2.31) and (2.32) define a minimal factorization of U.

Without loss of generality, we first suppose $D = D_1 = D_2 = I_m$. The factor U_1 may then be written as

$$U_1(\lambda) = I_m + C_1 (\lambda I - A_1)^{-1} B_1$$

where

$$A_1 = A\big|_M \qquad (A \quad \text{restricted to} \quad M)$$

and

$$B_1 = (I_n - \pi) B, \quad C_1 = C\big|_M .$$

Let H_1 denote $PH\big|_M$, where P is the orthogonal projection from $\mathbf{C}_n$ onto M in the usual metric of $\mathbf{C}_n$. We note first that H_1 is invertible. Indeed, let m in M be such that $H_1 m = 0$. Then, for any m' in M, $\langle H_1 m, m' \rangle = 0$ and thus $[m, m']_H = 0$. Hence m is equal to the zero vector since M is non-degenerate.

In view of Theorem 2.1, in order to prove that U is J-unitary on the line it is sufficient to check that

$$A_1^* H_1 + H_1 A = -C_1^* J C_1 \tag{2.33}$$

$$B_1 = -H_1^{-1} C_1^* J \tag{2.34}$$

We check (2.34). We already know that

$$HB = -C^* J$$

since U is J-unitary on the line. Hence,

$$PHB = -PC^*J \tag{2.35}$$

The projection π is selfadjoint with respect to $[\ ,\ ]_H$, $\pi = \pi^{[*]}$, and hence

$$(I_n - \pi^*)\, H = H\,(I_n - \pi) \tag{2.36}$$

Moreover, $P = (I_n - \pi)\, P$ and thus

$$P\,(I_n - \pi^*) = P \tag{2.37}$$

Using (2.35), (2.36) and (2.37), we have

$$P\,(I_n - \pi^*)\, HB = -PC^*J$$

and

$$PH\,(I_n - \pi)\, B = -PC^*J$$

and hence

$$H_1B_1 = -C_1^*J \tag{2.38}$$

This last equation coincides with equation (2.34).

The equality (2.33) is proved in the same way. This concludes the proof of the theorem when $D = D_1 = D_2 = I_m$, and the general case of arbitrary J-unitary constants is easily adapted.

We now focus on the converse statement. Let $U = U_1U_2$ be a minimal J-unitary factorization of U and let $U_i(\lambda) = D_i + C_i\,(\lambda I_{n_i} - A_i)^{-1}\, B_i$ be minimal realizations of $U_i, i = 1, 2$, with associated matrices $H_i, i = 1, 2$. Theorem 2.5 gives a minimal realization of U, $U(\lambda) = D + C\,(\lambda I_n - A)^{-1}\, B$, with associated hermitian matrix H given by (2.26). The subspace $M = \left\{ \binom{f}{0}; f \in C_{n_1} \right\}$ (where 0 is the zero-vector of C_{n_2}) is then an A invariant subspace non-degenerate in $[\ ,\ ]_H$, and, as is easily checked, generates the factorization $U = U_1U_2$.

Finally, the one to one correspondence mentioned in the statement of the theorem follows from the general theorem about minimal factorizations from [BGK]. □

We note that J-unitary factorization is studied in the paper of L.A. Sakhnovich (chapter 2, [S]). Our results intersect with some of his. However, our aims are different.

THEOREM 2.7. *Let U be a rational function J-unitary on the imaginary line. Let $U(\lambda) = D + C\,(\lambda I_n - A)^{-1}\, B$ be a minimal realization of U, with associated* *hermitian matrix H. Let f be an eigenvector of A, corresponding to the eigenvalue ω, and suppose $[f,f]_H \neq 0$. Then U admits a minimal J-unitary factorization*

$$U = U_1 U_2$$

where U_1 has the form

$$U_1(\lambda) = I_m - PJ + \frac{\lambda + \omega^*}{\lambda - \omega} PJ \tag{2.39}$$

with

$$P = \frac{Cff^*C^*J}{f^*C^*JCf}$$

for the case $Re\omega \neq 0$, and U_1 has the form

$$U_1(\lambda) = I_m - \frac{Cff^*C^*}{(\lambda - \omega)[f,f]_H} \tag{2.40}$$

for the case $Re\omega = 0$.

PROOF. The existence of the factorization is a consequence of Theorem 2.6. The factor U_1 is given by formula (2.31), where we suppose $D_1 = I_m$ and where π is defined by

$$\text{Ker } \pi = \text{span}\{f\} \quad , \quad \text{Im } \pi = (\text{span}\{f\})^{[\perp]}$$

Thus, for g in C_n,

$$(I - \pi)g = f\frac{[g,f]_H}{[f,f]_H} \tag{2.41}$$

and hence formula (2.31) becomes (with $D_1 = D_2 = I_m$)

$$U_1(\lambda) = I_m + \frac{Cff^*HB}{[f,f]_H \cdot (\lambda - \omega)}$$

Using equation (2.3), we obtain

$$U_1(\lambda) = I_m - \frac{Cff^*C^*J}{[f,f]_H \cdot (\lambda - \omega)}$$

which is (2.40) when $Re\omega = 0$. For $Af = \omega f$, equation (2.2) implies

$$(\omega^* + \omega)[f,f]_H = -\langle JCf, Cf\rangle \tag{2.42}$$

and thus, if $Re\omega = 0$, the vector Cf is J-neutral. Equation (2.42) implies, when $Re\omega \neq 0$,

$$[f,f]_H = -\frac{\langle JCf, Cf\rangle}{\omega^* + \omega} = -\frac{f^*C^*JCf}{\omega * + \omega}$$

from which formula (2.39) follows easily. □

In the electrical engineering literature, when $[f,f]_H > 0$, factors of the form (2.39) are called Blaschke-Potapov factors and factors of the form (2.40) are called Brune factors (see [DD]).

We conclude this subsection by showing that the function U exhibited in example 2.2 has no minimal J-unitary factorizations. Let us first obtain a minimal realization for U. We express U in the form $U = U_1U_2$, where

$$U_1(\lambda) = I_m + \left(\frac{\lambda - \omega_1}{\lambda + \omega_2^*} - 1\right) W_{12}$$

$$U_2(\lambda) = I_m + \left(\frac{\lambda - \omega_2}{\lambda + \omega_1^*} - 1\right) W_{21}$$

Then the following minimal realizations for U_1 and U_2 hold:

$$U_i(\lambda) = I_m + C_i\left(\lambda I_1 - A_i\right)^{-1} B_i \qquad i = 1,2$$

where

$$A_1 f = -\omega_1^* f,\ f \in C;\ B_1 = \frac{u_2^* J}{u_2^* J u_1}; \qquad C_1 = -(\omega_1 + \omega_2^*)u_1$$

and

$$A_2 f = \omega_2^* f\ ,\ f \in C\ ; \quad B_2 = \frac{u_1^* J}{u_1^* J u_2}; \qquad C_2 = -(\omega_1^* + \omega_2)u_2\ .$$

Thus, U admits the minimal realization $U(\lambda) = I_m + C\left(\lambda I_2 - A\right)^{-1} B$ (see formulas (2.27),(2.28)), where

$$A = \begin{pmatrix} A_1 & B_1C_2 \\ 0 & A_2 \end{pmatrix} = \begin{pmatrix} \omega_1^* & 0 \\ 0 & \omega_2^* \end{pmatrix},$$

$$B = \begin{pmatrix} B_1 \\ B_2 \end{pmatrix} = \begin{pmatrix} u_2^*J/u_2^*Ju_1 \\ u_1^*J/u_1^*Ju_2 \end{pmatrix}, \quad C = (C_1 \quad C_2) = (-(\omega_1 + \omega_2^*)u_1, -(\omega_1^* + \omega_2)u_2)$$

Two eigenvectors of A are $f_1 = \binom{1}{0}$ and $f_2 = \binom{0}{1}$. Both Cf_1 and Cf_2 are J-neutral vectors, and by equation (2.42), this forces $[f_1, f_1]_H = [f_2, f_2]_H = 0$, where H is the associated hermitian matrix associated with the present minimal realization of U. All invariant subspaces of A are degenerate and thus U lacks minimal J-unitary factorizations. Note that we did not compute the associated hermitian matrix H.

It is easy to check that the function defined in Example 2.3 also lacks minimal J-unitary factorizations.

2.4 Rational Matrix Functions Unitary on the Line

In this subsection we specialize some of the preceding results to the case in which J is the identity matrix. The corresponding functions are now unitary on the line.

THEOREM 2.8. *Let U be a matrix-valued rational function analytic at infinity and let $U(\lambda) = D + C(\lambda I_m - A)^{-1}B$ be a minimal realization of U. Then U is unitary on the imaginary line if and only if the following conditions hold:*

a) D is a unitary matrix

b) There exists an hermitian solution of the Lyapunov equation

$$A^*H + HA = -C^*C \tag{2.43}$$

and

$$C = -JB^*H$$

Condition b) is equivalent to

b') There exists an hermitian solution to the Lyapunov equation

$$GA^* + AG = -BB^* \tag{2.44}$$

and

$$B = -GC^*J$$

PROOF. To obtain Theorem 2.8 from Theorem 2.1, it suffices to show that any hermitian solution H to the Lyapunov equation (2.43) is invertible. Let f be an element in the kernel of H. Then

$$\langle HAf,\ f\rangle = \langle Af,\ Hf\rangle = 0$$

and thus, equation (2.43) leads to $Cf = 0$, that is, Ker $H \subset$ Ker C.

Let us decompose $C_n =$ Ker $H \oplus$ Ran H, and let $A = \begin{pmatrix} A_{11} & A_{12} \\ A_{21} & A_{22} \end{pmatrix}$, $H = \begin{pmatrix} 0 & 0 \\ 0 & H_{22} \end{pmatrix}$ and $C = \begin{pmatrix} 0 & C_2 \end{pmatrix}$ be the corresponding decompositions of A, H and C. Equation (2.43) leads to

$$A_{12}^*H_{22} = 0\ , \qquad H_{22}A_{21} = 0$$

In particular, $A_{21} = A_{12} = 0$, and $CA^n = (0, C_2A_{22}^n)$. Therefore the pair (C, A) is not observable, which contradicts the hypothesis. □

We note that, in a similar way, one can show that any hermitian solution of the Lyapunov equation (2.44) is invertible.

The counterpart of Theorem 2.2 in the present case is the following theorem.

THEOREM 2.9 *Let (C, A) be an observable pair of matrices. Then there exists a rational function unitary on the line with minimal realization $D + C(\lambda I_n - A)^{-1}B$ if and only if the Lyapunov equation*

$$A^*H + HA = -C^*C$$

has an hermitian solution H. If such a solution exists, it is invertible and possible choices of D and B are

$$D_0 = I_m \ , B_0 = -H^{-1}C^*$$

Finally, for a given H, all other choices of B and D differ from B_0 and D_0 by a right multiplicative unitary constant matrix.

The proof of Theorem 2.9 is a direct application of Theorem 2.2 and Theorem 2.10.

The next theorem concerns the multiplicative structure of a matrix-valued rational function unitary on the line.

THEOREM 2.10 *Let U be a rational function unitary on the line and analytic at infinity. Let $U(\lambda) = I_m + C(\lambda I_n - A)^{-1}B$ be a minimal realization of U, with associated hermitian matrix H and associated inner product $[,]_H$. Then*

a) *A has no spectrum on the imaginary line*

b) *$[f, f]_H \neq 0$ for any eigenvector f of A*

c) *U is the product of a unitary constant matrix and a minimal product of n rational functions of the form*

$$I_m - P + \frac{\lambda + \omega^*}{\lambda - \omega} P \tag{2.45}$$

where P is an orthogonal projection of rank one and $Re\omega \neq 0$.

PROOF. Let f be an eigenvector of A with eigenvalue ω. Equation (2.42) with $J = I_m$ becomes

$$(\omega^* + \omega)[f, f]_H = -\langle Cf, Cf\rangle$$

Suppose that either $Re\omega = 0$ or $[f, f]_H = 0$. In both cases, equation (2.44) forces $Cf = 0$ and thus $CA^n f = 0$ for all integers $n \geq 0$. Then the minimality of the realization leads to $f = 0$, and thus a) and b) are proved. We now turn to c). Let f be an eigenvector of A, corresponding to the eigenvalue ω. By Theorem 2.7, U admits the minimal factorization $U = U_1U_2$, where both factors are unitary on the line and where U_1 is of the form (2.43) with

$$P = \frac{Cff^*C}{f^*CC^*f}$$

The process can be iterated on U_2, and after n such iterations we get to the desired result. □

The matrix counterpart of Theorem 2.10 is the following theorem.

THEOREM 2.11. *Let (C, A) be an observable pair of matrices and suppose that the Lyapunov equation*

$$A^*H + HA = -C^*C$$

has an hermitian solution. Then this solution is invertible and there exists a basis $\varphi_0, \ldots, \varphi_{n-1}$ of C_n such that in this basis A has an upper triangular form, $H^{-1}A^*H$ has a lower triangular form, and all of the spaces span $\{\varphi_0, \ldots, \varphi_j\}$, $j = 0, \ldots, n-1$, are non-degenerate in $[\ ,\ ]_H$.

PROOF. Let U denote the function

$$U(\lambda) = I_m + iC(\lambda I_n - A)^{-1} H^{-1}C^*$$

By Theorem 2.1, U is unitary on the line. The theorem is then a direct consequence of the factorization of U into n factors of the form (2.45). □

A product of a constant unitary matrix and a finite number of factors of the form (2.45) with $\omega \in C_-$ will be called a Blaschke-Potapov product. By Theorem 2.10 a rational function unitary on the unit circle is a Blaschke-Potapov product iff it is analytic in C_+. The MacMillan degree of a Blaschke-Potapov product is equal to the total multiplicity of its zeros.

THEOREM 2.12. *Let U be a rational function unitary on the line. Then there exist Blaschke-Potapov products B_1, B_2, B_3, B_4 such that*

$$U = B_1 B_2^{-1} \tag{2.46}$$

$$U = B_3^{-1} B_4 \tag{2.47}$$

with

$$\deg U = \deg B_1 + \deg B_2 = \deg B_3 + \deg B_4$$

and

$$\nu(U) = \deg B_3 = \deg B_2$$

In particular, $\nu(U)$ is equal to the number of poles of U in C_+, and $n - \nu(U)$ is equal to the number of poles of U in C_-.

PROOF. The representations (2.46) and (2.47) can be deduced from the proof of Theorem 2.8.

From formula (2.45), we see that any associated hermitian matrix associated with a function of the form (2.44) with ω in C_- is a positive number. Similarly, when ω is in C_+, any associated matrix of the function (2.44) will then be a negative number, and the equality $\deg B_3 = \deg B_2 = \nu(U)$ follows from Theorem 2.5 and Theorem 2.4.The final statement follows from $degB_2 = \nu(U)$. □

This theorem enables us to obtain a simple proof of an inertia theorem.

THEOREM 2.13. *Let (C, A) be an observable pair of matrices and let H be an hermitian solution of the equation*

$$A^*H + HA = -C^*C$$

Then the matrix H is invertible, the matrix A has no spectrum on the imaginary line, and the number of negative (resp. positive) eigenvalues of H is equal to the number of eigenvalues of A in the open right (resp. left) half plane.

PROOF. The matrix H is invertible, as shown in the proof of Theorem 2.8. Then $U(\lambda) = I_m - C(\lambda I_n - A)^{-1}H^{-1}C^*$ is a minimal realization of a matrix-valued function U unitary on the line, and, by Theorem 2.10, the matrix A has no spectrum on the imaginary line. By Theorem 2.12, $\nu(U)$ is equal to the number of poles of U in C_+, and thus to the number of eigenvalues of A in C_+. By Theorem 2.4, $\nu(U)$ is also equal to the number of negative eigenvalues of H, which concludes the proof. □

We refer to the paper [C] for further references and details about the inertia theorems.

2.5. More on Factorizations.

We now present criteria for the existence of minimal J-unitary factorizations. The first theorem is from [AD].

THEOREM 2.14. *Let U be a rational function J-unitary on the line and analytic at infinity, and suppose that U has only simple poles. Then there exists a minimal J-unitary factorization $U = U_1U_2 \ldots U_k$, where each U_i is of degree less than or equal to 2.*

PROOF. Let $U(\lambda) = D + C\left(\lambda I_n - A\right)^{-1}B$ be a minimal realization of U, with associated hermitian matrix H. Let $f_1, \ldots, f_n$ be eigenvectors of A, corresponding to the poles $\omega_1, \ldots, \omega_n$ of U. If $[f_i, f_i]_H \neq 0$ for some i, we get a minimal J-unitary factorization $U = U_1V_1$ with a factor U_1 of degree 1 and J-unitary on the line. Let us suppose now that $[f_i, f_i]_H = 0$ for $i = 1, \ldots, n$. One of the inner products $[f_i, f_j]_H$ has to be non-zero, and the span of f_i and f_j for this choice of i, j will generate a minimal J-unitary factorization $U = U_1V_1$ with $\deg U_1 = 2$.

The process can be iterated on V_1 since V_1 still has only simple poles, and thus the theorem is proved. □

A simple modification of this argument leads to

THEOREM 2.15. *Let U be a rational function J-unitary on the line and analytic at infinity, and let N be the number of generalized eigenspaces corresponding to the poles of U. Then, if $N > \nu(U)$, there exists a minimal J-unitary factorization $U = U_1 U_2$ with $\deg U_1 \leq 2$.*

PROOF. Let $D + C(\lambda I_n - A)^{-1} B$ be a minimal realization of U with associated hermitian matrix H. Let $f_1, \ldots, f_N$ be the eigenvectors in the generalized eigenspaces. If for all i, j, $[f_i, f_j]_H = 0$, then the span of these eigenvectors would be a neutral subspace, which contradicts the hypothesis $N > \nu(U)$. The claim then follows as in Theorem 2.14. □

A matrix-valued function U that is J-unitary on the line is J-inner if

$$U(\lambda) J U^*(\lambda) \leq J \tag{2.48}$$

at those points of the right open half plane at which it is defined.

In the next theorem we study the factorization of rational J-inner functions and get a special case of a theorem of Potapov [P]. This result appears also in [GVKDM]. In the latter paper J-inner functions are called J-lossless functions.

THEOREM 2.16. *Let U be a rational matrix-valued function J-unitary on the line and analytic at infinity, and let $U(\lambda) = D + C(\lambda I_n - A)^{-1} B$ be a minimal realization of U. Then*

a) The function U is J-inner if and only if its associated matrix is strictly positive

b) Any J-inner rational matrix function is a minimal product of n J-inner rational matrix functions of degree 1 of the form (2.39) or (2.40).

PROOF. Formula (2.10) specialized to $\lambda = \omega$ in C_+ leads to

$$c^* \frac{J - U(\lambda) J U^*(\lambda)}{\lambda + \lambda^*} c = c^* C(\lambda I_n - A)^{-1} H^{-1} (\lambda^* I_n - A)^{-1} C^* c \tag{2.49}$$

for any c in C_m. The pair (C, A) being observable, the span of the vectors $(\lambda^* I_n - A^*)^{-1} C^* c$, where λ is in the resolvent set of A and c is in C_m, is all of C_n and the statement follows from (2.49).

Any eigenvector f of A satisfies $[f, f]_H \neq 0$, since $H > 0$, and, consequently, proposition b) follows from Theorem 2.7 iterated n times. □

3. RATIONAL FUNCTIONS J-UNITARY ON THE UNIT CIRCLE

In this section we study rational functions which are J-unitary on the unit circle. The results which we obtain are parallel to the ones of Section 2. In general, a clear analogy with the line case is easy to follow, and hence, in many places, the proofs are omitted or shortened.

3.1. Realization Theorems and Examples

In this subsection we study the minimal realizations of rational functions which are J-unitary on the unit circle. Let U be such a function. For any z on the unit circle at which U is analytic, we have $U(z)JU\left(\frac{1}{z^*}\right)^* = J$, and, by analytic continuation, the identity extends to all points z at which $U(z)$ is analytic and invertible by

$$U(z)JU\left(\frac{1}{z^*}\right)^* = J \tag{3.1}$$

Equality (3.1) implies the following lemma, the proof of which is easy and will be omitted.

LEMMA 3.1. *Let U be a rational matrix-valued function J- unitary on the unit circle. Then U is analytic and invertible at the point z (including $z = \infty$) iff U is analytic and invertible at the point $1/z^*$. Moreover, if z is a singular point of U with partial multiplicities*

$$k_1 \geq k_2 \geq \cdots \geq k_r > 0 \geq k_{r+1} \geq \cdots \geq k_s ,$$

then $1/z^*$ is a singular point of U with partial multiplicities

$$-k_s \geq \cdots \geq -k_{r+1} \geq 0 > -k_r \geq \cdots \geq -k_2 \geq -k_1 .$$

After this preliminary lemma we turn to the main topic of this subsection, and begin with the following theorem.

THEOREM 3.1. *Let U be a rational function, analytic and invertible at infinity, and let $U(z) = D + C\left(zI_n - A\right)^{-1} B$ be a minimal realization of U. Then U is J-unitary on the unit circle if and only if*

a) U is analytic and invertible at the origin

b) There exists an invertible hermitian matrix H such that

$$\begin{pmatrix} A & B \\ C & D \end{pmatrix}^* \begin{pmatrix} H & 0 \\ 0 & -J \end{pmatrix} \begin{pmatrix} A & B \\ C & D \end{pmatrix} = \begin{pmatrix} H & 0 \\ 0 & -J \end{pmatrix} \tag{3.2}$$

PROOF. We first prove the necessity of conditions a) and b). By Lemma 3.1, the function U is also analytic and invertible at the origin and thus A is invertible since the given realization of U is minimal. Equation (3.1) may be rewritten as

$$U^{-1}(z) = JU\left(\frac{1}{z^*}\right)^* J$$

and thus,

$$D^{-1} - D^{-1}C\left(zI_n - A^\times\right)^{-1} BD^{-1} = J\left(D^* + zB^*\left(I_n - zA^*\right)^{-1} C^*\right) J \tag{3.3}$$

Letting $D_1 = JD^*J - JB^*(A^*)^{-1}C^*J$, we may rewrite equation (3.3) as

$$D^{-1} - D^{-1}C\left(zI_n - A^\times\right)^{-1} BD^{-1} = D_1 - JB^*(A^*)^{-1}\left(zI_n - (A^*)^{-1}\right)^{-1}(A^*)^{-1}C^*J \tag{3.4}$$

where $A^\times = A - BD^{-1}C$. Letting z go to infinity, we find that $D_1 = D^{-1}$, that is,

$$D^{-1} = JD^*J - JB^*(A^*)^{-1}C^*J \tag{3.5}$$

Moreover, (3.4) is an equality between two minimal realizations of a given rational function, and thus there exists a unique invertible matrix H such that

$$D^{-1}C = JB^*(A^*)^{-1}H \tag{3.6}$$

$$(A^*)^{-1} = HA^\times H^{-1} \tag{3.7}$$

$$H^{-1}(A^*)^{-1}C^*J = BD^{-1} \tag{3.8}$$

These three equations are also satisfied by H^*, as is easily checked; by the uniqueness of the similarity matrix we get $H = H^*$. Moreover, equations (3.6) and (3.8) in conjunction with equation (3.5) lead easily to (3.2).

Conversely, define $U(z) = D + C\left(zI_n - A\right)^{-1} B$, where A, B, C, D satisfy equation (3.2) for some signature matrix J and some hermitian invertible matrix H. Computing $U(\omega)^*JU(z)$ for z and ω in the resolvent set of A, we find that

$$U(\omega)^*JU(z) = \left(D^* + B^*(\omega^*I_n - A^*)^{-1}C^*\right) J\left(D + C\left(zI_n - A\right)^{-1} B\right)$$

$$= D^*JD + B^*(\omega^*I_n - A^*)C^*JD + D^*JC\left(zI_n - A\right)^{-1} B$$

$$+ B^*(\omega^*I_n - A^*)^{-1}C^*JC\left(zI_n - A\right)^{-1} B$$

Using (3.2), we replace D^*JD by $J + B^*HB$, C^*JD by A^*HB and C^*JC by $H - A^*HA$, and, after some algebra, we obtain

$$U(\omega)^*JU(z) = J - (1 - z\omega^*)B^* \left(\omega^* I_n - A^*\right)^{-1} H \left(zI_n - A\right)^{-1} B \tag{3.9}$$

Therefore U is J-unitary on the unit circle. □

As in Section 2, it follows from the proof of this theorem that the matrix H is uniquely defined. This allows us to introduce the following definition: the matrix H will be called the *associated hermitian matrix* associated with the minimal realization $U(z) = D + C(zI_n - A)^{-1}B$ of a matrix-valued rational function U that is J-unitary on the circle.

For H we can also give the following formulas.

$$H = \left(\mathrm{col}(JB^*((A^*)^{-1})^{(j-1)})_0^{n-1}\right)^+ \left(\mathrm{col}(D^{-1}C(A^\times)^j)_0^{n-1}\right)$$

$$H = \left(\mathrm{row}(((A^*)^{-1})^{(j-1)}C^*J)_0^{n-1}\right) \left(\mathrm{row}((A^\times)^j BD^{-1})_0^{n-1}\right)^\dagger$$

We note that equation (3.2) is equivalent to

$$\begin{pmatrix} A & B \\ C & D \end{pmatrix} \begin{pmatrix} H^{-1} & 0 \\ 0 & -J \end{pmatrix} \begin{pmatrix} A & B \\ C & D \end{pmatrix}^* = \begin{pmatrix} H^{-1} & 0 \\ 0 & -J \end{pmatrix} \tag{3.10}$$

Moreover, (3.2) may be replaced by the following three equalities

$$H - A^*HA = -C^*JC \tag{3.11}$$

$$C^*JD = A^*HB \tag{3.12}$$

$$J - D^*JD = -B^*HB \tag{3.13}$$

while (3.10) may be written as

$$H^{-1} - AH^{-1}A^* = -BJB^* \tag{3.14}$$

$$CH^{-1}A^* = DJB^* \tag{3.15}$$

$$J - DJD^* = -CH^{-1}C^* \tag{3.16}$$

Thus, a rational function $U(z) = D + C\left(zI_n - A\right)^{-1} B$ which is J-unitary on the unit circle can be represented in the form

$$U(z) = D\left(I_m + JB^*(A^*)^{-1}H^{-1}\left(zI_n - A\right)^{-1} B\right)$$

$$= \left(I_m + C\left(zI_n - A\right)^{-1} H(A^*)^{-1}C^*J\right) D$$

For the case $H = I_n$, a small computation shows that $U^*(z^*)$ is the characteristic operator function of A^{-1}, defined in [BrGK].

Let A, B, C, D be matrices satisfying equation (3.2) for some signature matrix J and some (not necessarily invertible) hermitian matrix H. Then $U(z) = D + C(zI_n - A)^{-1}B$ defines a rational matrix function J-unitary on the unit circle. This follows from equation (3.9), still valid in this case. This realization may not be minimal in general.

Similarly, if A, B, C, D satisfy (3.10), where H^{-1} is replaced by some hermitian matrix Y, then $U(z) = D + C(zI_n - A)^{-1}B$ is a rational matrix function J-unitary on the unit circle, as follows from the formula

$$U(z)JU(\omega)^* = J - (1 - z\omega^*)C(zI_n - A)^{-1}Y(\omega^* I_n - A^*)^{-1}C^* \tag{3.17}$$

valid for z and ω in the resolvent set of A.

The examples of Section 2, adapted to the case of the unit circle, are as follows.

EXAMPLE 3.1. *With the notation of Example 2.1, and with ω not on the unit circle, the function*

$$U(z) = I_m - PJ + \left(\frac{z - \omega}{1 - z\omega^*}\right) PJ \tag{3.18}$$

is J-unitary on the unit circle.

EXAMPLE 3.2. *The function*

$$U(z) = I_m + \left(\frac{z - \omega_2}{1 - z\omega_1^*} - 1\right) W_{12} + \left(\frac{z - \omega_1}{1 - z\omega_2^*} - 1\right) W_{21} \tag{3.19}$$

is J-unitary on the unit circle, where W_{ij} is defined in Example 2.2 and where ω_1 and ω_2 are of modulus different from 1.

EXAMPLE 3.3. *Let α be on the unit circle and let u be a J-neutral vector. Then*

$$U(z) = I_m + i\left(\frac{z\alpha}{1 - z\alpha}\right)^n uu^*J$$

is J-unitary on the unit circle.

Theorem 3.1. has corollaries about the existence of solutions of Stein equations just as Theorem 2.1 has corollaries about Lyapunov equations. In the present case, computations are somewhat more involved, and we begin with two lemmas.

LEMMA 3.3. *Let A and C be given matrices in $C_{n\times n}$ and $C_{m\times n}$, respectively, with A invertible, and let H and J be such that*

$$H - A^*HA = -C^*JC$$

where H is an invertible hermitian matrix and J is a signature matrix. Let α be in the resolvent set of A with $|\alpha| = 1$, and define

$$D_\alpha = I_m + CH^{-1}\left(I_n - \alpha A^*\right)^{-1} C^* J \tag{3.20}$$

$$B_\alpha = H^{-1}(A^*)^{-1}C^* J D_\alpha \tag{3.21}$$

Then

$$\begin{pmatrix} A & B_\alpha \\ C & D_\alpha \end{pmatrix}^* \begin{pmatrix} H & 0 \\ 0 & -J \end{pmatrix} \begin{pmatrix} A & B_\alpha \\ C & D_\alpha \end{pmatrix} = \begin{pmatrix} H & 0 \\ 0 & -J \end{pmatrix} \tag{3.22}$$

LEMMA 3.4. *Let A and B be given matrices in $C_{n\times n}$ and $C_{n\times m}$, respectively, with A invertible, let H be an invertible hermitian matrix, and let J be a signature matrix. Suppose that*

$$H^{-1} - AH^{-1}A^* = BJB^*$$

For α in the resolvent set of A, with $|\alpha| = 1$, define

$$D'_\alpha = I_m + JB^*\left(I_n - \alpha A^*\right)^{-1} HB \tag{3.23}$$

$$C'_\alpha = D'_\alpha J B^*(A^*)^{-1} H \tag{3.24}$$

Then

$$\begin{pmatrix} A & B \\ C'_\alpha & D'_\alpha \end{pmatrix} \begin{pmatrix} H^{-1} & 0 \\ 0 & -J \end{pmatrix} \begin{pmatrix} A & B \\ C'_\alpha & D'_\alpha \end{pmatrix}^* = \begin{pmatrix} H^{-1} & 0 \\ 0 & -J \end{pmatrix} \tag{3.25}$$

The proofs of these two lemmas are straightforward computations and will be omitted.

The counterpart of Theorem 2.2 is the following theorem.

THEOREM 3.2. *Let (C, A) be an observable pair of matrices, with A invertible and let J be a signature matrix. Then there exists a rational function J-unitary on the unit circle with minimal realization $D + C\left(zI_n - A\right)^{-1} B$ if and only if the Stein equation*

$$H - A^*HA = -C^*JC \tag{3.11}$$

has a solution which is both hermitian and invertible. Moreover, when such a solution exists, possible choices of B and D are $B_\alpha andD_\alpha$ defined in (3.20) and (3.21) and, for a given H, any other choices of B and D differ from B_α and D_α by a right J-unitary multiplicative constant matrix.

PROOF. Let H be a solution of the Stein equation (3.11) that is both hermitian and invertible, and build B_α and D_α as in (3.20) and (3.21) from some α, with $|\alpha| = 1$, in the resolvent set of A. It is easy to see that $\det D_\alpha \neq 0$. Let us consider the function $U_\alpha(z) = D_\alpha + C\left(zI_n - A\right)^{-1} B_\alpha$. It is J-unitary on the unit circle by Lemma 3.3 and Theorem 3.1. Moreover, the pair (A, B_α) is controllable; indeed, it is equivalent to show that the pair (B_α^*, A^*) is observable, or equivalently, that the pair $\left(CA^{-1}H^{-1}, A^*\right)$ is observable. Using the Stein equation (3.11), we see that for every k there exist matrices $K_0, \ldots, K_{k-1}$ such that

$$CA^{-1}H^{-1}(A^*)^k = C(A^{-1})^{k+1}H + K_0C(A^{-1})^kH + \cdots + K_{k-1}CA^{-1}H$$

Thus, if f is in the intersection $\bigcap_0^\infty \operatorname{Ker} \left(CA^{-1}H^{-1}(A^*)^k\right)$, then Hf will be in the intersection $\bigcap_1^\infty \operatorname{Ker} C(A^{-1})^k$, and hence Hf is equal to zero, since

$$\bigcap_{k=1}^{\infty} \operatorname{Ker} C(A^{-1})^k = \bigcap_{k=1}^{\infty} \operatorname{Ker} CA^k$$

Hence $f = 0$, and the given realization of U_α is minimal.

The other claims of this theorem are proved as in Theorem 2.2. □

Similarly the following theorem can be proved:

THEOREM 3.3 *Let (A, B) be a controllable pair of matrices, with A invertible and let J be a signature matrix. Then there exists a rational function J-unitary on the unit circle with minimal realization $D + C\left(zI_n - A\right)^{-1} B$ if and only if the Stein equation*

$$X - AXA^* = -BJB^* \tag{3.26}$$

has a solution which is both hermitian and invertible. Moreover, when such a solution exists, possible choices of C and D are given by $C'_\alpha andD'_\alpha$ defined in (3.23) and (3.24), where H^{-1} is the given solution of (3.26), and, for this given H, any other choices of C and D differ from C'_α and D'_α by a left J-unitary multiplicative constant.

3.2. The Associated Hermitian Matrix

In this section, we give the analogue of the results of Section 2.2.

LEMMA 3.2. *Let U be a rational function J-unitary on the unit circle and analytic and invertible at infinity. Let $D_i + C_i\left(zI_{n_i} - A\right)^{-1} B_i$, $i = 1, 2$, be two minimal realizations of U with associated hermitian matrices $H_i, i = 1, 2$. Then the two minimal realizations are linked by means of a unique similarity matrix S by equations (2.18), and H_1 and H_2 are linked by equation (2.19).*

We omit the proof since it is the same as the proof in the line case.

THEOREM 3.4. *Let U be a rational function analytic and invertible at infinity and J-unitary on the unit circle. Let $D + C\left(zI_n - A\right)^{-1} B$ be a minimal realization of U, with associated hermitian matrix H. Then the number of negative eigenvalues of H is equal to the number of negative squares of each of the functions*

$$K_U(z,\omega) = \frac{J - U(z)JU(\omega)^*}{1 - z\omega^*} \quad \textit{and} \quad K_{\widehat{U}}(\omega, z) = \frac{J - U(\omega)^*JU(z)}{1 - z\omega^*} \tag{3.27}$$

Finally, let $\mathcal{K}(U)$ be the span of the functions $z \to K_U(z,\omega)c$, where ω spans the points of analyticity of U and c spans C_m. Then $\mathcal{K}(U)$ is a finite-dimensional vector space of rational functions and its dimension is equal to the MacMillan degree of U.

PROOF. Formula (3.9) leads to the equality

$$\frac{J - U(z)JU(\omega)^*}{1 - z\omega^*} = C\left(zI_n - A\right)^{-1} H^{-1}\left(\omega^* I_n - A^*\right)^{-1} \tag{3.28}$$

valid for z and ω in the resolvent set of A. The statement about the number of negative eigenvalues of H is then proved exactly as in Theorem 2.4.

From (3.28), we see that any linear combination of functions $\frac{J-U(z)JU(\omega)^*}{1-z\omega^*}c$ is of the form

$$C\left(zI_n - A\right)^{-1} f$$

for some vector f in C_n, and thus $\dim \mathcal{K}(U) \leq n$. The observability of the pair (C, A) leads to

$$C\left(zI_n - A\right)^{-1} f \equiv 0 \quad \Rightarrow \quad f = 0$$

and then to $\dim \mathcal{K}(U) = n$. □

As in the line case, $\nu(U)$ will denote the number of negative squares of the functions defined in (3.27).

THEOREM 3.5. *Let $U_i, i = 1,2$ be two rational functions J unitary on the unit circle and analytic and invertible at infinity, with minimal realizations $D_i + C_i\left(zI_{n_i} - A_i\right)^{-1} B_i, i = 1,2$, and associated hermitian matrices $H_i, i = 1,2$. Then the matrix $\begin{pmatrix} H_1 & 0 \\ 0 & H_2 \end{pmatrix}$ is the associated hermitian matrix corresponding to the minimal realization $D + C\left(zI_n - A\right)^{-1} B$ of the product U_1U_2, where $n = n_1 + n_2$ and A, B, C, D are as in (2.27) and (2.28).*

We conclude this section with the following theorem.

THEOREM 3.6. *Let U be a rational matrix-valued function J-unitary on the unit circle and analytic at infinity. Let $U(z) = D + C(zI_n - A)^{-1}B$ be a minimal* *realization of U with associated hermitian matrix H and let α be a regular point of A with $|\alpha| = 1$. Then the function $V(\lambda) = U\left(\alpha \cdot \frac{\lambda-1}{\lambda+1}\right)$ is J-unitary on the imaginary line and $V(\lambda) = N_\alpha + M_\alpha(\lambda I_n(-K_\alpha)^{-1}L_\alpha$ with*

$$\begin{aligned} K_\alpha &= (A + \alpha I_n)(\alpha I_n - A)^{-1} \\ L_\alpha &= \sqrt{2}(\alpha I_n - A)^{-1}B \\ M_\alpha &= \sqrt{2}\alpha C(\alpha I_n - A)^{-1} \\ N_\alpha &= D + C(\alpha I_n - A)^{-1}B \end{aligned}$$

is a minimal realization of V with the same associated hermitian matrix H.

Conversely, let U be a rational matrix-valued function J-unitary on the line and analytic at infinity, with minimal realization $U(\lambda) = D + C(\lambda I_n - A)^{-1}B$ and associated hermitian matrix H. Let t be a regular point of $-A$ with $t > 0$. Then the function $V(z) = U\left(t \cdot \frac{1-z}{1+z}\right)$ is J-unitary on the unit circle and with

$$\begin{aligned} V(z) &= N_t' + M_t'(zI_n - K_t')^{-1}L_t' \\ K_t' &= (tI_n - A)(tI_n + A)^{-1} \\ L_t' &= -\sqrt{2t}(tI_n + A)^{-1}B \\ M_t' &= \sqrt{2t}C(tI_n + A)^{-1} \\ N_t' &= D - C(tI_n + A)^{-1}B \end{aligned}$$

is a minimal realization of V with the same associated hermitian matrix H.

PROOF. For α_1 with $|\alpha_1| = 1$, the map $z = \alpha_1\frac{\lambda-1}{\lambda+1}$ maps the right open half plane onto the open unit circle. The realization for the function $V(\lambda)$ is obtained via formula (1.36) from [BGK]. By easy computations we obtain that N_α is J-unitary, H satisfies the equation

$$K_\alpha^* H + HK_\alpha = -M_\alpha^* J M_\alpha$$

and

$$L_\alpha = -H^{-1}M_\alpha^* J N_\alpha \ .$$

The converse statement is proved in the same manner since the map $\lambda = t\frac{1-z}{z+1}$ maps the open unit disk onto the right open half plane for $t > 0$. □

3.3. Factorizations of Rational Functions J-Unitary on the Unit circle

This subsection deals with minimal factorizations of a rational function J-unitary on the unit circle into factors which are themselves J-unitary on the unit circle. Here, $[,]_H$ has the same significance as in Subsection 2.4.

THEOREM 3.7. *Let U be a rational function J-unitary on the unit circle and analytic and invertible at infinity. Let $U(z) = D + C(zI_n - A)^{-1}B$ be a minimal realization of U with associated hermitian matrix H. Finally, let M be an invariant subspace of A, non-degenerate in the associated inner product $[,]_H$, and let π be the projection defined by*

$$\text{Ker } \pi = M \qquad \text{Im } \pi = M^{[\perp]}$$

Then

$$U_1(z) = \left[I_m + C(zI_n - A)^{-1}(I_n - \pi)BD^{-1}\right] D_1 \tag{3.29}$$

$$U_2(z) = D_2 \left[I_m + D^{-1}C\pi(zI_n - A)^{-1}B\right] \tag{3.30}$$

with

$$D = D_1 D_2 \ ,$$

$$D_1 = I_m + C_1 H_1^{-1}(I - \alpha A_1^*)^{-1} C_1^* J \tag{3.31}$$

where α is of modulus 1 and belongs to the resolvent set of A_1 and where

$$C_1 = C\big|_M, A_1 = A\big|_m, H_1 + PH\big|_M \qquad \text{and} \qquad B_1 = (I - \pi)B$$

is a minimal J-unitary factorization of U (P being the orthogonal projection onto M in the usual metric of C_n).

Conversely, any minimal J-unitary factorization of U can be obtained in such a way, and the correspondence between J-unitary factorizations of U with $U_1(\alpha) = I_m$ and non-degenerate subspaces of A is one to one.

PROOF. Equation (3.7) reads

$$A^\times = \left(A^{[*]}\right)^{-1}$$

Thus, if M is A invariant, then $M^{[\perp]}$ is $A^\times$ invariant and when M is non-degenerate, π defines a supporting projection. The factorization $U = U_1 U_2$ is thus minimal ([BGK]). By Theorem 3.1, U may be written as

$$U(z) = \left(I_m + C\left(zI_n - A\right)^{-1} H^{-1} C^* J\right) D$$

and so,

$$U_1(z) = \left(I_m + C\left(zI_n - A\right)^{-1}(I_n - \pi)H^{-1}(A^*)^{-1}C^* J\right) D_1 \ .$$

We want to show that the choice of D_1 given in (3.31) leads to a factor U_1 J-unitary on the unit circle. To that purpose, we first remark that H_1 is invertible and satisfies the Stein equation

$$H_1 - A_1^* H_1 A_1 = -C_1^* J C_1 \tag{3.32}$$

The arguments follow those of Theorem 2.6 and are omitted. The function U_1 can also be written as

$$U_1(z) = \left[I_m + C_1 (zI - A_1)^{-1} H_1^{-1} (A_1^*)^{-1} C_1^* J\right] D_1 \tag{3.33}$$

and an application of Theorem 3.1 permits us therefore to conclude that U_1 is J-unitary on the unit circle.

The other statements of the theorem are proved in much the same way as in Theorem 2.6. □

The counterpart of Theorem 2.7 is the following theorem.

THEOREM 3.8. *Let U be a rational function J-unitary on the unit circle and analytic and invertible at infinity. Let $U(z) = D + C(zI_n - A)^{-1} B$ be a minimal realization of U, with associated hermitian matrix H. Let f be an eigenvector of A, corresponding to the eigenvalue ω, and suppose $[f, f]_H \neq 0$. Then U admits a minimal J-unitary factorization*

$$U = U_1 U_2$$

where U_1 has the form

$$U_1(z) = I_m - P + \frac{1 - z\omega^*}{z - \omega} P \tag{3.34}$$

with P a one-dimensional projection subject to

$$PJP^* = J$$

for the case $|\omega| \neq 1$, and where U_1 has the form

$$U_1(z) = I_m + ik \cdot uu^* J \cdot \frac{z + \omega}{z - \omega} \tag{3.35}$$

for some real number k and some J-neutral vector u for the case $|\omega| = 1$.

The proof is similar to the proof of Theorem 2.7 and will be omitted. It relies on Theorem 3.6 and on the equality

$$[f, f]_H (1 - |\mu|^2) \; = \; \langle JCf, Cf \rangle \tag{3.36}$$

for $Af = \mu f$, which follows from the Stein equation (3.11).

3.4. Rational Functions Unitary on the Unit Circle.

Here we study the special case $J = I_m$, i.e. functions which are unitary on the unit circle. The proofs of the results are close to the proofs of results in Subsection 2.4 and will be either omitted or merely outlined. We begin with the following theorem.

THEOREM 3.9. *Let U be a matrix-valued rational function analytic and invertible at infinity, and let $U(z) = D + C(zI_n - A)^{-1}B$ be a minimal realization of U. Then U is unitary on the unit circle if and only if the following conditions hold.*

a) *U is analytic and invertible at the origin*

b) *There exists an hermitian matrix H such that*

$$\begin{pmatrix} A & B \\ C & D \end{pmatrix}^* \begin{pmatrix} H & 0 \\ 0 & -I_m \end{pmatrix} \begin{pmatrix} A & B \\ C & D \end{pmatrix} = \begin{pmatrix} H & 0 \\ 0 & -I_m \end{pmatrix} \tag{3.37}$$

Condition b) is equivalent to

b') There exists an hermitian matrix such that

$$\begin{pmatrix} A & B \\ C & D \end{pmatrix} \begin{pmatrix} G & 0 \\ 0 & -I_m \end{pmatrix} \begin{pmatrix} A & B^* \\ C & D \end{pmatrix} = \begin{pmatrix} G & 0 \\ 0 & -I_m \end{pmatrix} \tag{3.38}$$

PROOF. The hermitian matrix H which satisfies (3.37) satisfies the equation

$$H - A^*HA = -C^*C$$

and thus H is invertible by Theorem 3.6. Hence this theorem follows from Theorem 3.1.

The counterpart of Theorem 3.2 in the present case is the following theorem.

THEOREM 3.10. *Let (C, A) be an observable pair of matrices, with A invertible. Then there exists a rational function unitary on the unit circle with minimal realization $U(z) = D + C(zI_n - A)^{-1}B$ if and only if the Stein equation*

$$H - A^*HA = -C^*C$$

has an hermitian solution. If such a solution exists, it is invertible and possible choices of B and D are given by B_α and D_α, defined in (3.20) and (3.21) (with $J = I_m$), and, for a given H, any other choices of B and D differ from B_α and D_α by a right unitary multiplicative constant matrix.

The analogue of Theorem 2.10 is the following.

THEOREM 3.11. *Let U be a rational function unitary on the unit circle and analytic and invertible at infinity. Let $U(z) = D + C(zI_n - A)^{-1}B$ be a minimal realization of U, with associated matrix H and associated inner product $[,]_H$. Then*

a) *A has no spectrum on the unit circle*

b) $[f,f]_H \neq 0$ *for any eigenvector* f *of* A

c) U *is the product of a unitary constant matrix and a minimal product of* n *rational functions of the form*

$$I_m - P + \frac{z-\omega}{1-z\omega^*}P \tag{3.39}$$

where P *is an orthogonal projection of rank one and* $|\omega| \neq 1$.

PROOF. The proof of this theorem relies on equation (3.36), with $J = I_m$, which implies conditions a) and b), and on Theorem 3.8, which, iterated n times, proves assertion c). □

A suitable Moebius transformation permits us to get the multiplicative structure even when U is not analytic at infinity.

THEOREM 3.12. *Let* U *be a rational function unitary on the unit circle, of Mac Millan degree* n. *Then* U *is a product of* n *factors of the form (3.39).*

The matrix counterpart of Theorem 3.11 is the following theorem.

THEOREM 3.13. *Let* (C, A) *be an observable pair of matrices, with* A *invertible, and suppose that the Stein equation*

$$H - A^*HA = -C^*C$$

has an hermitian solution H. *Then* H *is invertible and there exists a basis* $\varphi_0, \ldots, \varphi_{n-1}$ *of* C_n *such that, in this basis,* A *has an upper triangular form,* $A^{[*]}$ *has a lower triangular form, and all the spaces span* $\{\varphi_0, \ldots, \varphi_j\}, j = 0, \ldots, n-1$, *are non-degenerate in* $[,]_H$.

In an analogous way we can also introduce the Blaschke-Potapov product for the unit circle, and then Theorem 2.12 has the following analogue.

THEOREM 3.14. *Let* U *be a rational function unitary on the unit circle and let* $\nu(U)$ *be the number of negative squares of the function* $\frac{I_m - U(z)U^*(\omega)}{1-z\omega^*}$. *Then there exist Blaschke-Potapov products* B_1, B_2, B_3, B_4 *such that*

$$U = B_1B_2^{-1} = B_3^{-1}B_4$$

with

$$\deg\ U = \deg\ B_1 + \deg\ B_2 = \deg\ B_3 + \deg\ B_4$$

and

$$\deg\ B_3 = \deg\ B_2 = \nu(U)\ .$$

In particular, $\nu(U)$ is equal to the number of poles of U in the open unit circle $\mathbb{D}$, and $n - \nu(U)$ is equal to the number of poles of U outside the open unit disk.

PROOF. If the function U is analytic and invertible at infinity, the assertion follows from Theorem 3.8. We start with the eigenvalues outside $\mathbb{D}$ to get the first representation, and the eigenvalues inside $\mathbb{D}$ to get the second one. When the function fails to be analytic and invertible at infinity, we consider first $V(z) = U(\phi(z))$, where ϕ is a Moebius transformation mapping $\mathbb{D}$ onto $\mathbb{D}$ and such that V is analytic and invertible at infinity. The assertion follows then for U. □

We conclude this subsection with the following inertia theorem.

THEOREM 3.15. *Let (C, A) be an observable pair of matrices with A invertible, and let H be an hermitian solution of the Stein equation*

$$H - A^*HA = -C^*C$$

Then the matrix H is invertible, the matrix A has no spectrum on the unit circle, and the number of negative (resp. positive) eigenvalues of H is equal to the number of eigenvalues of A in the unit disk (resp. outside the unit disk).

PROOF. The matrix H is invertible by Theorem 3.6. We then build a matrix-valued function U unitary on the unit circle with minimal realization $U(z) = D+C(zI_n - A)^{-1}B$, with B and D as in (3.20) and (3.21) with $J = I_m$. From Theorem 3.11 it follows that A has no spectrum on the unit circle and the statements about the eigenvalues follow from Theorem 3.14. □

Theorem 3.14 is the finite-dimensional version of a more general theorem of Krein and Langer (see [KL]). The latter paper [KL] contains a general theory of the operator-valued functions f analytic at the origin, and such that $\frac{I-f(\lambda)f^*(\omega)}{1-\lambda\omega^*}$ has a finite number of negative squares. We also mention the connections with the paper [DLS], in which functions unitary on the unit circle are studied in the framework of characteristic operator functions.

Finally, results of Subsection 2.5 have obvious counterparts in the circle case and we omit their formulation.

4. RATIONAL FUNCTIONS SELFADJOINT ON THE LINE.

A $C_{p\times p}$ valued rational function ϕ is called selfadjoint on the line if at all purely imaginary points where it is defined we have

$$\phi(\lambda) = \phi(\lambda)^* \qquad \lambda \in \mathbb{R} \tag{4.1}$$

The aim of this section is to study the minimal selfadjoint additive decompositions of ϕ, i.e., decompositions of ϕ of the form

$$\phi = \phi_1 + \phi_2$$

where both ϕ_1 and ϕ_2 are selfadjoint on the line and

$$\deg \phi = \deg \phi_1 + \deg \phi_2$$

Rational functions selfadjoint on the line are closely related to certain rational functions that are J_1-unitary on the line, where

$$J_1 = \begin{pmatrix} 0 & I_p \\ I_p & 0 \end{pmatrix} \tag{4.2}$$

Indeed, if ϕ is selfadjoint on the line, then the function

$$U(\lambda) = \begin{pmatrix} I_p & i\phi(\lambda) \\ 0 & I_p \end{pmatrix} \tag{4.3}$$

is J_1-unitary on the line.

This section is modelled after Section 2 and is divided into three subsections. It focuses on functions analytic at infinity.

4.1. Realization Theorems

We first study realizations of rational functions selfadjoint on the real line. Theorem 4.1 with a complete proof can be found in [GLR1] and [GLR2].

THEOREM 4.1. *Let ϕ be a $C_{p\times p}$ valued function analytic at infinity, and let $\phi(\lambda) = D + C(\lambda I_n - A)^{-1} B$ be a minimal realization of ϕ. Then ϕ is selfadjoint on the imaginary line if and only if the following conditions hold.*

a) *D is selfadjoint*

b) *There exists an invertible hermitian matrix H such that*

$$A^*H + HA = 0 \tag{4.4}$$

and

$$C = iB^*H \tag{4.5}$$

PROOF. The function U defined in (4.3) is J_1-unitary on the line if and only if ϕ is selfadjoint on the line; moreover, it admits the minimal realization

$$U(\lambda) = \begin{pmatrix} I_p & iD \\ 0 & I_p \end{pmatrix} + i\begin{pmatrix} C \\ 0 \end{pmatrix} (\lambda I_n - A)^{-1} \begin{pmatrix} 0 & B \end{pmatrix} \tag{4.6}$$

Using Theorem 2.1, we thus see that ϕ is selfadjoint on the line if and only if

1) $\begin{pmatrix} I_p & iD \\ 0 & I_p \end{pmatrix}$ is J_1-unitary

2) There exists an invertible hermitian matrix H such that

$$A^*H + HA = -\begin{pmatrix} iC \\ 0 \end{pmatrix} J_1 \begin{pmatrix} iC \\ 0 \end{pmatrix}^* \tag{4.7}$$

and

$$\begin{pmatrix} 0 & B \end{pmatrix} = -H^{-1}\begin{pmatrix} iC \\ 0 \end{pmatrix}^* J_1 \begin{pmatrix} I_p & iD \\ 0 & I_p \end{pmatrix} \tag{4.8}$$

These conditions are easily seen to be equivalent to a) and b) in the statement of the theorem. □

From Theorem 2.1 it follows that the matrix H is uniquely determined from the given minimal realization of ϕ. It will be called the *associated hermitian matrix* associated with the given minimal realization of ϕ. The matrix H is also given by the formulas

$$H = -\left[\operatorname{col}\left(B^*(A^*)^j\right)_0^{n-1}\right]^+ \left[\operatorname{col}\left(C(A)^j\right)_0^{n-1}\right] \tag{4.9}$$

and

$$H = -\left[\operatorname{row}\left((A^*)^j C^*\right)_0^{n-1}\right]\left[\operatorname{row}\left(A^j B\right)_0^{n-1}\right]^\dagger \tag{4.10}$$

Equations (4.4) and (4.5) easily lead to

$$\phi(\lambda) - \phi(\omega)^* = i(\lambda + \omega^*)C\left(\lambda I_n - A\right)^{-1} H^{-1}\left(\omega^* I_n - A^*\right)^{-1} C^* \tag{4.11}$$

$$\phi(\lambda) - \phi(\omega)^* = i(\lambda + \omega^*)B^*\left(\omega^* I_n - A^*\right)^{-1} H\left(\lambda I_n - A\right)^{-1} B \tag{4.12}$$

We note that if A, B, C are matrices that satisfy (4.4) and (4.5) for some not necessarily invertible hermitian matrix H, and if D is a selfadjoint matrix, then $\phi(\lambda) = D + C\left(\lambda I_n - A\right)^{-1} B$ is a rational function selfadjoint on the line. This follows from the fact that (4.12) is still valid in this case. The realization of ϕ is in general not minimal.

If A, B, C satisfy the equalities

$$GA^* + AG = 0 \tag{4.13}$$

$$B = iGC^* \tag{4.14}$$

for some not necessarily invertible hermitian matrix G, then equation (4.11) is valid with H^{-1} replaced by G, and hence ϕ is selfadjoint on the line.

As in Section 2, we can solve inverse problems using Theorem 4.1.

THEOREM 4.2. *Let (C, A) be an observable pair of matrices. Then there exists a rational function selfadjoint on the line with minimal realization $D+C\left(\lambda I_n - A\right)^{-1} B$ if and only if the equation*

$$A^*H + HA = 0$$

has a solution which is both invertible and hermitian. When such a solution H exists, D can be any hermitian matrix and B is equal to $iH^{-1}C^*$.

THEOREM 4.3. *Let (A, B) be a controllable pair of matrices. Then there exists a rational function selfadjoint on the line with minimal realization $D+C\left(\lambda I_n - A\right)^{-1} B$ if and only if the equation*

$$GA^* + AG = 0$$

has a solution which is both hermitian and invertible. When such a solution G exists, C is equal to $-B^*G^{-1}$ and D may be any hermitian matrix.

The proofs are easy and omitted.

4.2. The Associated Hermitian Matrix

We now present the analogues of the results of Subsection 2.2 in the framework of rational functions selfadjoint on the line.

LEMMA 4.1. *Let ϕ be a $C_{p\times p}$ valued rational function analytic at infinity and selfadjoint on the real line, and let $\phi(\lambda) = D + C_i\left(\lambda I_n - A_i\right)^{-1} B_i, i = 1, 2$, be two minimal realizations of ϕ, with associated hermitian matrices $H_i, i = 1, 2$. Then the two minimal realizations and the matrices H_i are linked by equations (2.18)-(2.19).*

The proof follows from Lemma 3.2 by considering the function U defined in (4,3), which is J_1-unitary on the line.

For λ and ω points of analyticity of ϕ we have

$$\frac{J_1 - U(\lambda)J_1U(\omega)^*}{(\lambda + \omega^*)} = \begin{pmatrix} \frac{\phi(\lambda)-\phi(\omega)^*}{i(\lambda+\omega^*)} & 0 \\ 0 & 0 \end{pmatrix} \tag{4.15}$$

Combining this equation with (4.11) and (4.12), we have the following analogue of Theorem 2.4.

THEOREM 4.4. *Let ϕ be a $C_{p\times p}$ valued function analytic at infinity and selfadjoint on the real line and let $\phi(\lambda) = D + C\left(\lambda I_n - A\right)^{-1} B$ be a minimal realization* *of ϕ, with associated hermitian matrix H. Then the number of negative eigenvalues of H is equal to the number of negative squares of the function*

$$\frac{\phi(\lambda)-\phi(\omega)^*}{i(\lambda+\omega^*)} \tag{4.16}$$

and the linear span of the functions $\lambda \to \frac{\phi(\lambda)-\phi(\omega)^*}{i(\lambda+\omega^*)}c$, where ω spans the points of analyticity of ϕ and c spans C_p, is a finite-dimensional vector space of rational functions of dimension $\deg\phi$.

To state the analogue of Theorem 2.5, we need the following definition.

DEFINITION 4.1. *Let ϕ_1 and ϕ_2 be two $C_{p\times p}$ valued rational functions. The sum $\phi_1+\phi_2$ is minimal if*

$$\deg(\phi_1+\phi_2) = \deg\phi_1 + \deg\phi_2\ . \tag{4.17}$$

THEOREM 4.5. *Let $\phi_i, i=1,2$, be two $C_{p\times p}$ valued rational functions, analytic at infinity and selfadjoint on the line, let $\phi_i(\lambda) = D_i + C_i\left(\lambda I_{n_i} - A_i\right)^{-1} B_i$, $i=1,2$, be minimal realizations of $\phi_i, i=1,2$, with associated hermitian matrices $H_i, i=1,2$, and suppose the sum $\phi=\phi_1+\phi_2$ is minimal. Then $\phi(\lambda) = D + C\left(\lambda I_n - A\right)^{-1} B$, where*

$$n = n_1+n_2, \quad D = D_1+D_2$$

$$C = (C_1\ C_2), \qquad B = \binom{B_1}{B_2}$$

and

$$A = \begin{pmatrix} A_1 & 0 \\ 0 & A_2 \end{pmatrix}$$

is a minimal realization of ϕ, with associated hermitian matrix $\begin{pmatrix} H_1 & 0 \\ 0 & H_2 \end{pmatrix}$

4.3 Minimal Additive Selfadjoint Decompositions.

The set of all minimal selfadjoint decompositions of a given matrix-valued rational function selfadjoint on the line is given by the following theorem.

THEOREM 4.6. *Let ϕ be a $C_{p\times p}$ valued rational function analytic at infinity and selfadjoint on the line. Let $\phi(\lambda) = D + C\left(\lambda I_n - A\right)^{-1} B$ be a minimal realization of U with associated hermitian matrix H. Finally, let M be an invariant subspace of A, non-degenerate in the associated inner product $[\ ,\ ]_H$, let π be the projection defined by*

$$\text{Ker}\,\pi = M \qquad \text{Im}\,\pi = M^{[\perp]}$$

and let $D = D_1 + D_2$ *be a decomposition of* D *into two hermitian matrices. Then the decomposition* $\phi = \phi_1 + \phi_2$, *where*

$$\phi_1(\lambda) = D_1 + C(\lambda I_n - A)^{-1}(I_n - \pi)B \tag{4.18}$$

$$\phi_2(\lambda) = D_2 + C\pi(\lambda I_n - A)^{-1}B \tag{4.19}$$

is a minimal selfadjoint decomposition of ϕ.

Conversely, any minimal selfadjoint decomposition of ϕ *can be obtained in such a way and, for a fixed decomposition* $D = D_1 + D_2$, *the correspondence between minimal selfadjoint decompositions of* ϕ *and non-degenerate invariant subspaces of* A *is one to one.*

PROOF. Let U be the function J_1-unitary on the line defined in (4.2). A minimal realization of U is given by equation (4.6), and then

$$A^{\times} = A - (0\ B)\begin{pmatrix} I_p & iD \\ 0 & I_p \end{pmatrix}^{-1}\begin{pmatrix} iC \\ 0 \end{pmatrix} = A \tag{4.20}$$

Equation (4.4) implies that the main operator A in (4.6) is hermitian in the associated inner product $[\ ,\ ]_H$ and hence, since $A = A^{\times}$, for any non-degenerate A-invariant subspace M, π is a supporting projection and generates minimal factorizations of U by

$$U_1(\lambda) = T_1 + i\begin{pmatrix} C \\ 0 \end{pmatrix}(\lambda I_n - A)^{-1}(I_n - \pi)(0\ B)T_2^{-1}$$

$$U_2(\lambda) = T_2 + iT_1^{-1}\begin{pmatrix} C \\ 0 \end{pmatrix}\pi(\lambda I_n - A)^{-1}(0\ B)$$

where T_1 and T_2 are $C_{2p\times 2p}$ elements such that

$$T_1T_2 = \begin{pmatrix} I_p & iD \\ 0 & I_p \end{pmatrix}$$

We choose

$$T_1 = \begin{pmatrix} I_p & iD_1 \\ 0 & I_p \end{pmatrix}$$

Then the fact that $U = U_1U_2$ leads to the additive decomposition

$$\phi = \phi_1 + \phi_2$$

where ϕ_1 and ϕ_2 are defined by (4.16) and (4.17). It is clear that this decomposition is a selfadjoint minimal decomposition when D_1 (and thus D_2) is selfadjoint. The other statements of the theorem are proved as in Theorem 2.6. □

It is clear that a rational function selfadjoint on the line may lack minimal selfadjoint decompositions, as the example $f(\lambda) = \frac{1}{(\lambda-\gamma_0)^2}$, $\gamma_0 \in i\mathbb{R}$, shows.

We conclude this section with the following analogue of Theorem 2.7.

THEOREM 4.7. *Let ϕ be a $C_{p\times p}$ valued rational function analytic at infinity and selfadjoint on the line. Let $D + C(\lambda I_n - A)^{-1} B$ be a minimal realization of U, with associated hermitian matrix H. Let f be an eigenvector of A, corresponding to the eigenvalue ω, and suppose $[f, f]_H \neq 0$. Then ω is purely imaginary and the decomposition $\phi = \phi_1 + \phi_2$, where*

$$\phi_1(\lambda) = \frac{Cff^*C^*}{i(\lambda + \omega^*)[f, f]_H}$$

is a minimal selfadjoint decompositon of ϕ.

The proof is a simple consequence of Theorem 2.7 and will be omitted.

5. RATIONAL FUNCTIONS SELFADJOINT ON THE UNIT CIRCLE

In this section we briefly review analogues of theorems of section 4 for the unit circle case. We begin with the following theorem.

THEOREM 5.1 *Let ϕ be a $C_{p\times p}$ valued rational function analytic at infinity and let $\phi(z) = D + C(zI_n - A)^{-1} B$ be a minimal realization of ϕ. Then ϕ is selfadjoint on the unit circle if and only if*

a) ϕ is analytic at the origin

b) There exists an hermitian invertible matrix H such that

$$A = H(A^*)^{-1}H^{-1} \tag{5.1}$$

$$D - D^* = iCHC^* \tag{5.2}$$

$$B = iAH^{-1}C^* \tag{5.3}$$

PROOF. Let us consider the function $U(z) = \begin{pmatrix} I_p & i\phi(z) \\ 0 & I_p \end{pmatrix}$. It is analytic and invertible at infinity, since ϕ is analytic at infinity. Using Theorem 3.1, we see that U is J_1 unitary on the unit circle if and only if

1) U is analytic and invertible at the origin

2) There exists a hermitian invertible matrix H such that

$$\begin{pmatrix} A & 0 & B \\ iC & I_p & iD \\ 0 & 0 & I_p \end{pmatrix}^* \begin{pmatrix} H & 0 & 0 \\ 0 & 0 & I_p \\ 0 & I_p & 0 \end{pmatrix} \begin{pmatrix} A & 0 & B \\ iC & I_p & iD \\ 0 & 0 & I_p \end{pmatrix} = \begin{pmatrix} H & 0 & 0 \\ 0 & 0 & I_p \\ 0 & I_p & 0 \end{pmatrix} \tag{5.4}$$

and these two conditions are easily seen to be equivalent to the conditions stated in the theorem. □

For a given minimal realization of ϕ the matrix H is unique, as follows from Theorem 3.1. It will be called the associated hermitian matrix of ϕ (for the given minimal realization).

The associated hermitian matrix H permits us to describe the minimal selfadjoint decomposition of ϕ.

THEOREM 5.2. *Let ϕ be a $C_{p\times p}$ valued rational function selfadjoint on the unit circle and analytic at infinity. Let $\phi(z) = D + C\left(zI_n - A\right)^{-1} B$ be a minimal realization of ϕ, with associated hermitian matrix H. Finally, let M be an invariant subspace of A, non-degenerate in $[\ ,\ ]_H$, and let π be the projection defined by*

$$\operatorname{Ker} \pi = M \qquad \operatorname{Im} \pi = M^{[\perp]}$$

Then

$$\phi = \phi_1 + \phi_2$$

where

$$\phi_1(z) = D + C\left(zI_n - A\right)^{-1}(I_n - \pi)B$$

$$\phi_2(z) = D_2 + C\pi\left(zI_n - A\right)^{-1} B$$

with $D_1 = \frac{i}{2}C_1H_1^{-1}C_1^* + S$, S being an arbitrary selfadjoint matrix and

$$C_1 = C\big|_M, H_1 = PH\big|_M$$

is a minimal selfadjoint decomposition of ϕ (P denoting the orthogonal projection onto M in the usual metric of C_n).

Conversely, any minimal selfadjoint decomposition of ϕ is obtained in such a way, and for fixed S the correspondence between minimal selfadjoint decompositions of ϕ and non-degenerate invariant subspaces of A is one to one.

The proof follows the proof of Theorem 3.6, specialized to the case $J = J_1, U(z) = \begin{pmatrix} I_p & i\phi(z) \\ 0 & I_p \end{pmatrix}$, and is omitted.

Finally, we note that for ϕ selfadjoint on the unit circle the function $i\phi$ is called pseudo-Carathéodory; such functions are investigated in [DGK].

6. FINITE-DIMENSIONAL DE BRANGES SPACES AND SPECIAL REALIZATIONS

Finite-dimensional de Branges spaces could be used to generate a special realization of rational J-unitary functions. In [A] and [AD] an approach was developed for divisibility of such functions based on this realization. In this section we explain the connections between this approach and the approach presented in the previous sections.

Let U be a $C_{m\times m}$ valued rational function J-unitary on the unit circle or on the imaginary line. Recall that $\mathcal{K}(U)$ was defined to be the linear span of the functions $K_U(\cdot,\omega)c$, where ω is a point of analyticity of U and c is in C_m, and where K_U is defined by

$$K_U(\lambda,\omega)=\frac{J-U(\lambda)JU(\omega)^*}{(\lambda+\omega^*)} \tag{6.1a}$$

if U is J-unitary on the imaginary line and by

$$K_U(z,\omega)=\frac{J-U(z)JU(\omega)^*}{1-z\omega^*} \tag{6.1b}$$

if U is J-unitary on the unit circle. In either case, we define an hermitian form $[\ ,\ ]_U$ by

$$[K_U(\cdot,\omega)c,K_U(\cdot,\nu)d]_U=d^*K(\nu,\omega)c$$

This form is easily seen to be well defined on the linear span of the functions $\lambda\to K_U(\lambda,\omega)c$, that is, if f and g are in this linear span and

$$f(\lambda)=\sum K_U(\lambda,\omega_i)c_i=\sum K_U(\lambda,\omega_i')c_i'$$

$$g(\lambda)=\sum K_U(\lambda,\nu_i)d_i=\sum K_U(\lambda,\nu_i')d_i'$$

then,

$$[f,g]_U=\left[\sum K_U(\cdot,\omega_i)c_i,\sum K_U(\cdot,\nu_i)d_i\right]_U=\left[\sum K_U(\cdot,\omega_i')c_i',\sum K_U(\cdot,\nu_i')d_i'\right]_U$$

DEFINITION 6.1. *Let U be a $C_{m\times m}$ valued rational function J-unitary on the line or on the unit circle. The space $\mathcal{K}(U)$ endowed with the hermitian form $[\ ,\ ]_U$ is called the de Branges space associated with U.*

The de Branges space $\mathcal{K}(U)$ provides a realization in which the minimal J-unitary factorizations of U may be described. To see this, we need some properties of the spaces $\mathcal{K}(U)$. We first define the operator R_α by

$$(R_\alpha f)(z)=\frac{f(z)-f(\alpha)}{z-\alpha} \tag{6.2}$$

where f is a vector-valued function analytic at the point α. The family (R_α) satisfies the resolvent identity

$$R_\alpha-R_\beta=(\alpha-\beta)R_\alpha R_\beta \tag{6.3}$$

and a space M of functions analytic in some open set Ω will be resolvent invariant if $R_\omega M \subset M$ for ω in Ω. When the space M is finite-dimensional, equation (6.3) implies that $R_\omega M \subset M$ for all ω in Ω if and only if the inclusion holds for some ω in Ω.

·We now gather the main properties of the de Branges space which will be needed in the sequel.

THEOREM 6.1. *Let U be a $C_{m\times m}$ valued rational function J-unitary on the imaginary line or the unit circle. Then the associated de Branges space is a finite-dimensional resolvent invariant space of C_m valued rational functions analytic everywhere except at the poles of U, and the dimension of $\mathcal{K}(U)$ is equal to the MacMillan degree of U. The dimension of any maximal negative subspace of $\mathcal{K}(U)$ is $\nu(U)$. Finally, for any point α of analyticity of U and any c in C_m, the function $(R_\alpha U)c$ belongs to $\mathcal{K}(U)$.*

The definition of $\nu(U)$ is in Subsection 2.2. A proof of this theorem appears in [AD]. The theorem can also be proved directly using formulas (2.24) for the line case and (3.28) for the unit circle case.

Let us now consider the case where U is analytic and invertible at the origin. Then the equation

$$U(z) = U(0) + z\frac{U(z) - U(0)}{z}$$

may also be written as

$$U(z) = D + zC(I - zA)^{-1}B \tag{6.4}$$

or

$$U(z) = D - CA^{-1}B - CA^{-1}\left(zI - A^{-1}\right)^{-1}A^{-1}B \tag{6.5}$$

where A, B, C, D are defined as follows:

$$A: \quad f \in \mathcal{K}(U) \to R_0 f \in \mathcal{K}(U)$$

$$B: c \in C_m \to (R_0 U)c \in \mathcal{K}(U)$$

$$C: f \in \mathcal{K}(U) \to f(0) \in C_m$$

$$D: c \in C_m \to U(0)c \in C_m\ .$$

The minimality of (6.5) follows from the fact that $\dim \mathcal{K}(U) = \deg U$.

We now focus on the analogue of the *associated hermitian matrix* in the present context. We first recall (see [dB], [B]) that, when U is analytic at the origin, the following identity holds in $\mathcal{K}(U)$:

$$[Af,g]_U + [f,Ag]_U = -g^*(0)Jf(0) \tag{6.6a}$$

for any f,g in $\mathcal{K}(U)$ if U is J-unitary on the line, and

$$[f,g]_U - [Af,Ag]_U = g^*(0)Jf(0) \tag{6.6b}$$

if U is J-unitary on the unit circle.

Let $\langle\ ,\ \rangle_U$ be some inner product with respect to which $\mathcal{K}(U)$ is a Hilbert space. Then there exists a unique operator from $\mathcal{K}(U)$ into itself such that

$$[f,g]_U = \langle f, Hg\rangle_U$$

The operator H is hermitian in $\mathcal{K}(U)$ endowed with $\langle\ ,\ \rangle_U$, and equation (6.6) becomes

$$HA + A^*H = -C^*JC \qquad \text{(imaginary line case)} \tag{6.7a}$$

$$H - A^*HA = C^*JC \qquad \text{(unit circle case)} \tag{6.7b}$$

where A^* denotes the adjoint of A with respect to the inner product $\langle\ ,\ \rangle$, and C^* denotes the adjoint of the operator C from $\mathcal{K}(U)$ endowed with $\langle\ ,\ \rangle_U$ into $\mathbf{C}_m$ endowed with the usual inner product. The equalities (6.7) can also be written in the form

$$(A^*)^{-1}H + H(A^{-1}) = -\left(CA^{-1}\right)^* JCA^{-1} \tag{6.8a}$$

and

$$H - \left(CA^{-1}\right)^* H\left(CA^{-1}\right) = -\left(CA^{-1}\right)^* J\left(CA^{-1}\right) \tag{6.8b}$$

which present H as a solution of the Lyapunov equation (2.2) (imaginary line case) or the Stein equation (3.11) (circle case) associated with the minimal realization (6.5).

In fact, it is easy to see that, in the line case, H satisfies equations (2.3) for the minimal realization (6.5), i.e.,

$$A^{-1}B = -H^{-1}(A^*)^{-1}C^*J\left(D - CA^{-1}B\right) \tag{6.9}$$

and in the circle case, H satisfies equation (3.2), which is related to the minimal realization (6.5). Thus, using the factorization theorems of Sections 2 and 3, we get the following result.

THEOREM 6.2. *Let U be a $C_{m\times m}$ valued rational function analytic and invertible at the origin and J-unitary on the unit circle or on the imaginary line. Then there is a one to one correspondence between minimal J-unitary factorizations of U (up to J-unitary constants) and non-degenerate resolvent invariant subspaces of $\mathcal{K}(U)$.*

This theorem appears in [AD].

When $J = J_1$ and $U = \begin{pmatrix} I_p & i\phi \\ 0 & I_p \end{pmatrix}$, where ϕ is a $C_{p\times p}$ valued selfadjoint rational function, the function defined in (6.1) becomes

$$K_U(\lambda,\omega) = \begin{pmatrix} k_\phi(\lambda,\omega) & 0 \\ 0 & 0 \end{pmatrix}$$

where

$$k_\phi(z,\omega) = -i\frac{\phi(z)-\phi(\omega)^*}{1-z\omega^*} \tag{6.10a}$$

if ϕ is selfadjoint on the unit circle while

$$k_\phi(\lambda,\omega) = \frac{\phi(\lambda)-\phi(\omega)^*}{(\lambda-\omega^*)} \tag{6.10b}$$

if ϕ is selfadjoint on the line.

We will denote by $\mathcal{L}(\phi)$ the linear span of the functions $k_\phi(\cdot,\omega)c$, where ω spans the points of analyticity of ϕ and c spans C_p. On $\mathcal{L}(\phi)$ one defines the analogue of $[\ ,\ ]_U$ by

$$[k_\phi(\cdot,\omega)c, \quad k_\phi(\cdot,\nu)d]_\phi = d^* k_\phi(\nu,\omega)c$$

The space $\mathcal{L}(\phi)$ endowed with the hermitian form $[\ ,\ ]_\phi$ will be called the de Branges space associated with ϕ. Properties of $\mathcal{L}(\phi)$ can be read from the associated $\mathcal{K}(U)$, where $U = \begin{pmatrix} I_p & i\phi \\ 0 & I_p \end{pmatrix}$, and when analytic at 0, ϕ admits the representation

$$\phi(z) = D + zC(I-zA)^{-1}B$$

where $D = \phi(0), A = R_0$ in $\mathcal{L}(\phi)$, $Cf = f(0)$, and $Bc = \frac{\phi(z)-\phi(0)}{z}c$.

Finally, there is a one to one correspondence between non-degenerate resolvent invariant subspaces of $\mathcal{L}(\phi)$ and minimal selfadjoint decompositions of ϕ (up to additive selfadjoint constants).

REFERENCES

[A] D. Alpay, Reproducing kernel Krein spaces of analytic functions and inverse scattering, Ph.D. Thesis, The Weizman Institute of Science, (Rehovot, Israel), October 1985.

[AD] D. Alpay and H. Dym, On applications of reproducing kernel spaces to the Schur algorithm and rational J-unitary factorizations, in: I. Schur Methods in Operator Theory and Signal Processing, Edited by I. Gohberg, OT18: Operator Theory: Advances and Applications, Vol. 18, Birkhäuser Verlag, Basel, 1986.

[AG] D. Alpay and I. Gohberg, Unitary rational matrix functions and orthogonal matrix polynomials, preprint, 1987.

[B] J. Ball, Models for non-contractions, *J. Math. Analysis and Applications*, **52** (1975), 235-254.

[BR] J. Ball and A. Ran, Local inverse spectral problems for rational matrix functions, *Integral Equation and Operator Theory*, **10** (1987), 350-415.

[BGK] H. Bart, I. Gohberg and M.A. Kaashoek, *Minimal factorizations of matrix and Operator Functions*, OT1: Operator Theory: Advances and Applications, Vol 1, Birkhäuser Verlag, Basel, 1979.

[dB] L. de Branges, *Hilbert spaces of entire functions*, Prentice Hall, Englewood Cliffs, N.J., 1968.

[Br] M.S. Brodskii, *Triangular and Jordan Representations of Linear Operators*, Trans. Math. Monographs, Vol 32, Amer. Math. Soc. Providence, R.I., 1971.

[BrGK] V.M. Brodskii, I. Gohberg and M.G. Krein, On characteristic functions of an invertible operator, *Acta Sc. Math. Szeged*, **32** (1971), 140-164.

[C] B. Cain, Inertia Theory, *Linear algebra and its applications*, **30** (1980), 211-240.

[D] J. Doyle, Advances in multivariable control, ONR/Honeywell workshop, 1984.

[DGK] P. Delsarte, Y. Genin and Y. Kamp, Pseudo Lossless functions with applications to the problem of locating the zeros of a polynomial, Manuscript M74, January 1984, Phillips Research Laboratory, Brussels, Av. Van Beceleare 2, Box 8, B1170, Brussels, Belgium.

[DD] P. Dewilde and H. Dym, Lossless inverse scattering for digital filters, *IEEE Trans. Inf. Theory*, **30** (1984), 644-662.

[DLS] A.Dijksma, H. Langer and H.S.V. de Snoo, Characteristic functions of unitary operator colligations in π_k spaces, in Operator Theory and Systems, edited by H. Bart, I. Gohberg and M.A. Kaashoek, OT 19: Operator Theory: Advances and Applications, Vol 19, Birkhäuser Verlag, Basel, 1986.

[G] K. Glover, All optimal Hankel-norm approximations of linear multivariable systems and their L^{∞} error bounds, *Int. J. Control*, **39** (1984), 115-1193.

[GKLR] I. Gohberg, M.A. Kaashoek, L. Lerer and L. Rodman, Minimal divisors of rational matrix functions with prescribed zero and pole structures, in Topics in Operator Theory, Systems and Networks, edited by H. Dym and I. Gohberg, OT12: Operator Theory: Advances and Applications, Vol 12, Birkhäuser Verlag, Basel, 1984.

[GLR1] I. Gohberg, P. Lancaster and L. Rodman, A sign characteristic for selfadjoint rational matrix functions, in Mathematical Theory of Networks and Systems, Proceeding of the MTNS-83, Beer-Sheva, Israel, Lecture Notes in Control and Information Sciences, 58, Springer Verlag (1983), 363-369.

[GLR2] I. Gohberg, P. Lancaster and L. Rodman, *Matrices and indefinite scalar products*, OT8: Operator Theory: Advances and Applications Vol 8, Birkhäuser Verlag, Basel, 1983.

[GVKDM] Y. Genin, P. Van Dooren, T. Kailath, J.M. Delosme and M. Morf, On $\sum$-lossless transfer functions and related questions, *Linear algebra and its applications*, **50** (1983), 251-275.

[KL] M.G. Krein and H. Langer, Über die verallgemeinerten Resolventen und die charakteristische Funktion eines isometrisches Operators in Raume π_k, *colloquia Mathematica Societatis Janos Bolyai* 5. *Hilbert space operators*, Tihany, (Hungary), 1970, 353-399.

[P] V.P. Potapov, The multiplicative structure of J-contractive matrix functions. *Amer. Math. Soc. Transl.*, **15** (1960), 131-243.

[S] L.A. Sakhnovich, Factorization problems and operator identities, *Russian Mathematical Surveys* **41**:1, (1986), 1-64.

Daniel Alpay
Department of Electronic Systems
Tel-Aviv University
Tel-Aviv, 69978 Israel

and

Israel Gohberg
Raymond and Beverly Sackler
Faculty of Exact Sciences
School of Mathematical Sciences
Tel-Aviv University
Tel-Aviv, 69978 Israel

Present address:

Daniel Alpay
Department of Mathematics
Groningen University
P.O.B. 800, 9700AV
Groningen, Holland

Operator Theory:
Advances and Applications, Vol. 33

PROPER CONTRACTIONS AND THEIR UNITARY MINIMAL COMPLETIONS

Israel Gohberg and Sorin Rubinstein

The paper contains characterizations of all rational matrix functions $W(\lambda)$ which are contractions for every real λ and strict contractions at infinity. The results are obtained in terms of minimal realizations. A description of all minimal, unitary on the line completions of such functions is given. This description is used in the study of linear fractional decompositions of proper contractions $W(\lambda)$.

1. INTRODUCTION

This paper deals with proper contractions, by which we mean rational matrix functions of a complex variable $W(\lambda)$: $\mathbb{C}^m \to \mathbb{C}^p$ which satisfy the following two properties:

a) $I - W(\lambda)W^*(\lambda) \geq 0$ for $\lambda \in \mathbb{R}$.

b) $W(\infty) = \lim_{\lambda\to\infty} W(\lambda)$ exists and $I - W(\infty)W^*(\infty) > 0$.

We start with a characterization of all proper contractions in terms of their realizations. Let $W(\lambda)$ be a rational matrix function with property b). This function admits a realization of the form

$$W(\lambda) = D + C(\lambda I - A)^{-1}B \tag{1.1}$$

with D: $\mathbb{C}^m \to \mathbb{C}^p$, C: $\mathbb{C}^n \to \mathbb{C}^p$, B: $\mathbb{C}^m \to \mathbb{C}^n$ and A: $\mathbb{C}^n \to \mathbb{C}^n$ for some natural number n. Moreover, we take a realization (1.1) for which n is minimal among all possible realizations of $W(\lambda)$. With the realization (1.1) we associate the following matrix which we call the state characteristic matrix of $W(\lambda)$

$$\mathcal{A} = \begin{bmatrix} \alpha & \beta \\ \gamma & \alpha^* \end{bmatrix} = \begin{bmatrix} A + BD^*(I-DD^*)^{-1}C & B(I-D^*D)^{-1}B^* \\ C^*(I-DD^*)^{-1}C & A^* + C^*(I-DD^*)^{-1}DB^* \end{bmatrix}.$$

One way to describe the state characteristic matrix of $W(\lambda)$ associated with the realization (1.1) is:

$$\begin{bmatrix} \gamma & \alpha^* \\ \alpha & \beta \end{bmatrix} = \begin{bmatrix} C & D \\ A & B \end{bmatrix} o_g \begin{bmatrix} C & D \\ A & B \end{bmatrix}^*$$

where "o_g" stands for the general cascade composition [6] first studied by Redheffer [9] (where it is called the star product). Other characterization of the state characteristic matrix is given in section 3 of this paper. We prove that the rational matrix function

$W(\lambda)$, of which the value at infinity is a strict contraction, is a proper contraction if and only if the real eigenvalues of the state characteristic matrix have even partial multiplicities. As a consequence we obtain the following second characterization of all proper contractions: $W(\lambda)$ is a proper contraction if and only if the following Riccati equation:

$$X\gamma X - iX\alpha^* + i\alpha X + \beta = 0 \tag{1.2}$$

has a hermitian solution. Here the coefficients γ, α^*, α, β are the blocks of the characteristic matrix $\mathcal{A}$. We call equation (1.2) the state characteristic equation of $W(\lambda)$ associated with the realization (1.1). It turns out that any hermitian solution of (1.2) is invertible. Moreover the number of negative (resp. positive) eigenvalues of any such a solution is equal to the number of poles of $W(\lambda)$ in the upper (resp. lower) half plane.

This results we apply for the study of all minimal, unitary on the line completions of a proper contraction $W(\lambda)$. This completions are defined by

$$\tilde{W}(\lambda) = \begin{bmatrix} W_{11}(\lambda) & W(\lambda) \\ W_{21}(\lambda) & W_{22}(\lambda) \end{bmatrix} : \mathbb{C}^p \oplus \mathbb{C}^m \to \mathbb{C}^p \oplus \mathbb{C}^m$$

where $\tilde{W}(\lambda)\cdot(\tilde{W}(\lambda))^* = I$ for every $\lambda \in \mathbb{R}$ and $\deg \tilde{W}(\lambda) = \deg W(\lambda)$. Here "deg" stands for the McMillan degree. This kind of completions are important in network theory and especially in Darlington Synthesis and were studied in this context in a series of papers of which we mention [3] and [4]. We prove that this kind of completions which have a given value at infinity are in a one to one correspondence with the hermitian solutions of the state characteristic equation of $W(\lambda)$. This correspondence is presented in an explicit form. We also suggest a method for finding the number of minimal completions of $W(\lambda)$ which have a given value at infinity. This result is based on the state characteristic matrix and is not using solutions of the state characteristic equation.

The last section of the paper deals with regular minimal fractional decompositions of rational matrix functions. Let $W(\lambda)$ be a rational matrix function analytic at infinity. A decomposition of the form:

$$W(\lambda) = \begin{bmatrix} \Omega_{11}(\lambda) & \Omega_{12}(\lambda) \\ \Omega_{21}(\lambda) & \Omega_{22}(\lambda) \end{bmatrix} o_f \omega(\lambda) \stackrel{\text{def}}{=}$$
$$\stackrel{\text{def}}{=} \Omega_{21}(\lambda) + \Omega_{22}(\lambda)\omega(\lambda)(I - \Omega_{12}(\lambda)\omega(\lambda))^{-1}\Omega_{11}(\lambda)$$

for which $[\Omega_{ij}(\lambda)]_{i,j=1,2}$ and $\omega(\lambda)$ are rational matrix functions analytic at infinity is called a fractional decomposition of $W(\lambda)$. This decomposition is called minimal if $\deg W(\lambda) = \deg[\Omega_{ij}(\lambda)]_{i,j=1,2} + \deg \omega(\lambda)$, and regular if $\Omega_{11}(\infty)$ and $\Omega_{22}(\infty)$ are invertible. In [8] Ball and Helton showed that any minimal regular fractional decomposition may be obtained from a minimal realization $\Sigma = (A, B, C, D; X, U, Y)$ of $W(\lambda)$ by means of a set of two subspaces X_1 and X_2 and two operators $F: X \to U$ and $G: Y \to X$ which satisfy:

a) $X = X_1 \oplus X_2$.

b) X_1 is $(A + BF)$-invariant.

c) X_2 is $(A + GC)$-invariant.

Explicit formulae for the components $[\Omega_{ij}(\lambda)]_{i,j=1,2}$ and $\omega(\lambda)$ are also given there. We consider the case that $W(\lambda)$ is a proper contraction and find the parameters X_1, X_2, F and G which define decompositions of $W(\lambda)$ for which $[\Omega_{ij}(\lambda)]_{i,j=1,2}$ is unitary on the line, and (therefore) $\omega(\lambda)$ is a proper contraction.

The paper is divided in five sections. After the introduction follows the second section which is of a preliminary character and is purely expository. For the proof of the results stated there we refer the reader to [2], [5] and [6]. The third section is concerned with the characterizations of the proper contractions. The fourth section deals with the minimal, unitary on the line completions of a proper contraction. The fifth section is concerned with the study of the linear fractional decompositions of a proper contraction which have a unitary first component.

2. PRELIMINARIES

In this sections we recall some definitions, notations and results about systems and realizations.

A system:

(2.1) $$\Sigma = (A, B, C, D; X, U, Y)$$

is a set of three finite dimensional spaces and four operators:

$$A: X \to X, \qquad B: U \to X, \qquad C: X \to Y, \qquad D: U \to Y$$

X, U and Y are called the state, input and output spaces respectively; A, B and C are called the state, input and output operator respectively. If

$$\Sigma' = (A', B', C', D'; X', U, Y)$$

is another system which has the same input and output spaces, Σ' is said to be *similar* to Σ if $D = D'$ and there exists an invertible operator $S: X \to X'$ called a system *similarity* such that

$$A = S^{-1}A'S, \qquad B = S^{-1}B', \qquad C = C'S.$$

Associated with any system (2.1) is its transfer function

$$W_\Sigma(\lambda) = D + C(\lambda - A)^{-1}B, \qquad \lambda \in \rho(A)$$

which is a rational function from the resolvent set of A, $\rho(A)$ to $L(U, Y)$, analytic at infinity. Conversely, for any rational matrix function which is analytic at infinity $W(\lambda)$, a system Σ may be found such that $W(\lambda) = W_\Sigma(\lambda)$.

The system Σ is called in this case a realization of $W(\lambda)$. A *realization* Σ of $W(\lambda)$ for which the dimension of the state space X is minimal among all the possible realizations is called a *minimal realization*, and the system Σ is also called a minimal system. The dimension of the state space in a minimal realization of $W(\lambda)$ is known by the name of *McMillan degree* a $W(\lambda)$ and written throughout the paper $\deg W(\lambda)$. A rational matrix function analytic at infinity has an infinite set of minimal realizations, but any two of them are similar. It is well known that a system (2.1) is minimal if and only if the following two conditions are satisfied

$$\operatorname{rank}[\ \lambda - A \quad B\ ] = \dim X, \qquad \lambda \in \mathbb{C} \tag{2.2}$$

$$\operatorname{rank}\begin{bmatrix} \lambda - A \\ C \end{bmatrix} = \dim X, \qquad \lambda \in \mathbb{C}. \tag{2.3}$$

Equivalent to (2.2) and (2.3) respectively are the conditions

$$\operatorname{span}_{i=0,1,2,\ldots,\dim X-1} \operatorname{Im}(A^i B) = X \tag{2.2'}$$

$$\bigcap_{i=0}^{\dim X-1} \ker(CA^i) = 0. \tag{2.3'}$$

A system which satisfies (2.2) (or (2.2′)) is called *controllable* and a system which satisfies (2.3) (or (2.3′)) is called *observable*. A system (2.1) is said to be *minimal at a point* $\lambda_0 \in \mathbb{C}$ if the system

$$\Sigma_{\lambda_0} = (A_{|\operatorname{Im} Q}, QB, C_{|\operatorname{Im} Q}, D; \operatorname{Im} Q, U, Y)$$

is minimal, where Q is the Riesz projection of A corresponding to λ_0. In this case, $\dim \operatorname{Im} Q$ is named the local McMillan degree of $W_\Sigma(\lambda)$ at λ_0 and denoted by $\deg_{\lambda_0} W_\Sigma(\lambda)$. A system is said to be minimal on a set if it is minimal at every point of the set.

We shall describe now some transformation in the set of systems. The associated system of a system (2.1) for which D is invertible is defined (and denoted) by

$$\Sigma^\times = (A - BD^{-1}C, BD^{-1}, -D^{-1}C, D^{-1}; X, Y, U).$$

By the above notation it is often written $A^\times$ for $A - BD^{-1}C$. $A^\times$ is called the associated state operator of Σ. In terms of transfer functions the following formula holds:

$$W_{\Sigma^\times}(\lambda) = W_\Sigma^{-1}(\lambda).$$

A system is minimal on a set if and only if its associated system is minimal on the same set.

Let the system (2.1) have its input and output spaces decomposed: $U = U_1 \oplus U_2$, $Y = Y_1 \oplus Y_2$. With respect to these decompositions the system may be written in the form

$$\Sigma = \left(A, [\ B_1 \quad B_2\ ], \begin{bmatrix} C_1 \\ C_2 \end{bmatrix}, \begin{bmatrix} D_{11} & D_{12} \\ D_{21} & D_{22} \end{bmatrix}; X, U_1 \oplus U_2, Y_1 \oplus Y_2 \right).$$

Assume that D_{11} is invertible. Then, the following system, called the partial inverse of Σ is defined:

$$\Sigma^{<} = \Big(A - B_1 D_{11}^{-1} C_1, [\, B_1 D_{11}^{-1} \quad B_2 - B_1 D_{11}^{-1} D_{12} \,],$$

$$\begin{bmatrix} -D_{11}^{-1} C_1 \\ C_2 - D_{21} D_{11}^{-1} C_1 \end{bmatrix}, \begin{bmatrix} D_{11}^{-1} & -D_{11}^{-1} D_{12} \\ D_{21} D_{11}^{-1} & D_{22} - D_{21} D_{11}^{-1} D_{12} \end{bmatrix}, X, Y_1 \oplus U_2, U_1 \oplus Y_2 \Big).$$

We shall denote also by $A^{<}$, $[\, B_1 \quad B_2 \,]^{<}$, $\begin{bmatrix} C_1 \\ C_2 \end{bmatrix}^{<}$ and $\begin{bmatrix} D_{11} & D_{12} \\ D_{21} & D_{22} \end{bmatrix}^{<}$ respectively the operators of $\Sigma^{<}$. If

$$W(\lambda) = W_{\Sigma}(\lambda) = \begin{bmatrix} W_{11}(\lambda) & W_{12}(\lambda) \\ W_{21}(\lambda) & W_{22}(\lambda) \end{bmatrix}$$

is the transfer function of Σ, its partial inverse is by definition:

$$W^{<}(\lambda) = W_{\Sigma^{<}}(\lambda) = \begin{bmatrix} W_{11}^{-1}(\lambda) & -W_{11}^{-1}(\lambda) W_{12}(\lambda) \\ W_{21}(\lambda) W_{11}^{-1}(\lambda) & W_{22}(\lambda) - W_{21}(\lambda) W_{11}^{-1}(\lambda) W_{12}(\lambda) \end{bmatrix}.$$

Finally, given a rational matrix function analytic at infinity $W(\lambda)$, the notation $W_*(\lambda)$ will be used throughout this paper to denote the rational matrix function which coincides with the adjoint of the matrix $W(\lambda)$ for every $\lambda \in \mathbb{R}$. If $W(\lambda)$ has a minimal realization (2.1) then $W_*(\lambda)$ has the minimal realization

$$\Sigma_* = (A^*, C^*, B^*, D^*; X, Y, U).$$

Moreover, any other minimal realization of $W_*(\lambda)$ may be obtained from this one by means of a hermitian similarity.

3. PROPER CONTRACTIONS

Let $W(\lambda)$ be a rational matrix function analytic at infinity. We call the function $W(\lambda)$ a *proper contraction* iff all its values on the real line are contractions and $W(\infty)$ is a strict contraction. The aim of this section is to characterize all the proper contractions in terms of their realizations.

Let us start by considering a rational matrix function $W(\lambda)$ analytic at infinity and with $W(\infty)$ a strict contraction, and let

$$\Sigma = (A, B, C, D;\ \mathbb{C}^n,\ \mathbb{C}^m,\ \mathbb{C}^p)$$

be a minimal realization of $W(\lambda)$. The function $(I - W(\lambda)W_*(\lambda))^{-1}$ is defined in a neighbourhood of the infinity and has the not necessarily minimal realization:

$$(I-\Sigma\Sigma_*)^\times =$$

$$= \left(\begin{bmatrix} A & BB^* \\ 0 & A^* \end{bmatrix}, \begin{bmatrix} BD^* \\ C^* \end{bmatrix}, [\,-C \quad -DB^*\,], I-DD^*;\ \mathbb{C}^n \oplus \mathbb{C}^n,\ \mathbb{C}^p,\ \mathbb{C}^p \right)^\times =$$

$$= \left(\begin{bmatrix} A+BD^*(I-DD^*)^{-1}C & B(I-D^*D)^{-1}B^* \\ C^*(I-DD^*)^{-1}C & A^*+C^*(I-DD^*)^{-1}DB^* \end{bmatrix}, \begin{bmatrix} BD^*(I-DD^*)^{-1} \\ C^*(I-DD^*)^{-1} \end{bmatrix}, \right.$$

$$\left. [\,(I-DD^*)^{-1}C \quad (I-DD^*)^{-1}DB^*\,], (I-DD^*)^{-1};\ \mathbb{C}^n \oplus \mathbb{C}^n,\ \mathbb{C}^p,\ \mathbb{C}^p \right)$$

We shall call the state operator of $(I-\Sigma\Sigma_*)^\times$

$$\mathcal{A} = \begin{bmatrix} \alpha & \beta \\ \gamma & \alpha^* \end{bmatrix} = \begin{bmatrix} A+BD^*(I-DD^*)^{-1}C & B(I-D^*D)^{-1}B^* \\ C^*(I-DD^*)^{-1}C & A^*+C^*(I-DD^*)^{-1}DB^* \end{bmatrix}$$

the *state characteristic matrix* of $W(\lambda)$ associated with the minimal realization Σ.

Let

$$\Sigma' = (A', B', C', D;\ \mathbb{C}^n,\ \mathbb{C}^m,\ \mathbb{C}^p)$$

be another minimal realization of $W(\lambda)$. Then, there exists a unique invertible matrix S such that:

$$A' = SAS^{-1}, \qquad B' = SB, \qquad C' = CS^{-1}.$$

One easily sees that the state characteristic matrix $\begin{bmatrix} \alpha' & \beta' \\ \gamma' & \alpha'^* \end{bmatrix}$ associated with the system Σ' is similar to the state characteristic matrix $\begin{bmatrix} \alpha & \beta \\ \gamma & \alpha^* \end{bmatrix}$ associated with the system Σ and the similarity is given by:

$$\begin{bmatrix} \alpha' & \beta' \\ \gamma' & \alpha'^* \end{bmatrix} = \begin{bmatrix} S & 0 \\ 0 & S^{*-1} \end{bmatrix} \begin{bmatrix} \alpha & \beta \\ \gamma & \alpha^* \end{bmatrix} \begin{bmatrix} S^{-1} & 0 \\ 0 & S^* \end{bmatrix}. \tag{3.1}$$

Conversely, if $\begin{bmatrix} \alpha & \beta \\ \gamma & \alpha^* \end{bmatrix}$ is the state characteristic matrix of $W(\lambda)$ associated with the system Σ, then, for every invertible S, the matrix $\begin{bmatrix} \alpha' & \beta' \\ \gamma' & \alpha'^* \end{bmatrix}$ given by (3.1) is the state characteristic matrix of $W(\lambda)$ associated with the system $(SAS^{-1}, SB, CS^{-1}, D;\ \mathbb{C}^n,\ \mathbb{C}^m,\ \mathbb{C}^p)$. The relation (3.1) is a relation of equivalence in the set of all state characteristic matrices, which we shall call a *special similarity*. As we have shown, the set of all the state characteristic matrices of a function $W(\lambda)$ associated with all its minimal realizations is a class of equivalence with respect to this relation. We call this class the state characteristic matrix of $W(\lambda)$.

The following proposition gives a characterization of all the matrices which are state characteristic matrices for some function $W(\lambda)$.

PROPOSITION 3.1. *Let $W(\lambda)$ be a rational matrix function analytic at infinity and for which $W(\infty)$ is a strict contraction and let $\begin{bmatrix} \alpha & \beta \\ \gamma & \alpha^* \end{bmatrix}$ be its state characteristic matrix. Then*

(i) *β and γ are nonnegative.*

(ii) *The pairs (α, β) and (α^*, γ) are controllable.*

Conversely, any 2×2 block matrix $\begin{bmatrix} \alpha & \beta \\ \gamma & \alpha^* \end{bmatrix}$ with $n \times n$ blocks which has properties (i) *and* (ii) *is the state characteristic matrix of a rational matrix function $W(\lambda)$ for which $W(\infty)$ is a strict contraction.*

The set of all such functions $W(\lambda)$ for which $\begin{bmatrix} \alpha & \beta \\ \gamma & \alpha^* \end{bmatrix}$ is the state characteristic matrix is given by all $W(\lambda) = W_\Sigma(\lambda)$ with the minimal realization

$$\Sigma_{STD} = (\alpha - TD^*S, T(I - D^*D)^{1/2}, (I - DD^*)^{1/2}S, D;\ \mathbb{C}^n,\ \mathbb{C}^m,\ \mathbb{C}^p).$$

Here D is any $p \times m$ strict contraction and T and S are any $n \times m$ and $p \times n$ matrices respectively, which satisfy:

$$\beta = TT^* \quad \text{and} \quad \gamma = S^*S.$$

PROOF. One easily sees that the properties (i) and (ii) are independent of the choice of the matrix $\begin{bmatrix} \alpha & \beta \\ \gamma & \alpha^* \end{bmatrix}$ in its class of equivalence with respect to the relation of special similarity, and therefore on the choice of the minimal realization Σ of $W(\lambda)$. Let

$$\Sigma = (A, B, C, D;\ \mathbb{C}^n,\ \mathbb{C}^m,\ \mathbb{C}^p)$$

be such a realization. Property (i) is a direct consequence of the definition of the state characteristic matrix associated with Σ. In order to prove (ii) we shall remark that the pair:

$$(\alpha, \beta) = \big(A + BD^*(I - DD^*)^{-1}C, B(I - D^*D)^{-1}B^*\big)$$

is controllable if and only if the pair

$$\big((I - D^*D)^{-1/2}B^*, A^* + C^*(I - DD^*)^{-1}DB^*\big)$$

is observable. But this follows from the formula:

$$(I - D^*D)^{-1/2}B^*(A^* + C^*(I - DD^*)^{-1}DB^*)^k =$$
$$= (I - D^*D)^{-1/2}B^*A^{*k} + \sum_{\ell=0}^{k-1}(I - D^*D)^{-1/2} * B^*A^{*\ell}$$

and the observability of the pair (B^*, A^*).

The controllability of the pair (α^*, γ) may be proved the same way.

Let us prove the converse part of the proposition. The state characteristic matrix $\begin{bmatrix} \alpha' & \beta' \\ \gamma' & \alpha'^* \end{bmatrix}$ associated with the system Σ_{STD} has the block components:

$$\begin{aligned} \alpha' &= \alpha - TD^*S + T(I-D^*D)^{1/2}D^*(I-DD^*)^{-1}(I-DD^*)^{1/2}S = \\ &= \alpha - TD^*S + TD^*(I-DD^*)^{1/2}(I-DD^*)^{-1}(I-DD^*)^{1/2}S = \\ &= \alpha \\ \beta' &= T(I-D^*D)^{1/2}(I-D^*D)^{-1}(I-D^*D)^{1/2}T^* = TT^* = \beta \\ \gamma' &= S^*(I-DD^*)^{1/2}(I-DD^*)^{-1}(I-DD^*)^{1/2}S = S^*S = \gamma \end{aligned}$$

and thus is identical with $\begin{bmatrix} \alpha & \beta \\ \gamma & \alpha^* \end{bmatrix}$. Let us prove that Σ_{STD} is minimal. Indeed, it is easy to verify that in the chain of pairs: $(\alpha - TD^*S, T(I-D^*D)^{1/2})$, $(\alpha - TD^*S, T)$, (α, T), (α, T^*T), (α, β) any two consecutive pairs are controllable in the same time. On the other hand the pair $((I-DD^*)^{1/2}S, \alpha - TD^*S)$ is observable if and only if the pair $(\alpha^* - S^*DT^*, S^*(I-DD^*)^{1/2})$ is controllable, and one may show the same way that this happens if and only if the pair $(\alpha^*\gamma)$ is controllable. Thus $\begin{bmatrix} \alpha & \beta \\ \gamma & \alpha^* \end{bmatrix}$ is the state characteristic matrix of $W_{\Sigma_{STD}}$.

Now, assume $\begin{bmatrix} \alpha & \beta \\ \gamma & \alpha^* \end{bmatrix}$ is the state characteristic matrix of a function $W_\Sigma(\lambda)$ with the minimal realization:

$$\Sigma = (A, B, C, D;\ \mathbb{C}^n,\ \mathbb{C}^m,\ \mathbb{C}^p).$$

Then, one may choose $T = B(I-D^*D)^{-1/2}$ and $S = (I-DD^*)^{-1/2}C$ in order to write the system Σ in the form $\Sigma = \Sigma_{STD}$ which ends the proof. □

Let $W(\lambda)$ be a rational matrix function of which the value at infinity is a strict contraction and let Σ be a minimal realization of $W(\lambda)$, and $\begin{bmatrix} \alpha & \beta \\ \gamma & \alpha^* \end{bmatrix}$ the state characteristic matrix of $W(\lambda)$ associated to Σ. We define the following Riccati equation:

$$H\gamma H - iH\alpha^* + i\alpha H + \beta = 0 \tag{3.2}$$

which we call the *state characteristic equation* of $W(\lambda)$ associated with the minimal realization Σ. We say that two equations of the form (3.2)

$$H\gamma_i H_i - iH\alpha_i^* + i\alpha_i H + \beta_i = 0, \qquad i = 1, 2$$

are *similar* iff there exists an invertible matrix S such that

$$\gamma_1 = S^*\gamma_2 S, \qquad \alpha_1^* = S^*\alpha_2^* S^{*-1}, \qquad \alpha_1 = S^{-1}\alpha_2 S, \qquad \beta_1 = S^{-1}\beta_2 S^{*-1}.$$

One sees that the similarity of two Riccati equations is equivalent to the special similarity of the matrices $\begin{bmatrix} \alpha_i & \beta_i \\ \gamma_i & \alpha_i^* \end{bmatrix}$ of their coefficients. Thus the class of equivalence under similarity of the equation (3.2) is formed by the set of the state characteristic equation of $W(\lambda)$ associated with all its minimal realizations. Moreover let

$$HS^*\gamma SH - iHS^*\alpha^*S^{*-1} + iS^{-1}\alpha SH + S^{-1}\beta S^{*-1} = 0 \tag{3.3}$$

be an equation similar to (3.2) where S is any invertible matrix. One easily verifies that H_0 is a solution of (3.2) if and only if $S^{-1}H_0S^{*-1}$ is a solution of (3.3).

In what follows we shall often identify similar Riccati equations and speak about the state characteristic equation of a function $W(\lambda)$. We are now in a position to give the main theorem of this section:

THEOREM 3.2. *Let $W(\lambda)$ be a rational matrix function analytic at infinity and for which $W(\infty)$ is a strict contraction. The following statements are equivalent*

a) *$W(\lambda)$ is a proper contraction.*

b) *All the real eigenvalues of the state characteristic matrix of $W(\lambda)$ have even multiplicities.*

c) *The state characteristic equation of $W(\lambda)$ has a hermitian solution.*

The proof of this theorem will be preceded by the following lemma:

LEMMA 3.3. *Let $W(\lambda)$ be a rational matrix function analytic at infinity and let*

$$\Sigma = (A, B, C, D;\ \mathbb{C}^n,\ \mathbb{C}^m,\ \mathbb{C}^p)$$

be a minimal realization of $W(\lambda)$. Then:

a) *The realization $\Sigma\Sigma_*$ of $W(\lambda) \cdot W_*(\lambda)$ is minimal at every $\lambda \in \mathbb{R}$.*

b) *If $I - DD^*$ is invertible then the realization $(I - \Sigma\Sigma_*)^\times$ of $(I - W(\lambda)W_*(\lambda))^{-1}$ is minimal at every $\lambda \in \mathbb{R}$.*

PROOF. From the realization Σ of $W(\lambda)$ and

$$\Sigma_* = (A^*, C^*, B^*, D^*;\ \mathbb{C}^n,\ \mathbb{C}^p,\ \mathbb{C}^m)$$

of $W_*(\lambda)$ one obtains (cf. [2], p. 6) the realization

$$\Sigma \cdot \Sigma_* = \left(\begin{bmatrix} A & BB^* \\ 0 & A^* \end{bmatrix}, \begin{bmatrix} BD^* \\ C^* \end{bmatrix}, [\, C \quad DB^* \,], DD^*;\ \mathbb{C}^n \oplus \mathbb{C}^n,\ \mathbb{C}^p,\ \mathbb{C}^p \right)$$

of $W(\lambda)W_*(\lambda)$. For any fixed $\lambda \in \mathbb{R}$ there exists a decomposition $\mathbb{C}^n = \mathbb{C}^k \oplus \mathbb{C}^\ell$ of the state space $\mathbb{C}^n$ of Σ and an invertible operator P: $\mathbb{C}^n \to \mathbb{C}^n$ such that with respect to this decomposition one obtains:

$$P(\lambda - A)P^{-1} = \begin{bmatrix} 0 & 0 \\ A_1 & A_2 \end{bmatrix}, \qquad PB = \begin{bmatrix} B_1 \\ B_2 \end{bmatrix}, \qquad P^{*-1}C^* = \begin{bmatrix} C_1 \\ C_2 \end{bmatrix}$$

with $[\, A_1 \quad A_2 \,]$ of full rank. Since Σ is minimal one sees that:

$$\operatorname{rank}[\, P(\lambda - A)P^{-1} \quad PB \,] = \operatorname{rank} P[\, \lambda - A \quad B \,]\begin{bmatrix} P^{-1} & 0 \\ 0 & I \end{bmatrix} = n.$$

It follows that B_1 has full rank. Then

$$\operatorname{rank}\begin{bmatrix} P & 0 \\ 0 & P^{*-1} \end{bmatrix}\begin{bmatrix} \lambda - A & -BB^* & BD^* \\ 0 & \lambda - A^* & C^* \end{bmatrix}\begin{bmatrix} P^{-1} & 0 & 0 \\ 0 & P^* & 0 \\ 0 & 0 & I \end{bmatrix} =$$

$$= \operatorname{rank}\begin{bmatrix} P(\lambda - A)P^{-1} & -PBB^*P^* & PBD^* \\ 0 & P^{*-1}(\lambda - A^*)P^* & P^{*-1}C^* \end{bmatrix} =$$

$$= \operatorname{rank}\begin{bmatrix} 0 & 0 & -B_1B_1^* & * & * \\ A_1 & A_2 & * & * & * \\ 0 & 0 & 0 & A_1^* & C_1 \\ 0 & 0 & 0 & A_2^* & C_2 \end{bmatrix}.$$

But

$$\operatorname{rank}\begin{bmatrix} A_1^* & C_1 \\ A_2^* & C_2 \end{bmatrix} = \operatorname{rank} P^{*-1}[\, \lambda - A^* \quad C^* \,]\begin{bmatrix} P^* & 0 \\ 0 & I \end{bmatrix} = n$$

and $\operatorname{rank}[\, A_1 \quad A_2 \,] = n - \operatorname{rank} B_1$, such that

$$\operatorname{rank}\begin{bmatrix} \lambda - A & -BB^* & BD^* \\ 0 & \lambda - A^* & C^* \end{bmatrix} = 2n.$$

Also:

$$\operatorname{rank}\begin{bmatrix} \lambda - A & -BB^* \\ 0 & \lambda - A^* \\ C & DB^* \end{bmatrix} = \operatorname{rank}\begin{bmatrix} 0 & I & 0 \\ I & 0 & 0 \\ 0 & 0 & I \end{bmatrix}\begin{bmatrix} \lambda - A & -BB^* & BD^* \\ 0 & \lambda - A^* & C^* \end{bmatrix}^*\begin{bmatrix} 0 & I \\ I & 0 \end{bmatrix} = 2n.$$

Thus the realization $\Sigma \cdot \Sigma_*$ is minimal at every $\lambda \in \mathbb{R}$.

b) It is evident that if the realization $\Sigma\Sigma_*$ of $W(\lambda)W_*(\lambda)$ is minimal at every $\lambda \in \mathbb{R}$, then so is the realization

$$I - \Sigma\Sigma_* = \left(\begin{bmatrix} A & BB^* \\ 0 & A^* \end{bmatrix}, \begin{bmatrix} BD^* \\ C^* \end{bmatrix}, [\, -C \quad -DB^* \,], I - DD^*;\ \mathbb{C}^n,\ \mathbb{C}^p,\ \mathbb{C}^p \right)$$

of $I - W(\lambda)W_*(\lambda)$ and therefore (cf. [7], Lemma 1.1) its associate system $(I - \Sigma\Sigma_*)^\times$ as a realization of $\left(I - W(\lambda)W_*(\lambda)\right)^{-1}$. □

PROOF OF THEOREM 3.2. The proof is divided in three steps:

Step 1. *Property* a) *implies property* b). Assume $W(\lambda)$ is a contraction for every $\lambda \in \mathbb{R}$. Let

$$\Sigma = (A, B, C, D;\ \mathbb{C}^n,\ \mathbb{C}^m,\ \mathbb{C}^p)$$

be a minimal realization of $W(\lambda)$, and let $\mathcal{A}$ be the state characteristic matrix of $W(\lambda)$ associated with Σ. Then, $\Sigma_* = (A^*, C^*, B^*, D^*;\ \mathbb{C}^n,\ \mathbb{C}^p,\ \mathbb{C}^m)$ is a minimal realization of $W_*(\lambda)$ and $\mathcal{A}$ is the state operator of the realization $(I-\Sigma\Sigma_*)^\times$ of $\left(I-W(\lambda)W_*(\lambda)\right)^{-1}$. According to Lemma 3.3 the realization $(I-\Sigma\Sigma_*)^\times$ of $\left(I-W(\lambda)W_*(\lambda)\right)^{-1}$ is minimal at any real λ. The state space of $(I-\Sigma\Sigma_*)^\times$ admits (cf. Th. 3.2 in [2]) a decomposition, $\mathbb{C}^{2n} = X_1 \oplus X_0 \oplus X_2$ with respect to which the system admits the following representation:

$$(I-\Sigma\Sigma_*)^\times = \left(\begin{bmatrix} \mathcal{A}_{11} & * & * \\ 0 & \mathcal{A}_{00} & * \\ 0 & 0 & \mathcal{A}_{22} \end{bmatrix}, \begin{bmatrix} \mathcal{B}_1 \\ \mathcal{B}_0 \\ 0 \end{bmatrix}, [\,0 \quad \mathcal{C}_0 \quad \mathcal{C}_2\,], \mathcal{D};\right.$$
$$\left. X_1 \oplus X_0 \oplus X_2,\ \mathbb{C}^m,\ \mathbb{C}^p \right)$$

with $(\mathcal{A}_{00}, \mathcal{B}_0, \mathcal{C}_0, \mathcal{D}; X_0,\ \mathbb{C}^m,\ \mathbb{C}^p)$ minimal. It follows that $\mathcal{A}_{11}$ and $\mathcal{A}_{22}$ have no real eigenvalues. Otherwise, for some $\lambda \in \mathbb{R}$:

$$\operatorname{rank} \begin{bmatrix} \lambda I - \mathcal{A}_{11} & * & * & \mathcal{B}_1 \\ 0 & \lambda I - \mathcal{A}_{00} & * & \mathcal{B}_0 \\ 0 & 0 & \lambda I - \mathcal{A}_{22} & 0 \end{bmatrix} < 2n$$

or

$$\operatorname{rank} \begin{bmatrix} \lambda I - \mathcal{A}_{11} & * & * \\ 0 & \lambda I - \mathcal{A}_{00} & * \\ 0 & 0 & \lambda I - \mathcal{A}_{22} \\ 0 & \mathcal{C}_0 & \mathcal{C}_2 \end{bmatrix} < 2n$$

which contradicts the minimality of $(I-\Sigma\Sigma_*)^\times$ in λ.

Let $\lambda_0 \in \mathbb{R}$ be an eigenvalue of the state operator $\mathcal{A}$ of $(I-\Sigma\Sigma_*)^\times$. Then λ_0 is an eigenvalue of $\mathcal{A}_{00}$ and there exists an invertible matrix S such that $S^{-1}\mathcal{A}_{00}S$ has the form:

$$J = \begin{bmatrix} J_0 & 0 \\ 0 & J_1 \end{bmatrix}$$

where J_0 has only λ_0 as eigenvalue and λ_0 is not an eigenvalue of J_1. One may write the matrix

$$\begin{bmatrix} I & 0 & 0 \\ 0 & S^{-1} & 0 \\ 0 & 0 & I \end{bmatrix} \begin{bmatrix} \mathcal{A}_{11} & * & * \\ 0 & \mathcal{A}_{00} & * \\ 0 & 0 & \mathcal{A}_{22} \end{bmatrix} \begin{bmatrix} I & 0 & 0 \\ 0 & S & 0 \\ 0 & 0 & I \end{bmatrix} = \begin{bmatrix} \mathcal{A}_{11} & * & * \\ 0 & J & * \\ 0 & 0 & \mathcal{A}_{22} \end{bmatrix}$$

in the form: $\begin{bmatrix} \mathcal{A}_{11} & * & * \\ 0 & J_0 & \mathcal{A}'_{02} \\ 0 & 0 & \mathcal{A}'_{22} \end{bmatrix}$ where $\mathcal{A}'_{22} = \begin{bmatrix} J_1 & * \\ 0 & \mathcal{A}_{22} \end{bmatrix}$.

Let X be a solution of the equation $\mathcal{A}'_{02} = J_0 X - X\mathcal{A}'_{22}$. Such a solution does exist because J_0 and $\mathcal{A}'_{22}$ have no common eigenvalue. Then:

$$\begin{bmatrix} I & 0 & 0 \\ 0 & I & -X \\ 0 & 0 & I \end{bmatrix}^{-1} \begin{bmatrix} \mathcal{A}_{11} & * & * \\ 0 & J_0 & \mathcal{A}'_{02} \\ 0 & 0 & \mathcal{A}'_{22} \end{bmatrix} \begin{bmatrix} I & 0 & 0 \\ 0 & I & -X \\ 0 & 0 & I \end{bmatrix} = \begin{bmatrix} \mathcal{A}_{11} & \mathcal{A}'_{12} & \mathcal{A}'_{13} \\ 0 & J_0 & 0 \\ 0 & 0 & \mathcal{A}'_{22} \end{bmatrix}.$$

for some matrices $\mathcal{A}'_{12}$ and $\mathcal{A}'_{13}$.

Let Y be a solution of the equation $\mathcal{A}'_{12} = \mathcal{A}_{11}Y - YJ_0$. Then

$$\begin{bmatrix} I & -Y & 0 \\ 0 & I & 0 \\ 0 & 0 & I \end{bmatrix}^{-1} \begin{bmatrix} \mathcal{A}_{11} & \mathcal{A}'_{12} & \mathcal{A}'_{13} \\ 0 & J_0 & 0 \\ 0 & 0 & \mathcal{A}'_{22} \end{bmatrix} \begin{bmatrix} I & -Y & 0 \\ 0 & I & 0 \\ 0 & 0 & I \end{bmatrix} = \begin{bmatrix} \mathcal{A}_{11} & 0 & * \\ 0 & J_0 & 0 \\ 0 & 0 & \mathcal{A}'_{22} \end{bmatrix}.$$

It follows that the partial multiplicities of λ_0 in $\mathcal{A}$ and in $\mathcal{A}_{00}$ are the same. On the other hand the system $(\mathcal{A}_{00}, \mathcal{B}_0, \mathcal{C}_0, \mathcal{D}; X_0, \mathbb{C}^m, \mathbb{C}^p)$ is a minimal realization of $(I - W(\lambda)W_*(\lambda))^{-1}$. According to Theorem 3.3 of [2], the eigenvalues of $\mathcal{A}_{00}$ and the poles of $(I - W(\lambda)W_*(\lambda))^{-1}$ are the same and have the same partial multiplicities. Therefore, the real eigenvalues of $\mathcal{A}$ are exactly the real poles of $(I - W(\lambda)W_*(\lambda))^{-1}$ and have the same partial multiplicities. The matrix function $(I - W(\lambda)W_*(\lambda))^{-1}$ is nonnegative and by Theorem 3.11 of [5], Chap. II, the partial multiplicities of its real poles are even. Then, the partial multiplicities of the real eigenvalues of the state characteristic matrix $\mathcal{A}$ of $W(\lambda)$ are also even.

Step 2. *Property* b) *implies property* a). The matrix function $I - W(\lambda)W_*(\lambda)$ and the nonnegative matrix function $W(\lambda)W_*(\lambda)$ have minimal realizations with the same state operator. Hence, all the real poles of $I - W(\lambda)W_*(\lambda)$ have even partial multiplicities. On the other hand, it was proved in Step 1 that the real poles of $(I - W(\lambda)W_*(\lambda))^{-1}$ are exactly the real eigenvalues of the state characteristic matrix $\mathcal{A}$ and have the same partial multiplicities. According to property b), this partial multiplicities are even. The matrix function $I - W(\lambda)W_*(\lambda)$ admits therefore (cf. [5], p. 170) a representation of the form:

$$I - W(\lambda)W_*(\lambda) = U^{-1}(\lambda) \begin{bmatrix} \omega_1(\lambda) & & \\ & \ddots & \\ & & \omega_n(\lambda) \end{bmatrix} U(\lambda)$$

with $\omega_j(\lambda) = \mu_j(\lambda) - \frac{\prod_i (\lambda - \lambda_i)^{2k_{ij}}}{\prod_i (\lambda - \nu_i)^{2\ell_{ij}}}$ where $U(\lambda)$ is unitary and analytic on $\mathbb{R}$. μ_j is real and analytic on $\mathbb{R}$ and has no real zeros, λ_i and ν_i are the real zeros and poles of $I - W(\lambda)W_*(\lambda)$ respectively and $2k_{ij}$, $2\ell_{ij}$ their partial multiplicities. From the property $I - W(\infty)W_*(\infty) > 0$ there follows that, for a big enough λ, the matrix $I - W(\lambda)W_*(\lambda)$ is positive. Hence the functions $\mu_j(\lambda)$ which have a constant sign on the real line are all positive. Therefore the matrix $I - W(\lambda)W_*(\lambda)$ is nonnegative for every $\lambda \in \mathbb{R}$.

Step 3. *Property* b) *is equivalent to property* c). Proposition 3.1 shows that we are in the conditions of application of Theorem 4.3 in [5]. According to this theorem (applied to our case), the state characteristic equation has a hermitian solution if and only if the real eigenvalues of the matrix:

$$\mathcal{M} = i \begin{bmatrix} -i\alpha^* & \gamma \\ -\beta & -i\alpha \end{bmatrix}$$

have even multiplicities. The following identity:

$$\begin{bmatrix} 0 & -iI \\ -I & 0 \end{bmatrix} \begin{bmatrix} \alpha & \beta \\ \gamma & \alpha^* \end{bmatrix} \begin{bmatrix} 0 & -I \\ iI & 0 \end{bmatrix} = i \begin{bmatrix} -i\alpha^* & \gamma \\ -\beta & -i\alpha \end{bmatrix}$$

shows that $\mathcal{M}$ and $\mathcal{A}$ are unitary equivalent. This ends the proof of Step 3. □

The hermitian solutions of the state characteristic equation of a proper contraction will play an important role in the rest of the paper. The following theorem gives two important properties of this solutions.

THEOREM 3.4. *Let $W(\lambda)$ be a proper contraction and*

$$H\gamma H - iH\alpha^* + i\alpha H + \beta = 0 \tag{3.4}$$

its state characteristic equation. Then

a) *Any hermitian solution H of (3.4) is invertible.*

b) *The number of negative (resp. positive) eigenvalues of the solution H of (3.4) is equal to the number of poles of $W(\lambda)$ in the lower (resp. upper) half plane.*

PROOF. We start by proving property a). This property is independent of the particular choice of equation (3.4) in its class of similarity. We assume that (3.4) is associated to some minimal realization $\Sigma = (A, B, C, D;\ \mathbb{C}^n,\ \mathbb{C}^m,\ \mathbb{C}^p)$ of $W(\lambda)$. Let H be a hermitian solution of (3.4) and assume H is not invertible. There exists a decomposition $\mathbb{C}^n = X_1 \oplus X_2$ of the domain of H, and a unitary operator U such that:

$$UHU^* = \begin{bmatrix} 0 & 0 \\ 0 & \Lambda \end{bmatrix} : X_1 \oplus X_2 \to X_1 \oplus X_2$$

is the diagonal form of H. Let

$$\begin{aligned} U\alpha U^* &= [\alpha_{ij}]_{i,j=1,2}, \qquad & U\beta U^* &= [\beta_{ij}]^{i,j=1,2} \\ U\gamma U^* &= [\gamma_{ij}]_{i,j=1,2}, \qquad & U\alpha^* U^* &= [\alpha^*_{ji}]_{i,j=1,2} \end{aligned}$$

be written with respect to this decomposition of $\mathbb{C}^n$. Moreover, let us write $UB(I - D^*D)^{-1/2} = \begin{bmatrix} B_1 \\ B_2 \end{bmatrix}$ with respect to the same decomposition. It follows that $\beta_{ij} = B_i B_j^*$, $i, j = 1, 2$. The equation (3.4) becomes:

$$\begin{bmatrix} \beta_{11} & i\alpha_{12}\Lambda + \beta_{12} \\ -i\Lambda\alpha^*_{21} + \beta_{21} & \Lambda\gamma_{22}\Lambda - i\Lambda\alpha^*_{22} + \alpha_{22}\Lambda + \beta_{22} \end{bmatrix} = \begin{bmatrix} 0 & 0 \\ 0 & 0 \end{bmatrix}.$$

This implies $\beta_{11} = 0$. Therefore, $B_1 = 0$, $\beta_{12} = 0$ and $\alpha_{12} = 0$. Hence $U\beta U^*$ and $U\beta^k \alpha U^*$ have the form:

$$U\beta U^* = \begin{bmatrix} 0 & 0 \\ 0 & * \end{bmatrix} \quad \text{and} \quad U\alpha^k \beta U^* = \begin{bmatrix} 0 & 0 \\ * & * \end{bmatrix}.$$

It follows that:

$$\operatorname{rank}[\ \beta \quad \alpha\beta \cdots \alpha^{n-1}\beta\ ] < \deg W(\lambda)$$

which contradicts Proposition 3.1. This ends the proof of property a).

The proof of property b) will be given in the next section. □

4. UNITARY MINIMAL COMPLETIONS

Let $W(\lambda)$: $\mathbb{C}^m \to \mathbb{C}^p$ be a rational matrix function analytic at infinity. A rational matrix function

$$\tilde{W}(\lambda) = \begin{bmatrix} W_{11}(\lambda) & W_{12}(\lambda) \\ W_{21}(\lambda) & W_{22}(\lambda) \end{bmatrix} : \mathbb{C}^p \oplus \mathbb{C}^m \to \mathbb{C}^p \oplus \mathbb{C}^m \tag{4.1}$$

analitic at infinity, for which $W_{12}(\lambda) = W(\lambda)$ is called a completion of the function $W(\lambda)$. If moreover $\deg \tilde{W}(\lambda) = \deg W(\lambda)$ then $\tilde{W}(\lambda)$ is called a minimal completion of $W(\lambda)$.

In the following theorem we give a characterization of all the minimal completions of $W(\lambda)$ which are unitary on the line. The existence of such completions for every proper contraction is also implicitly proved.

THEOREM 4.1. *Let $W(\lambda)$: $\mathbb{C}^m \to \mathbb{C}^p$ be a proper contraction and let*

$$\Sigma = (A, B, C, D;\ \mathbb{C}^n,\ \mathbb{C}^m,\ \mathbb{C}^p)$$

be a minimal realization of $W(\lambda)$. Then

a) *For every hermitian solution H of the state characteristic equation of $W(\lambda)$ associated to the realization Σ, the function:*

$$\tilde{W}_H(\lambda) = W_{\tilde{\Sigma}}(\lambda)$$

with

$$\tilde{\Sigma} = \left(A, [\, X \quad B\,], \begin{bmatrix} C \\ Y \end{bmatrix}, \begin{bmatrix} M & D \\ N & P \end{bmatrix};\ \mathbb{C}^n,\ \mathbb{C}^{p+m},\ \mathbb{C}^{p+m} \right) \tag{4.2}$$

where:

$$N = -D^*, \qquad M = (I - DD^*)^{1/2}, \qquad P = (I - D^*D)^{1/2}$$
$$X = (iHC^* - BD^*)(I - DD^*)^{-1/2}$$
$$Y = (I - D^*D)^{-1/2}(iB^*H^{-1} - D^*C)$$

is a minimal completion of $W(\lambda)$ which is unitary on the line.

b) *Any minimal completion of $W(\lambda)$ which is unitary on the line has the form*

$$\tilde{W}(\lambda) = \begin{bmatrix} I & 0 \\ 0 & T \end{bmatrix} \tilde{W}_H(\lambda) \begin{bmatrix} S & 0 \\ 0 & I \end{bmatrix}$$

where T and S are unitary matrices of sizes $m \times m$ and $p \times p$ respectively, and H is a hermitian solution of the state characteristic equation of $W(\lambda)$. The unitary matrices S and T are uniquely determined by the value of $\tilde{W}(\lambda)$ at infinity and are given by

$$S = (I - W(\infty)W^*(\infty))^{-1/2}[\tilde{W}(\infty)]_{11}$$
$$T = [\tilde{W}(\infty)]_{22}(I - W^*(\infty)W(\infty))^{-1/2}.$$

c) *The correspondence*

$$H \to \begin{bmatrix} I & 0 \\ 0 & T \end{bmatrix} \tilde{W}_H(\lambda) \begin{bmatrix} S & 0 \\ 0 & I \end{bmatrix}$$

from the set of all hermitian solutions of the state characteristic equation of $W(\lambda)$ associated with Σ, onto the set of all minimal, unitary on the line completions of $W(\lambda)$ which have a given value at infinity is one to one.

PROOF. The system Σ is a minimal system. From this it easily follows that $\tilde{\Sigma}$ is also minimal. According to Theorem 2.1 from [1] applied to the system $\tilde{\Sigma}$, the function $W_{\tilde{\Sigma}}(\lambda)$ is unitary on the line if and only if the following equations have an invertible hermitian solution H:

$$AH - HA^* = i[\, X \quad B \,] \begin{bmatrix} X^* \\ B^* \end{bmatrix} \tag{4.3}$$

$$[\, X \quad B \,] = iH[\, C^* \quad Y^* \,] \begin{bmatrix} M & D \\ N & P \end{bmatrix} \tag{4.4}$$

$$\begin{bmatrix} M & D \\ N & P \end{bmatrix} \begin{bmatrix} M^* & N^* \\ D^* & P^* \end{bmatrix} = \begin{bmatrix} I & 0 \\ 0 & I \end{bmatrix}. \tag{4.5}$$

One easily verifies that, with M, N, D, P, X and Y defined in Theorem 4.1 a), the relations (4.3), (4.4) and (4.5) are satisfied for every hermitian solution H of the state characteristic equation of $W(\lambda)$ associated with Σ. This ends the proof of a).

We prove now b). Let $\tilde{W}(\lambda)$ be a minimal completion of $W(\lambda)$ which is unitary on the line. Then $\tilde{W}(\lambda)$ admits a realization $\tilde{\Sigma}$ of the form:

$$\tilde{\Sigma} = \left(A, [\, X \quad B \,], \begin{bmatrix} C \\ Y \end{bmatrix}, \begin{bmatrix} M & D \\ N & P \end{bmatrix}; \mathbb{C}^n, \mathbb{C}^{p+m}, \mathbb{C}^{p+m}\right). \tag{4.6}$$

Moreover, the operators of $\tilde{\Sigma}$ satisfy the relations (4.3), (4.4) and (4.5). The relation (4.5) leads to:

$$M = (I - DD^*)^{1/2} S \tag{4.7}$$

$$P = T(I - D^*D)^{1/2} \tag{4.8}$$

$$N = -TD^*S \tag{4.9}$$

for some unitary matrices S and T. From (4.4) follows

$$X = iH\big(C^*(I - DD^*)^{1/2} S - Y^*TD^*S\big)$$

$$B = iHC^*D + iHY^*T(I - D^*D)^{1/2}.$$

The second of this relations give:

$$Y = T(I - D^*D)^{-1/2}(iB^*H^{-1} - D^*C) \tag{4.10}$$

and

$$(4.11)\qquad \begin{aligned} X &= iH(C^*(I-DD^*)+iH^{-1}(B-iHC^*D)D^*)(I-DD^*)^{-1/2}S = \\ &= (iHC^*-BD^*)(I-DD^*)^{-1/2}S. \end{aligned}$$

Introducing the values (4.11) of X in (4.3) one obtains

$$H\gamma H - iH\alpha^* + i\alpha H + \beta = 0$$

where γ, α^*, α and β are the coefficients of the state characteristic equation of $W(\lambda)$ associated with Σ. The relation $\tilde{W}(\lambda) = W_{\tilde{\Sigma}}(\lambda)$ for M, P, N, X and Y defined by (4.7), (4.8), (4.9), (4.10) and (4.11) is equivalent to

$$\tilde{W}(\lambda) = \begin{bmatrix} I & 0 \\ 0 & T \end{bmatrix} \tilde{W}_H(\lambda) \begin{bmatrix} S & 0 \\ 0 & I \end{bmatrix}.$$

If the value of $\tilde{W}(\infty)$ is given then S and T may be computed from (4.7), (4.8)

$$S = (I-DD^*)^{-1/2}M = (I-DD^*)^{-1/2}[\tilde{W}(\infty)]_{11} \qquad (4.12)$$

$$T = P(I-D^*D)^{-1/2} = [\tilde{W}(\infty)]_{22}(I-D^*D)^{-1/2}. \qquad (4.13)$$

This ends the proof of property b).

Now let $\tilde{D}$ be a unitary completion of $W(\infty)$ and $\tilde{W}(\lambda)$ a minimal unitary on the line completion of $W(\lambda)$ with $\tilde{W}(\infty) = \tilde{D}$. Then according with the proof of property b), $\tilde{W}(\lambda)$ admits a realization of the form (4.6), with X and Y given by (4.10), (4.11), H is a hermitian solution of the state characteristic equation of $W(\lambda)$ associated with Σ and S and T defined by (4.12), (4.13). Moreover, in conformity with Lemma 4.3 of [6], X and Y are uniquely defined by the completion $\tilde{W}(\lambda)$ of $W(\lambda)$. The solution H of the state characteristic equation is also a solution for the system (4.3), (4.4), and in conformity with Theorem 2.1 in [1] and the remark that follows is uniquely defined. This proves property c). □

The following remark about the dependence of the minimal unitary on the line completions of $W(\lambda)$ on its realization Σ is in order. Let

$$\Sigma' = (A', B', C', D';\ \mathbb{C}^n,\ \mathbb{C}^m,\ \mathbb{C}^p)$$

be another minimal realization of $W(\lambda)$. There exists an invertible matrix Q such that $A' = QAQ^{-1}$, $B' = QB$, $C' = CQ^{-1}$. The state characteristic equation associated with Σ' may be obtained from the state characteristic equation associated with Σ by means of the change of variable $H' = QHQ^*$. Then, for every solution H of the state characteristic equation associated with Σ which leads to the completion $\tilde{\Sigma}$ of Σ, the solution QHQ^* of the corresponding characteristic equation associated with Σ' leads as one easily verifies to the completion

$$\tilde{\Sigma}' = \left(QAQ^{-1}, Q[\,X \quad B\,], \begin{bmatrix} C \\ Y \end{bmatrix} Q^{-1}, \begin{bmatrix} M & D \\ N & P \end{bmatrix};\ \mathbb{C}^n,\ \mathbb{C}^{m+p},\ \mathbb{C}^{m+p}\right)$$

of Σ' which is similar to $\tilde{\Sigma}$, and thus to the same unitary on the line minimal completion $\tilde{W}(\lambda)$ of $W(\lambda)$.

We are now in a position to prove the rest of Theorem 3.4.

PROOF OF THEOREM 3.4, PROPERTY b). Without loss of generality we may assume that equation (3.4) is associated with a minimal realization $\Sigma = (A, B, C, D; \mathbb{C}^n, \mathbb{C}^m, \mathbb{C}^p)$ of the proper contraction $W(\lambda)$. Let H be a hermitian solution of this equation. According to the proof of Theorem 4.1, H satisfies the equation (4.3), and therefore the equation:

$$(iA)H + H(iA)^* = -[\,X \quad B\,]\begin{bmatrix} X^* \\ B^* \end{bmatrix}.$$

The pair $\left(\begin{bmatrix} X^* \\ B^* \end{bmatrix}, (iA)^*\right)$ is observable. This allows us to apply Theorem 2.13 in [1]. To obtain that the number of negative (resp. positive) eigenvalues of H is equal to the number of eigenvalues of $(iA)^*$ in the open right (resp. left) half plane. On the other hand, the eigenvalues of A are known to be exactly the poles of $W(\lambda)$, and b) immediately follows. □

In Theorem 4.1 c) we proved that there is a one to one correspondence between the hermitian solutions of the state characteristic equation of a proper contraction and its minimal, unitary on the line completions which have a given value at infinity. This allows us to use a known result on the number of hermitian solutions of a symmetric Riccati equation in order to obtain the corresponding result on the number of minimal, unitary on the line completions of a proper contraction.

THEOREM 4.2. *Let $W(\lambda)$ be a proper contraction and $\mathcal{A}$ its state characteristic matrix. Then, for every unitary completion $\tilde{D} = \begin{bmatrix} M & W(\infty) \\ N & P \end{bmatrix}$ of $W(\infty)$,*

a) *$W(\lambda)$ admits a unique, unitary on the line, minimal completion $\tilde{W}(\lambda)$ with $\tilde{W}(\infty) = \tilde{D}$ if and only if all eigenvalues of $\mathcal{A}$ are real.*

b) *$W(\lambda)$ admits a finite number of unitary on the line minimal completions $\tilde{W}(\lambda)$ with $\tilde{W}(\infty) = \tilde{D}$ if and only if $\dim\ker(\lambda I - \mathcal{A}) = 1$ for every nonreal eigenvalue λ of $\mathcal{A}$. In this case, the number of such completions is exactly equal to $\prod_{j=1}^{\alpha}(k_j + 1)$ where k_j, $j = 1, 2, \ldots, \alpha$ are the multiplicities of all the distinct eigenvalues of $\mathcal{A}$ lying in the open upper half plane.*

c) *If $\dim\ker(\lambda I - \mathcal{A}) \geq 2$ for some nonreal eigenvalue λ of $\mathcal{A}$ then $W(\lambda)$ has a continuum of unitary on the line, minimal completions $\tilde{W}(\lambda)$ with $\tilde{W}(\infty) = \tilde{D}$.*

PROOF. Corollary 4.11 of Chap. II in [5] states that if an equation $XDX + XA + A^*X - C = 0$ for which $C^* = C$, D is nonnegative definite and the pair (A, D) is controllable has a hermitian solution then the following statements are true:

a) If $\dim\ker \lambda I - i\begin{bmatrix} A & D \\ C & -A^* \end{bmatrix} \leq 1$ for all λ in the open upper half plane, then the number of hermitian solutions of the given equation is $\prod_{j=1}^{\alpha}(k_j + 1)$, where $k_1, \ldots, k_\alpha$ are the multiplicities of all distinct eigenvalues $\lambda_1, \ldots, \lambda_\alpha$ of $i\begin{bmatrix} A & D \\ C & -A^* \end{bmatrix}$ lying in the open upper half plane.

b) If $\dim\ker(\lambda I - i\begin{bmatrix} A & D \\ C & -A^* \end{bmatrix}) \geq 2$ for some nonreal λ, then the given equation has a continuum of hermitian solution.

One easily sees that the state characteristic equation of $W(\lambda)$ satisfies the conditions of this corollary. The form of the matrix $i\begin{bmatrix} A & D \\ C & -A^* \end{bmatrix}$ in the case of the state characteristic equation of $W(\lambda)$ is

$$\mathcal{M} = i\begin{bmatrix} -i\alpha^* & \gamma \\ -\beta & -i\alpha \end{bmatrix}$$

which is similar to the state characteristic matrix of $W(\lambda)$

$$\mathcal{M} = \begin{bmatrix} 0 & -iI \\ -I & 0 \end{bmatrix}\begin{bmatrix} \alpha & \beta \\ \gamma & \alpha^* \end{bmatrix}\begin{bmatrix} 0 & -I \\ iI & 0 \end{bmatrix}.$$

This, together with Theorem 4.1 b) ends the proof of the theorem.

The following three examples illustrate each of the cases of Theorem 4.2.

EXAMPLE 1. Let $W(\lambda) = \mu\frac{\lambda}{\lambda - i}$, $-1 < \mu < 1$. Then $W(\lambda)$ has the minimal realization:

$$\Sigma = (i, i, \mu, \mu;\ \mathbb{C},\ \mathbb{C},\ \mathbb{C}).$$

The state characteristic matrix of $W(\lambda)$ is:

$$\mathcal{A} = \begin{bmatrix} \frac{i}{1-\mu^2} & \frac{1}{1-\mu^2} \\ \frac{\mu^2}{1-\mu^2} & -\frac{i}{1-\mu^2} \end{bmatrix}$$

and has two simple eigenvalues $\lambda_1 = \frac{i}{\sqrt{1-\mu^2}}$, $\lambda_2 = \frac{-i}{\sqrt{1-\mu^2}}$. According to the previous theorem, for every unitary on the line completion of $W(\infty)$, $W(\lambda)$ admits exactly two unitary on the line minimal completions. Let us compute them. The state characteristic equation of $W(\lambda)$ is

$$\frac{1}{1-\mu^2}\mu^2 h^2 - \frac{2}{1-\mu^2}h + \frac{1}{1-\mu^2} = 0.$$

This equation has two real solutions: $h_1 = 1 + \sqrt{1-\mu^2}$, $h_2 = 1 - \sqrt{1-\mu^2}$. Let

$$\begin{bmatrix} 1 & 0 \\ 0 & e^{it} \end{bmatrix}\begin{bmatrix} \sqrt{1-\mu^2} & \frac{\mu}{\sqrt{1-\mu^2}} \\ -\mu & \sqrt{1-\mu^2} \end{bmatrix}\begin{bmatrix} e^{is} & 0 \\ 0 & 1 \end{bmatrix}, \qquad s, t \in \mathbb{R}$$

be a unitary completion of $W(\infty)$. According to Theorem 4.1, $W(\lambda)$ admits the following unitary on the line minimal completions:

$W_{\tilde{\Sigma}_1}(\lambda)$ where

$$\tilde{\Sigma}_1 = \Bigg(i, [\ i\mu e^{is} \quad i\ ], \begin{bmatrix} \mu \\ \frac{1-\mu^4-\sqrt{1-\mu^2}}{\mu^2\sqrt{1-\mu^2}}e^{it} \end{bmatrix},$$

$$\begin{bmatrix} e^{is}\sqrt{1-\mu^2} & \mu \\ -\mu e^{i(t+s)} & e^{it}\sqrt{1-\mu^2} \end{bmatrix};\ \mathbb{C},\ \mathbb{C}^2,\ \mathbb{C}^2\Bigg).$$

$W_{\tilde{\Sigma}_2}(\lambda)$ where

$$\tilde{\Sigma}_2 = \left(i, [\ -i\mu e^{is} \quad i\ ], \begin{bmatrix} \mu \\ \frac{1-\mu^4+\sqrt{1-\mu^2}}{\mu^2\sqrt{1-\mu^2}} e^{it} \end{bmatrix}, \right.$$
$$\left. \begin{bmatrix} e^{is}\sqrt{1-\mu^2} & \mu \\ -\mu e^{i(t+s)} & e^{it}\sqrt{1-\mu^2} \end{bmatrix}; \mathbb{C}, \mathbb{C}^2, \mathbb{C}^2 \right).$$

EXAMPLE 2. Let $W(\lambda) = \frac{(\mu\lambda+1)(1-\mu^2)-\mu i}{\lambda(1-\mu^2)-i}$, $0 < \mu < 1$. $W(\lambda)$ has the minimal realization:

$$\Sigma = \left(\frac{i}{1-\mu^2}, 1, 1, \mu;\ \mathbb{C}, \mathbb{C}, \mathbb{C} \right).$$

The state characteristic equation of $W(\lambda)$ is:

$$\frac{1}{1-\mu^2}h_2 - 2h\frac{1}{1-\mu^2} + \frac{1}{1-\mu^2} = 0$$

and has the unique real solution $h = 1$. This means that $W(\lambda)$ is a contraction on the line. The state characteristic matrix of $W(\lambda)$ associated with Σ is

$$\mathcal{A} = \frac{1}{1-\mu^2}\begin{bmatrix} i+\mu & 1 \\ 1 & -i+\mu \end{bmatrix}$$

and has the only eigenvalue $\lambda = \frac{\mu}{1-\mu^2}$ which is real. According to the previous theorem $W(\lambda)$ admits a unique minimal completion which is unitary on the line. This completions may be constructed from the value of h and is:

$W_{\tilde{\Sigma}}$ where

$$\tilde{\Sigma} = \left(\frac{i}{1-\mu^2}, \begin{bmatrix} \frac{i-\mu}{\sqrt{1-\mu^2}} e^{is} & 1 \end{bmatrix}, \begin{bmatrix} 1 \\ \frac{i-\mu}{\sqrt{1-\mu^2}} e^{it} \end{bmatrix}, \right.$$
$$\left. \begin{bmatrix} e^{is}\sqrt{1-\mu^2} & \mu \\ -e^{i(t+s)}\mu & e^{it}\sqrt{1-\mu^2} \end{bmatrix}; \mathbb{C}, \mathbb{C}^2, \mathbb{C}^2 \right).$$

EXAMPLE 3.

$$W(\lambda) = \begin{bmatrix} \mu & 0 \\ 0 & \mu \end{bmatrix} + \begin{bmatrix} \lambda - \frac{i-\mu}{1-\mu^2} & -\frac{i}{1-\mu^2} \\ -\frac{i}{1-\mu^2} & \lambda + \frac{i+\mu}{1-\mu^2} \end{bmatrix}^{-1}, \qquad 0 < \mu < 1$$

has the realization

$$\Sigma = \left(\begin{bmatrix} \frac{i-\mu}{1-\mu^2} & \frac{i}{1-\mu^2} \\ \frac{i}{1-\mu^2} & \frac{-i-\mu}{1-\mu^2} \end{bmatrix}, \begin{bmatrix} 1 & 0 \\ 0 & 1 \end{bmatrix}, \begin{bmatrix} 1 & 0 \\ 0 & 1 \end{bmatrix}, \begin{bmatrix} \mu & 0 \\ 0 & \mu \end{bmatrix}; \mathbb{C}^2, \mathbb{C}^2, \mathbb{C}^2 \right)$$

and the state characteristic matrix is

$$\mathcal{A} = \begin{bmatrix} i & i & 1 & 0 \\ i & -i & 0 & 1 \\ 1 & 0 & -i & -i \\ 0 & 1 & -i & i \end{bmatrix} \frac{1}{1-\mu^2}.$$

The state characteristic equation of $W(\lambda)$ is

$$H^2 + iH \begin{bmatrix} i & i \\ i & -i \end{bmatrix} + i \begin{bmatrix} i & i \\ i & -i \end{bmatrix} H + \begin{bmatrix} 1 & 0 \\ 0 & 1 \end{bmatrix} = \begin{bmatrix} 0 & 0 \\ 0 & 0 \end{bmatrix}$$

and has the selfadjoint solution $H_1 = \begin{bmatrix} 1 & 0 \\ 0 & -1 \end{bmatrix}$. Thus $W(\lambda)$ is a contraction on the line. One sees that the eigenvalue $\lambda = \frac{-i}{1-\mu^2}$ of $\mathcal{A}$ has geometric multiplicity 2. Thus $W(\lambda)$ admits a continuum of minimal completions unitary on the line, for every given unitary completion of $W(\infty)$.

5. THE LINEAR FRACTIONAL DECOMPOSITION OF PROPER CONTRACTIONS

Let $W(\lambda) \in L(U,Y)$ be a rational matrix function analytic at infinity. A linear fractional decomposition of $W(\lambda)$ is a decomposition of the form:

$$W(\lambda) = \begin{bmatrix} \Omega_{11}(\lambda) & \Omega_{12}(\lambda) \\ \Omega_{21}(\lambda) & \Omega_{22}(\lambda) \end{bmatrix} o_f \omega(\lambda) \overset{\text{def}}{=} \tag{5.1}$$

$$\overset{\text{def}}{=} \Omega_{21}(\lambda) + \Omega_{22}(\lambda)\omega(\lambda)(I - \Omega_{12}(\lambda)\omega(\lambda))^{-1}\Omega_{11}(\lambda)$$

with $[\Omega_{ij}(\lambda)]_{i,j=1,2} \in L(U \oplus Y)$ and $\omega(\lambda) \in L(U,Y)$ rational matrix functions analytic at infinity and of appropriate sizes. The decomposition (6.1) is called minimal if:

$$\deg W(\lambda) = \deg \begin{bmatrix} \Omega_{11}(\lambda) & \Omega_{12}(\lambda) \\ \Omega_{21}(\lambda) & \Omega_{22}(\lambda) \end{bmatrix} + \deg \omega(\lambda).$$

A rewriting of this condition with "$\deg_{\lambda_0}$" instead of "deg" leads to the definition of a decomposition which is minimal at the point λ_0.

A decomposition (5.1) is called regular iff $\Omega_{11}(\infty)$ and $\Omega_{22}(\infty)$ are invertible.

In [8] (see also [6] of which the notations we use here) it is shown that any regular minimal linear fractional decomposition of the rational matrix function $W(\lambda)$ may be obtained from a minimal realization Σ of $W(\lambda)$:

$$\Sigma = (A, B, C, D; X, U, Y)$$

be means of two operators $F: X \to U$ and $G: U \to X$ and a decomposition $X = X_1 \oplus X_2$ of the state space of Σ which satisfies the conditions:

X_1 is $A + BF$ invariant

X_2 is $A + GC$ invariant.

Explicit formula for the components of the decomposition are given in realization form.

In what follows we shall find which are the operators F and G and the subspaces X_1 and X_2 which correspond to a decomposition of a proper contraction such that the first component to be unitary on the line, and therefore, the second a proper contraction. To this end we need the notion of general decomposition:

Let $\tilde{W}(\lambda)$ be a rational matrix function which operates on a direct sum of finite dimensional spaces

$$\tilde{W}(\lambda) = \begin{bmatrix} W_{11}(\lambda) & W_{12}(\lambda) \\ W_{21}(\lambda) & W_{22}(\lambda) \end{bmatrix} \in L(U \oplus Y).$$

A general cascade decomposition of $\tilde{W}(\lambda)$ is a decomposition of the form:

$$\tilde{W}(\lambda) = \begin{bmatrix} \Omega_{11}(\lambda) & \Omega_{12}(\lambda) \\ \Omega_{21}(\lambda) & \Omega_{22}(\lambda) \end{bmatrix} o_g \begin{bmatrix} \omega_{11}(\lambda) & \omega_{12}(\lambda) \\ \omega_{21}(\lambda) & \omega_{22}(\lambda) \end{bmatrix}$$

which means

$$\begin{aligned} W_{11}(\lambda) &= \omega_{11}(\lambda)(I - \Omega_{12}(\lambda)\omega_{21}(\lambda))^{-1}\Omega_{11}(\lambda) \\ W_{12}(\lambda) &= \omega_{12}(\lambda) + \omega_{11}(\lambda)\Omega_{12}(\lambda)(I - \omega_{21}(\lambda)\Omega_{12}(\lambda))^{-1}\omega_{22}(\lambda) \\ W_{21}(\lambda) &= \Omega_{21}(\lambda) + \Omega_{22}(\lambda)\omega_{21}(\lambda)(I - \Omega_{12}(\lambda)\omega_{21}(\lambda))^{-1}\Omega_{11}(\lambda) \\ W_{22}(\lambda) &= \Omega_{22}(\lambda)(I - \omega_{21}(\lambda)\Omega_{12}(\lambda))^{-1}\omega_{22}(\lambda). \end{aligned}$$

where the components $[\Omega_{ij}^{(\lambda)}]_{i,j=1,2}$ and $[\omega_{ij}(\lambda)]_{i,j=1,2}$ are rational matrix functions analytic at infinity with values in $L(U \oplus Y)$. The operation involved in this decomposition was first studied by Redheffer [9]. The set of all such decompositions for which $\deg \tilde{W}(\lambda) = \deg[\Omega_{ij}(\lambda)]_{i,j=1,2} + \deg[\omega_{ij}(\lambda)]_{i,j=1,2}$ is described in realization form in [6]. Two properties of this decomposition are of interest in what follows:

a) $\tilde{W}(\lambda) = \Omega(\lambda) o_g \omega(\lambda) \Rightarrow [\tilde{W}(\lambda)]_{21} = \Omega(\lambda) o_f [\omega(\lambda)]_{21}$

b) $\tilde{W}(\lambda) = \Omega(\lambda) o_g \omega(\lambda) \Rightarrow [\tilde{W}(\lambda)]^{<} = [\Omega(\lambda)]^{<} \cdot [\omega(\lambda)]^{<}$

provided the involved operations make sense at infinity. (Here the sign "<" stands for the operation of partial inversion defined in section 2.)

Once this recalled we can give the following theorem:

THEOREM 5.1. *Let $W(\lambda)$: $\mathbb{C}^m \to \mathbb{C}^p$ be a proper contraction. Let*

$$\Sigma(A, B, C, D;\ \mathbb{C}^n,\ \mathbb{C}^m,\ \mathbb{C}^p)$$

be a minimal realization of $W(\lambda)$, and

$$D = \begin{bmatrix} \Delta_{11} & \Delta_{12} \\ \Delta_{21} & \Delta_{22} \end{bmatrix} o_f d$$

be a fractional linear decomposition of D *for which* $[\Delta_{ij}]_{i,j=1,2}$ *is unitary and* Δ_{21} *and* d *are strict contractions. Then*

a) *To any hermitian solution of the state characteristic equation of* $W(\lambda)$ *associated with* Σ:

$$H\gamma H - iH\alpha^* + i\alpha H + \beta = 0$$

and any H^{-1}*-non degenerate*

$$(\alpha - i\beta H^{-1})\text{-invariant}$$

subspace M, *corresponds a minimal linear fractional decomposition*:

$$W(\lambda) = \begin{bmatrix} \Omega_{11}(\lambda) & \Omega_{12}(\lambda) \\ \Omega_{21}(\lambda) & \Omega_{22}(\lambda) \end{bmatrix} o_f \omega(\lambda)$$

of $W(\lambda)$ *for which* $[\Omega_{ij}(\infty)]_{i,j=1,2} = [\Delta_{ij}]_{i,j=1,2}$, $\omega(\infty) = d$, $[\Omega_{ij}(\lambda)]_{i,j=1,2}$ *is unitary on the line and* $\omega(\lambda)$ *is a proper contraction. The quadruple* (X_1, X_2, F, G) *which defines this decomposition is given by*:

$$X_1 = M, \qquad F = -(I - D^*D)^{-1}(iB^*H^{-1} - D^*C),$$
$$X_2 = M^{[\perp]_{H^{-1}}}, \qquad G = -(iHC^* - BD^*)(I - DD^*)^{-1}.$$

b) *All such decompositions may be obtained this way.*

PROOF. Let

$$\tilde{W}(\lambda) = \begin{bmatrix} W_{11}(\lambda) & W_{12}(\lambda) \\ W(\lambda) & W_{22}(\lambda) \end{bmatrix} \tag{5.2}$$

be a minimal completion of $W(\lambda)$ which is unitary on the line. Assume $\tilde{W}(\lambda)$ admits the general cascade decomposition:

$$\tilde{W}(\lambda) = \begin{bmatrix} \Omega_{11}(\lambda) & \Omega_{12}(\lambda) \\ \Omega_{21}(\lambda) & \Omega_{22}(\lambda) \end{bmatrix} o_g \begin{bmatrix} \omega_{11}(\lambda) & \omega_{12}(\lambda) \\ \omega_{21}(\lambda) & \omega_{22}(\lambda) \end{bmatrix} \tag{5.3}$$

of which the components $[\Omega_{ij}(\lambda)]_{i,j=1,2}$ and $[\omega_{ij}(\lambda)]_{i,j=1,2}$ are unitary on the line. Moreover, suppose that the decomposition (5.3) is minimal, by which we mean that $\deg \tilde{W}(\lambda) = \deg[\Omega_{ij}(\lambda)]_{i,j=1,2} + \deg[\omega_{ij}(\lambda)]_{i,j=1,2}$. Then $W(\lambda)$ admits the linear fractional decomposition:

$$W(\lambda) = \begin{bmatrix} \Omega_{11}(\lambda) & \Omega_{12}(\lambda) \\ \Omega_{21}(\lambda) & \Omega_{22}(\lambda) \end{bmatrix} o_f \omega_{21}(\lambda). \tag{5.4}$$

Moreover, $\deg W(\lambda) = \deg \tilde{W}(\lambda) = \deg[\Omega_{ij}(\lambda)]_{i,j=1,2} + \deg[\omega_{ij}(\lambda)]_{i,j=1,2} \geq \deg[\Omega_{ij}(\lambda)]_{i,j=1,2} + \deg \omega_{21}(\lambda)$. As it is always true that $\deg W(\lambda) \leq \deg[\Omega_{ij}(\lambda)]_{i,j=1,2} + \deg \omega_{21}(\lambda)$ it follows that the decomposition (5.4) is minimal. Let $W(\lambda)$, $[\Delta_{ij}]_{i,j=1,2}$ and d be as in the theorem. Then d admits a unitary completion of the form:

$$\tilde{d} = \begin{bmatrix} d_{11} & d_{12} \\ d & d_{22} \end{bmatrix}$$

and

$$\begin{bmatrix} D_{11} & D_{12} \\ D & D_{22} \end{bmatrix} \stackrel{\text{def}}{=} \begin{bmatrix} \Delta_{11} & \Delta_{12} \\ \Delta_{21} & \Delta_{22} \end{bmatrix} \circ_g \begin{bmatrix} d_{11} & d_{12} \\ d & d_{22} \end{bmatrix}$$

is a unitary completion of D. Since D, Δ_{21} and d are strict contractions, follows that D_{11}, D_{22}, Δ_{11}, Δ_{22}, d_{11} and d_{22} are invertible matrices. Let $\tilde{W}(\lambda)$, given by (5.2), be a minimal completion of $W(\lambda)$ which is unitary on the line and has the value $\begin{bmatrix} D_{11} & D_{12} \\ D & D_{22} \end{bmatrix}$ at infinity. $\tilde{W}(\lambda)$ admits a decomposition (5.3) if and only if the function $\tilde{W}(\lambda)^<$ admits the factorization:

$$\tilde{W}(\lambda)^< = \begin{bmatrix} \Omega_{11}(\lambda) & \Omega_{12}(\lambda) \\ \Omega_{22}(\lambda) & \Omega_{22}(\lambda) \end{bmatrix}^< \cdot \begin{bmatrix} \omega_{11}(\lambda) & \omega_{12}(\lambda) \\ \omega_{21}(\lambda) & \omega_{22}(\lambda) \end{bmatrix}^< . \tag{5.5}$$

On the other hand $\tilde{W}(\lambda)$ is unitary if and only if $\tilde{W}(\lambda)^<$ is J unitary for $J = \begin{bmatrix} -I & 0 \\ 0 & I \end{bmatrix}$. A slight modification of Theorem 4.1, for the case of a completion of form (5.2), shows that $\tilde{W}(\lambda)$ has the realization:

$$\tilde{\Sigma} = (A, [\, B \quad X \,], \begin{bmatrix} Y \\ C \end{bmatrix}, \begin{bmatrix} D_{11} & D_{12} \\ D & D_{22} \end{bmatrix}; \mathbb{C}^n, \mathbb{C}^{m+p}, \mathbb{C}^{m+p})$$

with

$$\begin{aligned} D_{12} &= -TD^*S \\ D_{22} &= (I - DD^*)^{1/2}S \\ D_{11} &= T(I - D^*D)^{1/2} \\ X &= (iHC^* - BD^*)(I - DD^*)^{-1/2}S \\ Y &= T(I - D^*D)^{-1/2}(iB^*H^{-1} - D^*C) \end{aligned}$$

for some unitary matrices S and T and a hermitian solution H of the state characteristic equation associated with Σ. Let us agree on the notation:

$$\tilde{\Sigma} = (\tilde{A}, \tilde{B}, \tilde{C}, \tilde{D}; \mathbb{C}^n, \mathbb{C}^{m+p}, \mathbb{C}^{m+p}).$$

An easy computation shows that the state operator $\tilde{A}^<$ of $\tilde{\Sigma}^<$ is given by:

$$\tilde{A}^< = \alpha - i\beta H^{-1}.$$

Theorem 2.6 of [1] applied to the system $\tilde{\Sigma}^<$ says that there exists an one-to-one correspondence between the J-unitary (product) factorization of $[\tilde{W}(\lambda)]^<$ and the subspaces M which are $\tilde{A}^<$-invariant and $\mathcal{H}^{-1}$-nondegenerate, where $\mathcal{H}$ is the only hermitian solution of the system:

$$\tilde{A}^<\mathcal{H} - \mathcal{H}[\tilde{A}^<]^* = i\tilde{B}^< J[\tilde{B}^<]^* \tag{5.6}$$

$$\tilde{B}^< = i\mathcal{H}\tilde{C}^* J\tilde{D}. \tag{5.7}$$

A very lengthy computation shows that the hermitian solution of the characteristic equation of associated with Σ that defines the completion $\tilde{\Sigma}$ is also a solution of the system (5.6), (5.7) and, therefore, $\mathcal{H} = H$. A shorter proof of this fact may be obtained from the remark that $-iH^{-1}$ is the only similarity between $\tilde{\Sigma}^{\times}$ and $\tilde{\Sigma}_*$ which means that it is the only similarity between $[\tilde{\Sigma}^{\times}]^{<}$ and $[\tilde{\Sigma}_*]^{<}$. On the other hand $-i\mathcal{H}^{-1}$ is the only similarity between $[\tilde{\Sigma}^{<}]^{\times}$ and $[\tilde{\Sigma}^{<}]_{*J}$, where

$$[\tilde{\Sigma}^{<}]_{*J} = ([\tilde{A}^{<}]^*, [\tilde{C}^{<}]^* J, J[\tilde{B}^{<}]^*, J[\tilde{D}^{<}]^* J;\ \mathbb{C}^n,\ \mathbb{C}^{m+p},\ \mathbb{C}^{m+p}).$$

Then, the equality $H = \mathcal{H}$ follows from:

$$[\tilde{\Sigma}^{\times}]^{<} = [\tilde{\Sigma}^{<}]^{\times} \quad \text{and} \quad [\tilde{\Sigma}_*]^{<} = [\tilde{\Sigma}^{<}]_{*J}.$$

In conclusion, for every subspace M of $\mathbb{C}^n$ which is H^{-1}-nondegenerate and $(\alpha - i\beta H^{-1})$-invariant one obtains a J unitary factorization of the form (5.5).

By an application of a partial inversion one obtains the minimal decomposition (5.3), and from it, the minimal decomposition (5.4). The martices F and G may be computed by means of the formulae in [6], p. 619. One obtains

$$F = -D_{11}^{-1} Y \quad \text{and} \quad G = -X D_{22}^{-1}.$$

Here D_{11}, D_{22}, X, Y are those in the definition of $\tilde{\Sigma}$. Hence

$$\begin{aligned} F &= -(I - D^*D)^{-1/2} T^* T (I - D^*D)^{-1/2} (iB^* H^{-1} - D^*C) = \\ &= -(I - D^*D)^{-1} (iB^* H^{-1} - D^*C) \end{aligned}$$

and

$$\begin{aligned} G &= -(iHC^* - BD^*)(I - DD^*)^{-1/2} S S^* (I - DD^*)^{-1/2} = \\ &= -(iHC^* - BD^*)(I - DD^*)^{-1}. \end{aligned}$$

b) Conversely, let

$$W(\lambda) = \begin{bmatrix} \Omega_{11}(\lambda) & \Omega_{12}(\lambda) \\ \Omega_{21}(\lambda) & \Omega_{22}(\lambda) \end{bmatrix} o_f \omega(\lambda)$$

be a minimal fractional linear decomposition of $W(\lambda)$ for which $[\Omega_{ij}(\lambda)]$ is analytic at infinity, unitary on the line, and such that $\Omega_{21}(\infty)$ is a strict contraction, and $\omega(\lambda)$ is a proper contraction. Let:

$$\tilde{\omega}(\lambda) = \begin{bmatrix} \omega_{11}(\lambda) & \omega_{12}(\lambda) \\ \omega(\lambda) & \omega_{22}(\lambda) \end{bmatrix}$$

be a unitary on the line minimal completion of $\omega(\lambda)$. Then

$$\tilde{W}(\lambda) \overset{\text{def}}{=} \begin{bmatrix} \Omega_{11}(\lambda) & \Omega_{12}(\lambda) \\ \Omega_{21}(\lambda) & \Omega_{22}(\lambda) \end{bmatrix} o_g \begin{bmatrix} \omega_{11}(\lambda) & \omega_{12}(\lambda) \\ \omega(\lambda) & \omega_{22}(\lambda) \end{bmatrix} \tag{5.8}$$

is a completion of $\tilde{W}(\lambda)$ which is unitary on the line. Moreover

$$\begin{aligned}\deg \tilde{W}(\lambda) \geq \deg W(\lambda) = \deg[\Omega_{ij}(\lambda)]_{i,j=1,2} + \deg \omega = \\ = \deg[\Omega_{ij}(\lambda)]_{i,j=1,2} + \deg \tilde{\omega}(\lambda).\end{aligned}$$

On the other hand one always has $\deg \tilde{W}(\lambda) \leq \deg[\Omega_{ij}(\lambda)]_{i,j=1,2} + \deg \tilde{\omega}(\lambda)$. Thus the completion $\tilde{W}(\lambda)$ and the decomposition (5.8) are minimal. It is also easy to verify that $\Omega_{11}(\lambda)$, $\Omega_{22}(\lambda)$, $\omega_{11}(\infty)$ and $\omega_{22}(\infty)$ are invertible. So, one may apply the method of a) to obtain the given decomposition. □

REFERENCES

[1] Alpay, D., Gohberg, I.: Unitary rational matrix functions, this volume.

[2] Bart, H., Gohberg, I., Kaashoek, M.A.: Minimal factorizations of matrix and operator functions, Birkhäuser Verlag, Basel, 1979.

[3] Dewilde, P.: Input-output descriptions of roomy systems, Siam J. Control and Optimization 14(4) (1976), 712–736.

[4] Glover, K.: Model reduction: A tutorial on Hankel-norm methods and lower bounds on L^2 errors, Reprint 288–293.

[5] Gohberg, I., Lancaster, P., Rodman, L.: Matrices and indefinite scalar products, Birkhäuser Verlag, Basel, 1983.

[6] Gohberg, I., Rubinstein, S.: Cascade decompositions of rational matrix functions and their stability, International J. of Control 46(2) (1987), 603–629.

[7] Gohberg, I., Kaashoek, M.A.: An inverse spectral problem for rational matrix functions and minimal divizibility, Integral Equations and Operation Theory 10 (1987), 437–465.

[8] Helton, J.W., Ball, J.A.: The cascade decomposition of a given system vs. the linear fractional decompositions of its transfer function, Integral Equation and Operator Theory 5 (1982), 341–385.

[9] Redheffer, R.M.: On a certain linear fractional transformation, J. Math. Phys. 39 (1960), 260–286.

Raymond and Beverly Sackler
Faculty of Exact Sciences
School of Mathematical Sciences
Tel-Aviv University
Ramat-Aviv, Israel